J. A. Gámez, S. Moral, A. Salmerón (Eds.)

Advances in Bayesian Networks

T0181271

Springer

Berlin
Heidelberg
New York
Hong Kong
London
Milano
Paris
Tokyo

Studies in Fuzziness and Soft Computing, Volume 146

Editor-in-chief
Prof. Janusz Kacprzyk
Systems Research Institute
Polish Academy of Sciences
ul. Newelska 6
01-447 Warsaw
Poland
E-mail: kacprzyk@ibspan.waw.pl

José A. Gámez
Serafín Moral
Antonio Salmerón (Eds.)

Advances in Bayesian Networks

 Springer

Dr. José A. Gámez
Universidad de Castilla-La Mancha
Escuela Politecnica Superior
Depto. Informatica
Campus Universitario s/n
02071 Albacete
Spain
E-mail: jose.gamez@uclm.es

Dr. Antonio Salmerón
Universidad de Almeria
ETS Ingenieria Informatica
Depto. Estadistica y
Matemática Aplicada
La Cañada de San Urbano s/n
04120 Almeria
Spain
E-mail: antonio.salmeron@ual.es

Professor Serafín Moral
Universidad de Granada
ETS Ingenieria Informatica
Depto. Ciencias Computacion
Campus Universitario Fuentenueva
18071 Granada
Spain
E-mail: smc@decsai.ugr.es

ISBN 978-3-642-05885-1   e-ISBN 978-3-540-39879-0

Library of Congress Cataloging-in-Publication-Data

A catalog record for this book is available from the Library of Congress.
Bibliographic information published by Die Deutsche Bibliothek.
Die Deutsche Bibliothek lists this publication in the Deutsche Nationalbibliographie;
detailed bibliographic data is available in the Internet at http://dnb.ddb.de

Springer-Verlag is a part of Springer Science+Business Media
springeronline.com

© Springer-Verlag Berlin Heidelberg 2010
Printed in Germany

Cover design: E. Kirchner, Springer-Verlag, Heidelberg
Printed on acid free paper   62/3020/M - 5 4 3 2 1 0

Preface

In the last years, probabilistic graphical models, specially Bayesian networks and decision graphs, have faced a significant theoretical development within areas such as Artificial Intelligence and Statistics. The sound theoretical foundation of these models as well as their powerful semantics have allowed their application to a wide range of practical problems which involve tasks such as data mining, information retrieval, diagnosis or classification. But several other applications based on Bayesian networks can be found, mainly in situations that require the modelisation of multivariate problems under uncertainty.

A *Bayesian network* is actually a representation of a multivariate probability distribution by means of a graph, where each node represents a random variable and the arcs indicate probabilistic dependencies between neighbour nodes. Associated with each node there is a conditional distribution for the variable in that node given its parents in the graph, so that the joint distribution factorises as the product of all the conditional distribution in the network. The induced factorisation allows to deal with models involving a large amount of variables, since there is no need to represent the joint distribution explicitly.

One of the most common tasks carried out over Bayesian networks is the inference or *probability propagation*, which consists in collecting the updated information over some variables of interest given that some other variables have been observed. Probability propagation is known to be an NP-hard problem. It means that it is always possible to find problems in which preforming the inference may become unfeasible. This fact motivates the development of sophisticated algorithms with the aim of expanding the class of tractable problems. When the response time or the available hardware resources are limited, inference must be carried out by means of approximate methods, which trade accuracy for economy of resources.

Influence Diagrams are graphical models in which apart from nodes representing random variables, we have decisions and utilities. In general, we may have several decisions that are made in a given sequence. The objective is to compute the optimal policy for them.

Bayesian network models can be constructed directly from an expert by means of Knowledge Engineering techniques, but when databases are available it is more usual to apply machine learning methods to construct the model without human intervention. The learning process involves two main aspects: deciding the network structure and estimating the parameters (the conditional distributions). The latest algorithms for structural learning are based on search strategies over the space of structures compatible with the problem variables, whose size is super-exponential. Regarding the parameters, they are usually estimated by maximum likelihood or by procedures of Bayesian Statistics.

This book is a compendium with three main types of chapters:

- Papers presenting some of the most recent advances in the area of probabilistic graphical models. There are chapters devoted to foundations (Daneshkhah and Smith; Xiang and Chen), decision graphs (Lu and Druzdzel), learning from data (Blanco, Inza, and Larrañaga; de Campos, Gámez and Puerta; Lucas; Kersting and Landwehr; Castelo and Perlman) and inference (Allen and Darwiche; Puch, Smith, and Bielza).
- Applications to different fields, as speech recognition (Deviren and Khalid Daoudi), Meteorology (Cano, Sordo, and Gutiérrez) and information retrieval (de Campos, Fernández-Luna and Huete).
- Survey papers describing the state of the art in some specific topics, as approximate propagation in Bayesian networks (Cano, Moral, and Salmerón), abductive inference (Gámez), decision graphs (Nielsen and Jensen), and applications of influence diagrams (Gómez).

Some of the contributions to this book (those from the two first groups above) have been selected from the papers presented at the First European Workshop on Probabilistic Graphical models (PGM'02), held in Cuenca (Spain) in November 2002.

Albacete,
Granada,
Almería,

José A. Gámez
Serafín Moral
Antonio Salmerón
September, 2003

Contents

Influence Diagrams

Learning

X

Applications

Applications of Bayesian Networks in Meteorology 309
Rafael Cano, Carmen Sordo, José M. Gutiérrez

Hypercausality, Randomisation, and Local and Global Independence

Alireza Daneshkhah and Jim.Q.Smith

Department of Statistics, The University of Warwick, Coventry CV4 7AL, UK

Abstract. In this paper we define the multicausal essential graph. Such graphical model demands further properties an equivalence class of Bayesian networks (BNs). In particular, each BN in an equivalence class is assumed to be causal in a stronger version of the manipulation causality of Spirtes et al (1993). In practice, the probabilities of any causal Bayesian network (CBN) will usually need to be estimated. To estimate the conditional probabilities in BNs it is common to assume local and global independence. The first result we prove in this paper is that if the BN is believed to be causal in a sufficiently strong sense i.e., is hypercausal, then it is necessary to make the assumption of local and global prior independence. Our second theorem develops this result. We give a characterisation of prior distributions sympathetic to causal hypotheses on all BNs in the equivalence class defined by an essential graph. We show that such strongly causally compatible priors satisfy a generalisation of the Geiger and Heckerman (1997) condition. In a special case when the essential graph is undirected, this family of prior distributions reduces to the Hyper-Dirichlet family, originally introduced by Dawid and Lauritzen (1993) as a prior family for decomposable graphical models.

1 Introduction

Causation in BNs is defined by Spirtes et al (1993), Pearl (2000) and Lauritzen (2001)), as a hypothesis of invariance of a collection of factorisations of a density to the manipulation of its nodes.

Let $p(x)$ represent a probability mass function on a set $\mathcal{X}$ of discrete random variables consistent with the conditional independence relations coded in a BN G in an idle system. Pearl defines a system to be in idle when its variables are observed and their values are not manipulated. Such data arise, for example, in cross sectional observational studies with no intervention. It is not unusual to want to make inferences about what will happen when specific variables in the system are manipulated to take certain values.

Let $p(x \mid do(V = v))$ denote the distribution resulting from the manipulation $do(V = v)$ that intervenes on a subset V of variables and sets them to values v. Denote by p_* the set of all distributions $p(x \mid do(V = v))$, $V \subseteq \mathcal{X}$ including $p(x)$, which represents no intervention (i.e., $V = \emptyset$). The following example makes clear the difference between *conditioning* and *manipulation* in the Bayesian networks.

Example 1 Consider the BN given in the following figure below. Here C represents *nuclear core activity* per hour; the *maximum temperature of cooling system* in an hour is represented by T; F indicates *failure of cooling system* in an hour. For simplicity, we will discretise the problem. So suppose, the possible values for C , T and F are respectively given by, (Normal, Critical, Meltdown), $(0^0 - 100^0, 100^0 - 200^0, 200^0 - 300^0, 400^0+)$ and (0, 1).
In the BN above, setting T to one of its values is just artificially increasing the

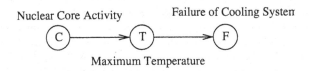

Fig. 1. The BN for the Nuclear Activity' Example

temperature. Clearly this will not affect core activity. However, conditioning on T (e.g., observing a high temperature at the value) is indicative of core activity. In both cases, if the failure of the cooling system is believed only to depend on T, then a natural extension of the unmanipulated BN is to make the obvious additional assumptions that doing T will have the same effect as observing T, i.e.,

$$p(F \mid do(T = t)) = p(F \mid T = t).$$

This motivates the following definition of a Causal Bayesian Network (CBN) on a vector of measurements, $x = (x_i, i \in \mathcal{I} = \{1, \ldots, k\})$ (see Pearl (1995)).

Definition 1 A BN G is said to be a *Causal Bayesian Network* (CBN) compatible p_* if and only if the following three conditions hold for every $p(. \mid do(V = v)) \in p_*$:

(i) $p(x \mid do(V = v))$ is Markov relative to G;

(ii) $p(x_i \mid do(V = v)) = 1$ for all $X_i \in V$ whenever x_i is consistent with $V = v$, and is otherwise zero;

(iii) $p(x_i \mid pa_i, do(V = v)) = p(x_i \mid pa_i)$ for all $X_i \notin V$ whenever pa_i is consistent with $V = v$, and is otherwise zero.

Causal Bayesian networks embody fierce assumptions. However there are many practical contexts, like the one above when this elaboration of a BN of an observational study is plausible. In such a context the CBN encodes a large set of assertions efficiently in the form of a single graph. The formula in Definition 1 imply that the distribution $p_v(x)$ resulting from an intervention $do(V = v)$ will result the truncated factorisation

$$p(x \mid do(V = v)) = \prod_{\{i \mid x_i \notin V\}} p(x_i \mid pa_i) \quad \forall \, x \text{ consistent with } v,$$

and

$$p(x \mid do(V = v)) = 0, \quad \text{x not consistent with v.}$$

When G is a CBN with respect to p_*, the following two properties must hold.

Property 1 For all i,

$$p(x_i \mid pa_i) = p(x_i \mid do(PA_i = pa_i))$$

Property 2 For all i and for every subset S of variables disjoint for $\{X_i, Pa(i)\}$, we have

$$p(x_i \mid do(S = s, PA_i = pa_i)) = p(x_i \mid PA_i = pa_i)$$

where PA_i denote to parent set of its child X_i.

Now although Pearl does not focus on this issue, in practice, the conditional probabilities in a BN usually need to be estimated (e.g., see Cowell et al (1999) and Jordan (1999)). It is therefore natural, from a Bayesian perspective, to ask what constraints, if any, need to be imposed on the prior probabilities on the idle system to ensure that its marginal mass function is consistent with the CBN hypothesis. One property that one might want to demand is that, whether we were to learn the value of a probability θ from some extraneous source or set this probability to a fixed values: e.g., by randomisation, it should be legitimate to 'plug-in' its value: i.e., to *manipulate* the value of θ to its known value and retain the integrity of the BN.

To make a connection between prior independence and causality, it is first necessary to strengthen slightly the assumptions of factorisation invariance under manipulation which induces randomised intervention. This set of assumptions is given in Daneshkhah and Smith (2003) (Hereafter by DS (2003)). In section 2, we introduce the Hypercausal Bayesian Network (HCBN) that asserts a set of factorisations of densities which are invariant to a class of "*do*" operations larger than those considered by Pearl. This requires us to develop the ideas of Koster (2000) and Lauritzen (2001) about randomised intervention. We show that if a BN is assumed to be Hypercausal and we wish to learn about the probabilities defining it, then the prior distributions on the probabilities of the idle system (i.e., the BN without intervention) must exhibit local and global parameter independence. On the other hand if we assume that the idle system exhibits local and global independence then it is extendable, in a natural way, to a HCBN and in particular to a CBN.

Two BNs are Markov equivalent if and only if their pattern (a mixed graph) (Verma and Pearl (1990, 1992)) or equivalently their essential graphs (also a mixed graph) (Anderson et al (1997)) agree. It follows that BNs with the same essential graphs will be indistinguishable from each other from an observational cross sectional study of an idle system. Some of the earliest algorithms for deducing causality from BNs (e.g., Verma and Pearl (1990, 1992) and Chickering (1995)) were based on the configurations of directed edges in essential graphs empirically fitted to exhaustive data sets from a cross sectional experiment. The directed edges of these mixed graphs deduced from cross sectional data allowed them to make causal assertions about what might happen were the system to be manipulated.

Now suppose, on the basis of observations of analogous idle system to the one under study, a Bayesian is confident in asserting a particular essential graph H as a valid hypothesis for another analogous unmanipulated system. The components of conditional probabilities of this new system are, however, uncertain and the researcher wants to make prior assumptions which are consistent with *every* HCBN consistent with the given essential graph. How can this be achieved?

In section 3 we define a concept of multicausality which asserts hypercausality for all BNs in the equivalence class of an essential graph. We show that the prior density on these probability parameters must satisfy a generalisation of the Geiger and Heckerman (GH) condition (see Theorem 2 in Geiger and Heckerman (1997)). In the special case when the essential graph is undirected, this family degenerates into the Hyper-Dirichlet family (see Dawid and Lauritzen (1993)). We conclude the paper by discussing the interpretation and implication of using priors of this form in BNs.

2 Relationships Between Causality And Parameter Independence

Let the vector $X = (X_1, X_2, \ldots, X_k)$, of nodes of a BN G have its components X_i, $1 \leq i \leq k$, listed in an order compatible with G and their corresponding vectors of probabilities $\underline{\theta}_1, \ldots, \underline{\theta}_k$ compatibly with the partial order induced by the directed edges of the BN. Thus the parameters are listed: $\underline{\theta}_1, \ldots, \underline{\theta}_k$, where $\underline{\theta}_i = \{\theta_{i|pa_{i(l)}, 1 \leq l \leq m_i}\}$. The components of $\underline{\theta}_i$ are taken in some arbitrary but fixed order within the vector. From the familiar rules of probability, we can write the general prior distribution for our study as

$$p(\underline{\theta}) = \prod_{i=1}^{k} p(\underline{\theta}_i \mid \underline{\theta}^{i-1}), \quad \underline{\theta}^{i-1} = \{\underline{\theta}_1, \ldots, \underline{\theta}_{i-1}\}$$

Let $\underline{\theta}_A$ represent the subset of $\underline{\theta}$ whose indices $i \in A$, A is a subset of $\{1, \ldots, k\}$. Here k is the total number of conditional probabilities needed to define G, or equivalently the number of components of $\underline{\theta}$. Let us consider each component of $\underline{\theta}_i$ as

$$\theta_{i(j)|pa_{i(l)}} = p(X_i = x_{i(j)} \mid PA_i = pa_{i(l)})$$

where $\theta_{i(j)|pa_{i(l)}}$ denotes the parameter associated with the level j of ith variable and the level l of its parents. Thus,

$$\underline{\theta}_i = \{\theta_{i(j)|pa_{i(l)}}, \ 1 \le j \le n_i, \ 1 \le l \le m_i\}$$

where the each component of $\underline{\theta}_i$ is positive and for the fixed l (the fixed level of parent or for the components in the same strata), $\sum_{j=1}^{n_i} \theta_{i(j)|pa_{i(l)}} = 1$. Here n_i and m_i denote the numbers of the states of the ith variable and its parents set, respectively.

The vector

$$\underline{\theta} = \{\theta_{i|pa_{i(l)}} : 1 \le l \le m_i, 1 \le i \le k\}$$

is said to exhibit *local* and *global independence* if where

$$\underline{\theta}_{i|pa_{i(l)}} = (\theta_{i(1)|pa_{i(l)}}, \ldots, \theta_{i(n_i)|pa_{i(l)}})$$

are all mutually independent (see Spiegelhalter and Lauritzen (1990) for details).

By the example below, we clarify the notations and concepts described above.

Example 2 Consider again the CBN shown in Figure (1) where the mentioned categories of values (in Example 1) of maximum temperature of cooling system in an hour (T), i.e., $(0^0 - 100^0, 100^0 - 200^0, 200^0 - 300^0, 400^0+)$ can be coded to 0, 1, 2, 3, respectively. Similarly, the values of C, that is, normal, critical and Meltdown are coded to 0, 1, 2, 3 respectively. Finally, F is a binary variable that $F = 1$ indicates the failure of cooling system in an hour. Let us rename C, T and F by X_1, X_2 and X_3 respectively. Then, the parameters associated with F is given by $\underline{\theta}_3 = (\theta_{3(0)|0}, \ldots, \theta_{3(0)|3}, \theta_{3(1)|0}, \ldots, \theta_{3(0)|3})$ (such that $\sum_{i=0}^{1} \theta_{3(i)|l} = 1$), where, for example, $\theta_{3(1)|3} = p(F = 1 \mid T = 3)$ and so on. If $\underline{\theta}_1, \underline{\theta}_2, \underline{\theta}_3$ are mutually independent of each other, then we say the parameters associated with each node of the BN above are globally independent. Furthermore, if $(\theta_{3(0)|0}, \ldots, \theta_{3(0)|3}, \theta_{3(1)|0}, \ldots, \theta_{3(0)|3})$ are mutually independent of each other, then we say that the parameters associated with Y are locally independent.

2.1   Randomisation and Cause

Although Pearl focuses on deducing causal relationships from observational studies, traditionally *causal* effects have been more usually investigated using randomised trials in designed experiments. Thus, for example, to investigate the efficacy of a medical treatment A over an alternative treatment B, we would typically manipulate the system by randomising the allocation of the treatments. Similarly, in the example above, the failure of cooling system might be investigated by a series of randomised trials monitoring failures within a range of temperatures. It is therefore natural to include such manipulations in any discussion of causation.

Definition 2   The contingent randomised intervention, $do(\underline{\theta}_A = \underline{\theta}_A^*)$ on a BN G whenever the contingent $pa_{i(l)}$ arises, manipulates X_i to a value $x_{i(j)}$ according to the set randomising probabilities

$$\theta_{i(j)|pa_{i(l)}}^* = p(X_i = \hat{x}_{i(j)} \mid PA_i = pa_{i(l)}) \quad \text{for each } i \in A.$$

When several interventions are employed simultaneously the effect of the manipulation is calculated in an order compatible with G.

A default choice for predicting the effect of a contingent manipulation conditional on the probability vector $\underline{\theta}$ is to use the following definition which extends, in an obvious way the definition of Spirtes et al (1993) and Pearl (2000).

Definition 3   A Bayesian network is said to be *Contingently Causal* (CCBN), if under the contingent manipulation

$$do(X_i = \hat{x}_{i(j)} \mid PA_i = pa_{i(l)}) = do(\theta_{i(j)|pa_{i(l)}} = \theta_{i(j)|pa_{i(l)}}^*),$$

for all configurations of X consistent with $\{X_i = \hat{x}_{i(j)} \mid PA_i = pa_{i(l)}\}$ and $pa_{i(l)}$, the joint mass function after this manipulation of the other variables follow the formula,

$$p(x \mid \theta_{\hat{\imath}}, do(\theta_{i(j)|pa_{i(l)}} = \theta_{i(j)|pa_{i(l)}}^*)) = \{ \prod_{v=1,k,v\neq i} \underline{\theta}_v \} \times \theta_{i(j)|pa_{i(l)}}^*$$

For all configurations, $pa_{i(l)}$ not consistent with $\{X_i = \hat{x}_{i(j)} \mid PA_i = pa_{i(l)}\}$, then we set

$$p(x \mid \underline{\theta}_{\hat{\imath}}, do(\theta_{i(j)|pa_{i(l)}} = \theta_{i(j)|pa_{i(l)}}^*)) = 0.$$

Here we have let $\underline{\theta}_{\hat{\imath}}$ denotes the $\underline{\theta}$ vector with the ith component missing.

Clearly, if we plan to randomise over A using $\underline{\theta}_A^*$ then the randomisation should not influence other parameters in the system (See DS (2003)).

Definition 4   Call a contingent randomised intervention $do(\underline{\theta}_A = \underline{\theta}_A^*)$ on an uncertain BN, *Bayes faithful* if

$$p(\underline{\theta}_{\bar{A}} \mid do(\underline{\theta}_A = \underline{\theta}_A^*)) = p(\underline{\theta}_{\bar{A}})$$

where $\bar{A}$ stands for the complement of A in $\{1, \ldots, k\}$, and $p(\underline{\theta}_{\bar{A}})$ is the Bayesian's prior marginal density on $\underline{\theta}_{\bar{A}}$.

2.2   Hypercausality and Randomisation

We are now ready to define hypercausality. Effectively, this takes the *extended* BN -i.e., one that includes parameters in the BN as if they are random variables as in Figure 1-and demands causal consistently with this BN as well as the original BN. Let $A_u = \{1, \ldots, u\}$, $1 \leq u \leq k$. In terms of the constructions above we have

Definition 5   Say an uncertain BN is a *Hypercausal BN* (HCBN) if it is a CCBN and for all Bayes faithful contingent intervention $do(\underline{\theta}_{A_u} = \hat{\underline{\theta}}_{A_u})$, $1 \leq u \leq k$ defined above, $p(\underline{\theta}_{\bar{A}_u} \mid do(\underline{\theta}_{A_u} = \hat{\underline{\theta}}_{A_u})) = p(\underline{\theta}_{\bar{A}_u} \mid \hat{\underline{\theta}}_{A_u})$. Where $\bar{A}_u$ denotes the complement of A_u.

Pearl (2000) focused on the definition of intervention $do(X_i = \hat{x}_i)$ for BNs with known probabilities. This can be thought of as a degenerate form of contingent randomised intervention on $\hat{\theta}_i$ on a CCBN as

$$\hat{\theta}_{i(j)|pa_{i(l)}} = \begin{cases} 1 \text{ if } x_{i(j)} = \hat{x}_i \text{ for all j and l} \\ 0 \text{ otherwise} \end{cases}$$

Koster (2000) has given a generalisation of this definition, setting

$$\hat{\theta}_{i(j)|pa_{i(l)}} = \begin{cases} \theta_i^*(x_i) \text{ if } x_{i(j)} = \hat{x}_i \text{ for all j and l} \\ 0 \qquad \text{otherwise} \end{cases}$$

It is easily checked that his intervention formula under CCBN (and hence CBN) and the Bayes faithfulness assumptions coincides with ours, conditional on $\hat{\theta}_i$.

However there are examples (see DS (2003)) when we may want to use different randomisations for different configurations of parents of X_j and there is absolutely no reason within this framework not to extend his definition to include this case.

Before we can define an uncertain analogue of a CBN, we first need to define unambiguously what is meant by "observing the probability θ_i". It is most natural to follow Pearl and Example 3 and to define this conditioning as on the 'perfect' estimate of a particular conditional probability in the BN obtained as a limiting proportion in an auxiliary sample from the same sample space.

So assume a selection mechanism, acting on a random sequence $\{X_{j(i)}[t]\}_{t\geq 1}$ of observations respecting the idle BN, which records $\{X_{j(i)}[t]\}_{t\geq 1}$ only for the m parent $pa_{j(i)}$ configuration values associated with indices $i \in A$, where m is the number of the components of A.

Call the selected subsequence $\{w_A(s)\}_{s\geq 1}$, where $w_A(s) = \{w_i(s_i) : i \in A\}$, where s_i indexes the s^{st} observation of the ith variable. We can now define

$$p(\theta_{\bar{A}} \mid \theta_A) = \lim_{r \to \infty} p(\theta_{\bar{A}} \mid \{w_A(s)\}_{1\leq s\leq r})$$

where $inf s_i \to \infty$ as $r \to \infty$.

It is easily checked that from the Bayesian viewpoint this will give us the standard formula for the conditional probability $p(\theta_{\bar{A}} \mid \theta_A)$.

This rather technical definition above allows us to examine more closely what assumptions on HCBN model encodes. Consider the following example.

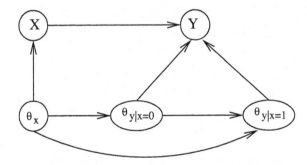

Fig. 2. The BN with the dependent parameters of Example 3

Example 3 Consider the extended BN above on two binary random variables X and Y, where the parameters θ_x, $\theta_{y|x=1}$ and $\theta_{y|x=0}$ are dependent. This dependency implies, for example, that if we learn the value of θ_x through, for example, by observing an enormous sample of units where we record the X-margin only - then or assessment of the probability $\theta_{y|x=1} = p(Y = 1 \mid X = 1)$ will, in general, change. This might occur if, for example, a higher than expected value for θ_x is associated with a 'bad scenario' which we believe would lead us to expect $\theta_{x|y=1}$ would be larger than expected as well.

On the other hand, if we were to manipulate that system to take a randomised sample on X so that $p(X = 1) = \theta_x$, then by definition this *randomisation* should leave $\theta_{y|x=1}$ unchanged in the idle system. So if there is a prior dependence between θ_x and $\theta_{y|x=1}$, then we should not expect to be able to identify a randomly manipulated system from one learnt from an idle system. This model would then not be an HCBN. This argument can be extended to all combinations of values of these parameters, and suggests a close relationship between the HCBN model assumption and local and global independence.

Note that the reason we needed to introduce contingent randomisation was to be able to consider the separate manipulation of all components θ_i of the probability vector $\underline{\theta}$ to values other than zero and one. In this way we are able to perform all necessary manipulations of each of the components of the vector of probabilities in the extended BN. This heuristic argument motivates the following theorem.

Theorem 1 G is an HCBN if and only if it exhibits local and global independence.

Proof By the definition of Bayes faithfulness on this CCBN

$$p(\underline{\theta}_{\bar{A}_u} \mid do(\underline{\theta}_{A_u} = \hat{\underline{\theta}}_{A_u})) = p(\hat{\underline{\theta}}_{A_u})$$

so by the definition of an HCBN, equivalently, the probabilities in the idle system satisfy

$$p(\underline{\theta}_{\bar{A}_u}) = p(\underline{\theta}_{\bar{A}_u} \mid \hat{\underline{\theta}}_{A_u}), \quad 1 \le u \le k$$

Hence by definition of A_u $\Longleftrightarrow$

$$\underline{\theta}_k \perp \underline{\theta}_1, \ldots, \underline{\theta}_{u-1}, \quad 1 \le u \le k-1 \iff \perp_{i=1}^k \underline{\theta}_i, \tag{1}$$

$\Longleftrightarrow$ G exhibits local and global independence.

Corollary 1 An HCBN exhibits the property that for all Bayes faithful contingent interventions on $\underline{\theta}_A, A \subseteq \{1, \ldots, k\}$,

$$p(\underline{\theta}_{\bar{A}} \mid do(\underline{\theta}_A = \hat{\underline{\theta}}_A)) = p(\underline{\theta}_{\bar{A}} \mid \hat{\underline{\theta}}_A))$$

Proof This follows trivially from equation (1).

When estimating models, local and global independence is assumed almost universally- see e.g., Geiger and Heckerman (1997), Cowell et al (1999) and Cooper and Yoo (1999). In so doing these implicitly assume such models are at least consistent with HCBNs. The causal interpretation above is therefore one way of checking whether this assumption is appropriate in a practical situation. For example, if the root node X in the BN of Figure 1 concerned the failure of a component I could ask myself whether taking a very large sample of such $X's$ in operational mode and observing their failure rate θ_x

could be expected to affect the system as a whole as if we simulated simply by randomising to failure with probability θ_x artificially. Only if the answer was 'yes' should I proceed with this assumption.

It is iteresting to note that whenever we move between actual and simulated scenarios we almost always implicitly make hypercausal assumptions.

3   The Multicausal Essential Graph

We often want to put prior distributions on *classes* of BNs, particularly those which are Markov equivalent. Having made a link between BN models whose factorisation are causally invariant, we are now able to investigate the causal implications of priors commonly used for such equivalence classes. We follow the suggestion of Anderson et al (1997) and use the essential graph to represent this equivalence class.

Definition 6   The *essential* graph G^* associated with G is a graph with the same skeleton as G, but where an edge is directed in G^* if and only if it occurs as a directed edge with the same orientation in every $G' \in [G]$ (we denote the equivalence class of BNs corresponding to G by [G]). Clearly, all other edges of G^* are undirected and the directed edges in G^* are called the essential arrows of G.

We now relate essential graphs to a class of causal hypotheses as follows:.

Definition 7   An uncertain essential graph, $\mathcal{P}$ is called *multicausal* if the prior on the uncertain probabilities of every BN in the equivalence class corresponding to $\mathcal{P}$, ([$\mathcal{P}$]), is consistent with an HCBN.

An important subclass of essential graphs is one where all its components are undirected. This class is called the *decomposable* equivalence class. It is natural to ask how priors might be set up on the uncertain probabilities in the system in a way which is invariant to equivalent BNs. In particular, what is the family of priors which exhibit local and global independence for every BN compatible with a given essential graph? In 1993, Dawid and Lauritzen proved that, for decomposable essential graphs, the Hyper-Dirichlet family of distributions preserved global independence. Later, Geiger and Hekerman (1997) proved a much stronger result for a two node essential graph. They demonstrated that the two BNs in this equivalence class both exhibited local and global independence if and only if the joint prior distribution on its two nodes was Dirichlet.

Explicitly, let $\{\psi_{ij}, 1 \leq i \leq k, 1 \leq j \leq n\}$, be positive random variables that sum to unity and denote the multinomial parameters in the two way probability table on the two equivalent BNs with two nodes and one edge.

Denote the Dirichlet, $\mathcal{D}(\alpha)$, density on $\underline{\theta}$ by

$$p(\underline{\theta}) = \frac{\Gamma(\alpha)}{\prod_{i=1}^{k} \Gamma(\alpha_i)} \prod_{i=1}^{k} \theta_i^{\alpha_i - 1}$$

where $\alpha = \Sigma_{i=1}^{k} \alpha_i$.

Geiger and Heckerman (1997) proved that for the priors on both BNs to exhibit local and global independence, it is necessary and sufficient that , using the obvious notation, the prior distributions on ψ_{ij}, $\theta_{i\cdot}$, $\theta_{j|i}$, $\theta_{\cdot j}$ and $\theta_{i|j}$ are all Dirichlet. Furthermore the parameters of their respective densities $\mathcal{D}(\alpha_{ij})$, $\mathcal{D}(\alpha_{1\cdot}, \dots, \alpha_{k\cdot})$, $\mathcal{D}(\alpha_{j|1}, \dots, \alpha_{j|k})$, $\mathcal{D}(\alpha_{\cdot 1}, \dots, \alpha_{\cdot n})$ and $\mathcal{D}(\alpha_{i|1}, \dots, \alpha_{i|n})$, would need to satisfy the further linear equations, $\alpha_{i\cdot} = \Sigma_{j=1}^{n} \alpha_{ij}$, $\alpha_{\cdot j} = \Sigma_{i=1}^{k} \alpha_{ij}$ and $\alpha_{i|j} = \alpha_{j|i} = \alpha_{ij}$.

We have seen above that there is a direct link between local and global independence and hypercausal hypotheses. Such classes of local and global independence statements must therefore correspond to several simultaneous causal hypotheses made by the Bayesian within the equivalence class of BNs in the posited essential graph.

Although the GH condition above is only valid for graphs with two variables, it is straightforward, using induction, to extend the GH condition to characterise complete (and hence undirected) essential graphs H, all of whose BNs exhibit local and global independence: see the Lemma below

Lemma 1 If the prior distributions of all BNs consistent with a complete essential graph H with n variables $(X_1, \dots, X_n)$ exhibit local and global independence, then the prior density on the joint probabilities $\underline{\theta}^{(n)}$ of $(X_1, \dots, X_n)$ in H must be Dirichlet, $n \geq 2$.

Proof Go by induction on the number k of variables in H. The assertion is clearly true for $k = 2$ by the GH condition. Suppose it is true for $k = n - 1$ and write $X^{(r)} = \{X_1, \dots, X_r\}$, $2 \leq r \leq n$. The essential graph H' of Figure 2 is valid, where, by the inductive hypothesis, the prior density on the joint probabilities $\underline{\theta}^{(n-1)}$ of $X^{(n-1)}$ is Dirichlet. Since the BN which introduces X_n first and the one that introduces X_n last both exhibit local and global independence, it follows that the two BNs associated with H' also exhibit local and global independence in their probabilities. The GH result now allows us to assert that $\underline{\theta}_n$ is Dirichlet. The well known properties of the Dirichlet now allow us to complete the inductive step. and so complete the inductive step.

Example 4 There are several BNs consistent with the following essential graph that is shown in the Figure 4. The edge between X_1 and X_2 could be in either directed, and there are 6 configurations of directions of arrows on the triangle of nodes (X_5, X_6, X_7). So this essential graph has an

Fig. 3. The essential graph H'

associated equivalence class of 12 BNs. For example , let us consider two BNs associated with undirected edge between nodes X_5 and X_6 regardless of the direction of other undirected edges in this essential graph. Since we assume multicausality, these two BNs (one with edge $X_5 \longrightarrow X_6$ and another one with edge $X_5 \longleftarrow X_6$) exhibit local and global independence. We now note that if we condition on any value of X_4 (the shared parent of X_5 and X_6), the conditional prior distributions on X_5 and X_6 exhibit local and global independence whichever way around we condition X_5 and X_6. Therefore, the GH condition must hold on X_5 and X_6 conditional on each value of X_4. Indeed if we have this condition then it is easily checked that with local and global independence on other nodes we must have multicausality.

We can claim the same thing on the BNs that are consistent with the essen-

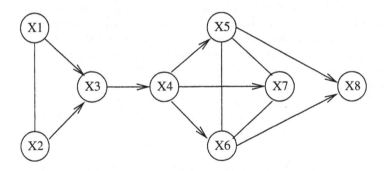

Fig. 4. The multicausal essential graph of Example 4

tial graph shown in the figure above one with edge $X_1 \longrightarrow X_2$ and another one with $X_1 \longleftarrow X_2$ (without any shared parents).

This motivates the following lemma and theorem.

Lemma 2   The nodes of an undirected component subgraph of an essential graph H all share the same (directed) parents in H.

Proof   This is a direct consequences of lemma 1 in Chickering (1995).

Definition 8   The *undirected cliques* of an essential graph H are the maximally connected subsets of nodes/variables of H, all connected to one another by undirected edges.

Theorem 2 let H be an essential graph. Then H is multicausal if and only if, for any particular BN G in the equivalence class defined by H,
(i) The probability vector $\underline{\theta}$ of G exhibits local and global independence.
(ii) The densities of the joint probabilities of the undirected cliques of H are Dirichlet conditional on each value of their (shared and directed) parent configurations in H.

Proof Suppose H is multicausal. Condition (i) is implied by the definition. Choose an undirected clique C in H and condition on a value of its (directed) parent set-which by lemma 2 is shared by nodes in C. Given this value of the parents, by the definition of H, the complete essential graph on the nodes in C is valid, and the probabilities on these nodes exhibit local and global independence for all BNs consistent with it. Condition (ii) now follows directly from lemma 1.

On the other hand, suppose H satisfies condition (i) and (ii) and let G' be any BN associated with H. According to lemma 3.2 in Andersson et al (1997), for any two BNs G and G' associated with H, there exists a finite sequence $G = D_1, \ldots, D_n = G'$ of BNs, all in the equivalence class described by H, and where D_i and D_{i+1}, $1 \leq i \leq n - 1$, differ in exactly one edge direction. It is therefore sufficient to prove that if G exhibits local and global independence and condition (ii) is met then G', differing by only one edge, also exhibits local and global independence.

So assume that G' is obtained from G by reversing the edge (X_i, X_j) in G. Since both G and G' are associated with H, (X_i, X_j) must be an undirected edge in H and so lie in one of its undirected cliques. By lemma 2, (X_i, X_j) share (directed) parents. Note the elementary property that if the joint probabilities of $X \in C$ have a Dirichlet density then the joint probabilities of $X \in C_1 \subseteq C$ also have a Dirichlet density. So by (ii) conditional on any value of their shared parents the density of the joint probabilities of (X_i, X_j) is Dirichlet. It follows from GH condition that G' must also exhibit local and global independence. This completes the theorem.

Consider a decomposable BN, M for a set of random variables $\mathcal{X} = (X_i, i \in \mathcal{I})$, i.e. one whose essential graph is undirected. Let $\mathcal{V} = \prod_{i \in \mathcal{I}} \mathcal{X}_i$ denote the set of possible configurations of $\mathcal{X}$. Denote by $\theta(v)$ the probability of a state $v \in \mathcal{V}$. Then $\theta(v)$ is determined by the clique marginal probability tables $\theta_C, C \in \mathcal{C}$ where $\mathcal{C}$ denotes the set of cliques of M:

$$\theta(v) = \frac{\prod_{C \in \mathcal{C}} \theta_C(v_C)}{\prod_{S \in \mathcal{S}} \theta_S(v_S)}, v \in \mathcal{V}$$

Here $\mathcal{S}$ denotes the system of clique separators in an arbitrary perfect ordering of $\mathcal{C}$.

For each clique $C \in \mathcal{C}$, let $\mathcal{D}(\alpha_C)$ denote the Dirichlet distribution for θ_C with density

$$\pi(\theta_C \mid \alpha_C) \propto \prod_{v_C \in \mathcal{V}_C} \theta_C(v_C)^{\alpha_C(v_C)-1}$$

where $\alpha_C(v_C) > 0$ for all $v_C \in \mathcal{V}_C$.

Now let us suppose that the collection of specifications $\mathcal{D}(\alpha_C)$, $C \in \mathcal{C}$ are constructed in such a way that for any two cliques C and D in $\mathcal{C}$ we have:

$$\alpha_C(v_{C \cap D}) = \alpha_D(v_{C \cap D}) \tag{2}$$

that is, if the cliques C and D overlap, then the parameters α_C and α_D are such that each implies the same marginal distribution for $\theta_{C \cap D}$.

Definition 9   For the prefect ordering of cliques $\alpha_{\mathcal{C}} = \{\alpha_C\}_{C \in \mathcal{C}}$ there exists a unique Dirichlet distribution for $\underline{\theta} = \{\theta_C\}_{C \in \mathcal{C}}$, that is called Hyper Dirichlet and denoted by $\mathcal{HD}(\alpha_C)$, which is hyper Markov over M and the distribution on each clique $C \in \mathcal{C}$ is $\mathcal{D}(\alpha_C)$.

Note that, it has been shown in Theorem 3.9 in Dawid and Lauritzen (1993) that there exists a unique "Hyper-Dirichlet" distribution for $\underline{\theta}$ over M such that θ_C has the marginal density $\mathcal{D}(\alpha_C)$ for all $C \in \mathcal{C}$ [1].

We can now state a corollary of Theorem 2.

Corollary 2   If H is undirected then it is multicausal if and only if the clique margins have a Hyper-Dirichlet distribution.

Proof   If H is undirected then the BN is decomposable. Because all the clique margins are (consistently) Dirichlet, the result now follows directly from the definition of the Hyper-Dirichlet distribution above.

4   Discussion

Because of the pioneering works of Robins (1986) and Spirtes et al (1993), recent studies of causality have mainly focussed on how to deduce causal

---

[1] In practice, one would construct a hyper-Dirichlet distribution by first identifying a perfect ordering of the cliques, for example, $\{C_1, \ldots, C_n\}$. Place a Dirichlet distribution $\mathcal{D}(\alpha_{C_1})$ on θ_{C_1}, next place a Dirichlet distribution $\mathcal{D}(\alpha_{C_2})$ on θ_{C_2}, with parameters constructed by (2) and realizations constrained so that $\theta_{C_1 \cap C_2}$ is identical for θ_{C_1} and θ_{C_2}. For each subsequent clique C_i, place a Dirichlet distribution on θ_{C_i} such that the parameters and the realizations of that distribution are consistent with those specified for the previous cliques.

relationships from observational studies. But, causal assertions have traditionally been studied through manipulating treatments in randomised trials
(see Smith (2002)). Components of a BN model are therefore usually most
strongly supported by the sort of randomised contingent experiments we discuss in this paper. From the subjectivist viewpoint, we contend that the bold
part of the hypercausal hypothesis is that the Bayesian asserts that the process in the field will behave in the same way as it did under (analogues)
carefully controlled randomised trials (not vice versa). This hypothesis will
usually fail when, extraneous dependencies get introduced through not being
able to control conditions in the field. Hypercausal priors (which demand
identity of these two cases) allows us to think about this problem the right
way round.

When data is not exhaustive it is common for idle systems to exhibit
extraneous dependencies not linked with science but with paucity of information. These dependencies in the idle system can be induced by misspecified
priors, trial sample data sets on non-ancestral subsets of variables, selection
variables and so on, quite spurious for any assessment of causality, but intrinsic for learning about the idle system.
In our opinion it is these issues which make causal deductions from uncertain
idle system so prone to mislead.

Moving to multicausal models, we have established that if a Bayesian is
prepared to make bold enough causal assertions within a single uncertain BN
then this not only introduces independence relationships between parameters,
but can also characterise prior families of distributions on these parameters.
We believe this is a very helpful way of thinking about this class of models.
Note that this multicausal essential graph characterisation concerns a single
hypothesised model. It is not an assertion about a common prior to be used
for CBN for the model selection as is more typical in, for example see Geiger
and Heckerman (1997), Cowell et al (1999) and Cooper and Yoo (1999) and
references therein.

How strong an assumption is multicausality, in learnt BNs? Although
this assumption is almost universally made in practice, we would argue that
this corresponds to a very unusual circumstance and demands very specific
structures on prior information before it is valid. To illustrate this, consider
supplementing the BN of Example 4 with experimental evidence observing Y
after we have randomised on X and conditioned on this value. The essential
graph is $X - Y$ so that the Hyperessential prior is just a Hyper-Dirichlet
prior which sets the joint density of $\psi = (\psi_{00}, \psi_{01}, \psi_{10}, \psi_{11})$,
where $\psi_{ij} = p(X = i, Y = j)$, $i, j = 0, 1$, having a Dirichlet density $p(\psi \mid \alpha)$
given by

$$p(\psi \mid \alpha) = K \psi_{00}^{\alpha_{00}-1} \psi_{01}^{\alpha_{01}-1} \psi_{10}^{\alpha_{10}-1} \psi_{11}^{\alpha_{11}-1}$$

where K denotes to the normalising constant. Furthermore, θ_x and $\underline{\theta}_{y|x}$ are distributed respectively as

$$p(\theta_x) = K_1(1 - \theta_x)^{\alpha_{0.} - 1}\theta_x^{\alpha_{1.} - 1}$$

and

$$p(\underline{\theta}_{y|x}) = K_2\theta_{y|x}^{\alpha_{11}-1}\theta_{y|\bar{x}}^{\alpha_{01}-1}\theta_{\bar{y}|x}^{\alpha_{10}-1}\theta_{\bar{y}|\bar{x}}^{\alpha_{00}-1}$$

similarly, we have

$$p(\theta_y) = K_3(1 - \theta_y)^{\alpha_{.0}-1}\theta_y^{\alpha_{.1}-1}$$

and

$$p(\underline{\theta}_{x|y}) = K_4\theta_{x|y}^{\alpha_{11}-1}\theta_{x|\bar{y}}^{\alpha_{10}-1}\theta_{\bar{x}|y}^{\alpha_{01}-1}\theta_{\bar{x}|\bar{y}}^{\alpha_{00}-1}$$

where $\alpha_{i.} = \alpha_{i0} + \alpha_{i1}$, $\alpha_{.j} = \alpha_{0j} + \alpha_{1j}$, $i, j = 0, 1$ and K_l, $l = 1, \ldots, 4$, denotes to the normalising constants. Note that, in the distributions above, $\theta_x \perp \theta_{y|x}$ and $\theta_y \perp \theta_{x|y}$. This supplementing experimental evidence updates $\theta_{y|x}$ retains θ_x and keeps $\theta_x \perp \theta_{y|x}$. However, because the GH condition is no longer a posteriori, θ_y is no longer independent of $\theta_{x|y}$, the posterior density is not Hyperessential and the new prior is not multicausal. So information other than data equivalent to the direct observation in the idle system of joint margins of (X, Y) will prevent us identifying the manipulated and the idle system, implicit in the multicausal assumptions.

Note that the data affects the estimation of the idle margin of (X, Y) and this prevents the model from being an HCBN. As we discussed above, it is therefore the idle system drifting away from the more scientific manipulated system- not vice versa !
Estimation of the probabilities in the idle model has introduced dependencies in the marginal BN due solely to the estimation process and spurious with respect to the mechanisms in the model. In particular, once the data set used to estimate the probabilities is large and exhaustive the model probabilities will be (almost) known and the model therefore (almost) hypercausal. These types of limiting results and approximate Hypercausality are discussed in more details in DS (2003).

This feature will be exacerbated by the fact that evidence about effects -perhaps symptoms- (in this case Y) often being more accessible than evidence about causes -perhaps diseases- (in this cause X). So a spurious causal directionality in the BN from the idle system might thus be deduced just because of the form of the data or information we have and in this case reversed!

To conclude: hypercausal BNs will tend to be rare and multicausal BNs even more so. In more structured problems observational studies they can be expected to give many spurious indications of causal direction when parameters are being estimated. However the causal framework developed in this paper at least provides a vehicle through which to seriously discuss some of the more basic inferential consequences of priors assuming local and global independence on the probabilities in a BN.

References

1. Andersson, S. A., Madigan, D., and Perlman, M. D. (1997) A characterisation of Markov equivalence classes for acyclic. *The Annals of Statistics,* **24**, 505–541.
2. Chickering, D. M. (1995) A transformational characterisation of Bayesian network structures. In P. Besnard and S. Hanks (eds.), *Uncertainty in Artificial Intelligence,* **11**, 87–98. San Francisco: Morgan Kaufmann.
3. Cooper, G. F., and Yoo, C. (1999) Causal Discovery from a Mixture of Experimental and Observational Data. In K.B. Laskey and H. Prade (eds.),*Proceeding of the Fifteenth Conference on Uncertainty in Artificial Interlligence.* Morgan Kaufmann Publishers, San Francisco.
4. Cowell, R. G., Dawid, A. P., Lauritzen, S. L., and Spiegelhalter, D. J. (1999) *Probabilistic Networks and Expert Systems.* Springer Verlag.
5. Daneshkhah, A. R., and Smith, J. Q. (2003) A Relationship between randomised Manipulation and Parameter Independenceks. To be appear in J. M. Bernardo, M. J. Bayarri, J. O. Berger, A. P. Dawid, D. Heckerman, A. F. M, Smith and M. West (eds.), *Bayesian Statistics 7.* Oxford University Press.
6. Dawid, A. P., and Lauritzen, S. L. (1993) Hyper Markov laws in the statistical analysis of decomposable graphical models. *The Annals of Statistics* ,**21**, 1272–1317.
7. Geiger, D., and Heckerman, D. (1997) A characterisation of the Dirichlet distribution through global and local parameter independence. *The Annals of Statistics,* **25**, 1344–1369.
8. Jordan, M. I. (1999) *Learning in Graphical Models.* The MIT Press.
9. Koster, J. T. A. (2000) Graphs, Causality and Structural Equation Models. The slides of the presentation is available from WWW URL: http://www.knaw.nl/09public/rm/ koster.pdf.
10. Lauritzen, S. L. (2001) Causal inference from graphical models. In O.E. Barndorff-Nielson, D.R. Cox and C. Kluppelberg (eds.), *Complex stochastic systems,* 63-108. Chapman and Hall/CRC.
11. Pearl, J. (1995) Causal diagrams for experimental research. *Biometrika,* **28**, 669–710.
12. Pearl, J. (2000)*Causality: Models, Reasoning, and Inference.*, Cambridge University Press.
13. Robins, J. M. (1986) A new approach to causal inference in mortality studies with a sustained exposure period-applications to control of the healthy worker survivor effect. *Mathematical Modelling,* **7**, 1393–1512.
14. Smith, J. Q. (2002) Discussion of ' Chain graph model and their causal interpretation', by S. L. Lauritzen and T. Richardson. *J. R. Statist. Soc. B,* **64**.
15. Spiegelhalter, D. J., and Lauritzen, S. L. (1990) Sequential updating of conditional probabilities on directed graphical structures. *Networks,* **20**, 579–605.
16. Spirtes, P., Glymour, C., and Scheines, R. (1993) *Causation, Prediction, and Search.* New York: Springer Verlag.

17. Verma, T., and Pearl, J. (1990) Equivalence and sythesis of causal models. In P. Bonissone, M. Henrion, L. N. Kanal, and J. F. Lemmer (eds.), *Uncertainty in Artificial Intelligence,* **6**, 255–68. Amsterdam: Elsevier.

Interface Verification for Multiagent Probabilistic Inference

Y. Xiang and X. Chen

University of Guelph, Guelph, Ontario, Canada

Abstract. Multiply sectioned Bayesian networks support representation of probabilistic knowledge in multiagent systems. To ensure exact, distributed reasoning, agent interfaces must satisfy the d-sepset condition. Otherwise, the system will behave incorrectly. We present a method that allows agents to verify cooperatively the d-sepset condition through message passing. Each message reveals only partial information on the adjacency of a shared node in an agent's local network. Hence, the method respects agent's privacy, protects agent vendors' know-how, and promotes integration of multiagent systems from independently developed agents.

1 Introduction

As the cost of computers and networking continues to drop and distributed systems are widely deployed, users are expecting more intelligent behaviors from such systems – multiagent systems (MAS) [14]. Agents in an MAS perform a set of tasks depending on the particular application domain. A common task is for a set of cooperative agents to determine what is the current state of the domain so that they can act accordingly. Agents monitoring a piece of equipment need to determine whether the equipment is functioning normally and, if not, which components have failed. Agents populating a smart house should recognize the current need of inhabitants and adjust the appliances accordingly. Similar situations arise in other domains such as cooperative design, battle field assessment, and surveillance. Often agents have only uncertain knowledge about the domain and must perform the task based on partial observations. Such a task has been termed *distributed interpretation* [7] by some authors. We shall refer to it as multiagent situation assessment.

Different approaches have been proposed to tackle multiagent situation assessment. *Blackboard* [10] offers a framework for multiagent inference and cooperation. It does not dictate how uncertain knowledge should be represented nor offers any guarantee of inference coherence. DATMS [8] and DTMS [4] offer inference frameworks based on default reasoning. Relation between BDI model and decision-tree is studied in [12]. Reasoning about the mental state of an agent from the received communication is considered by [2]. Monitoring whether a multiagent system is functioning normally by focusing on agent-relation is investigated in [5]. Emotions of agents are studied using decision theory in [3]. Proving hypotheses by agents with distributed knowledge using dialectical argumentation is proposed in [9]. Multiply sectioned Bayesian networks (MSBNs) [15] provide a framework where agents' knowledge can be encoded with graphical models and agent's belief can be updated by distributed, exact probabilistic reasoning. Multiagent MSBNs (MAMSBNs) are the focus of this work.

Distributed and exact inference requires that an MAMSBN observes a set of constraints [15]. When building an MAMSBN, these constraints on the knowledge representation need to be verified before inference for situation assessment takes place. Otherwise, garbage-in-garbage-out may occur and the resultant MAS will not reason correctly. When agents are autonomous and may be constructed by independent vendors (hence privacy of agents becomes an issue), verification of these constraints raises a challenge. In this work, we study verification of agent interface. We present a method that verifies the correctness of agent interfaces in an MAMSBN without compromising agent autonomy and privacy.

Section 2 briefly overviews MAMSBNs and introduces formal background necessary for the remainder of the paper.

2   Overview of MAMSBNs

A BN [11] S is a triplet (N, G, P), where N is a set of domain variables, G is a DAG whose nodes are labeled by elements of N, and P is a joint probability distribution (jpd) over N. In an MAMSBN, a set of $n > 1$ agents $A_0, ..., A_{n-1}$ populates a *total universe* V of variables. Each A_i has knowledge over a *subdomain* $V_i \subset V$ encoded as a Bayesian subnet (V_i, G_i, P_i). The collection of local DAGs $\{G_i\}$ encodes agents' knowledge of domain dependency. Distributed and exact reasoning requires these local DAGs to satisfy some constraints [15] described below:

Let $G_i = (V_i, E_i)$ $(i = 0, 1)$ be two graphs. The graph $G = (V_0 \cup V_1, E_0 \cup E_1)$ is referred to as the *union* of G_0 and G_1, denoted by $G = G_0 \sqcup G_1$. If each G_i is the subgraph of G spanned by V_i, we say that G is *sectioned* into G_i $(i = 0, 1)$. Local DAGs of an MAMSBN should overlap and be organized into a hypertree.

Definition 1 *Let* $G = (V, E)$ *be a connected graph sectioned into subgraphs* $\{G_i = (V_i, E_i)\}$. *Let the* G_is *be organized as a connected tree* Ψ, *where each node is labeled by a* G_i *and each link between* G_k *and* G_m *is labeled by the interface* $V_k \cap V_m$ *such that for each* i *and* j, $V_i \cap V_j$ *is contained in each subgraph on the path between* G_i *and* G_j *in* Ψ. *Then* Ψ *is a* **hypertree** *over* G. *Each* G_i *is a* **hypernode** *and each interface is a* **hyperlink**.

Each hyperlink serves as the information channel between agents connected and is referred to as an *agent interface*. To allow efficient and exact inference, each hyperlink should render the subdomains connected conditionally independent. It can be shown (by extending results in [15]) that this implies the following structural condition.

Definition 2 *Let G be a directed graph such that a hypertree over G exists. A node x contained in more than one subgraph with its parents $\pi(x)$ in G is a **d−sepnode** if there exists one subgraph that contains $\pi(x)$. An interface I is a **d−sepset** if every $x \in I$ is a d-sepnode.*

The overall structure of an MAMSBN is a hypertree MSDAG:

Definition 3 *A **hypertree MSDAG** $\mathcal{D} = \bigsqcup_i D_i$, where each D_i is a DAG, is a connected DAG such that (1) there exists a hypertree ψ over $\mathcal{D}$, and (2) each hyperlink in ψ is a d-sepset.*

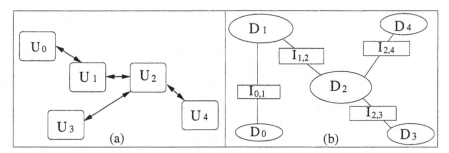

Fig. 1. (a) A digital system of five components. (b) The hypertree modeling.

Table 1. Agent communication interfaces.

Interface	Interface Composition
$I_{0,1}$	$\{a_0, b_0, c_0, e_0, f_0, g_1, g_2, x_3, z_2\}$
$I_{1,2}$	$\{g_7, g_8, g_9, i_0, k_0, n_0, o_0, p_0, q_0, r_0, t_2, y_2, z_4\}$
$I_{2,3}$	$\{a_2, b_2, d_1, d_2, d_3, s_0, u_0, w_0, x_0, y_0, z_0\}$
$I_{2,4}$	$\{e_2, h_2, i_2, j_2, t_4, t_5, t_7, w_2, x_4, y_4, z_5\}$

As a small example, Figure 1 (a) shows a digital system with five components U_i ($i = 0, ..., 4$). Although how components are interfaced, as shown in (a), and the set of interface variables, as shown in Table 1, are known to the system integrator, internal details of each component are proprietary. To give readers a concrete idea on the scenario, a centralized perspective of the digital system is shown in Figure 2.

The subnets for agents A_1 and A_2 are shown in Figures 3 and 4, where each node is labeled by the variable name and an index. The agent interface $I_{1,2}$ between them contains 13 variables and is a d-sepset. For instance, the

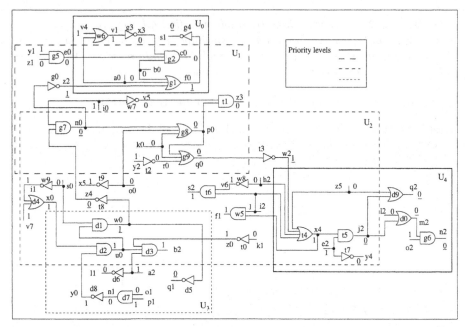

Fig. 2. A digital system.

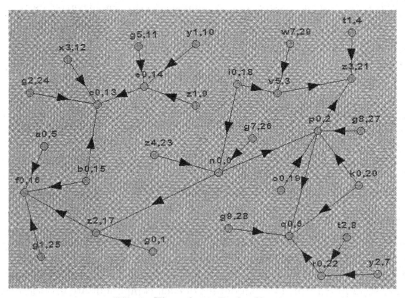

Fig. 3. The subnet D_1 for U_1.

parents of z_4 are all contained in D_2, while those of n_0 are contained in both D_1 and D_2.

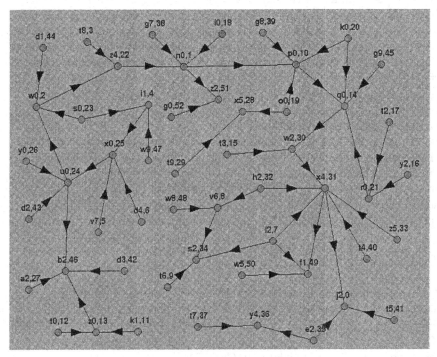

Fig. 4. The subnet D_2 for U_2.

In an MAMSBN integrated from agents from different vendors, no agent has the perspective of Figure 2, nor the simultaneous knowledge of D_1 and D_2. Only the nodes in an agent interface are *public*. All other nodes in a subnet are *private* and known to the corresponding agent only. This forms the constraint of many operations in an MAMSBN, e.g., triangulation [17] and communication [16]. Using these operations, agents can reason about their environment probabilistically based on local observations and limited communication. More formal details on MAMSBNs can be found in references noted above.

3   The Issue of Cooperative Verification

Each agent interface in an MAMSBN should be a d-sepset (Def. 2). When an MAS is integrated from independently developed agents, there is no guarantee that this is the case. Blindly performing MAMSBN operations on the MAS would result in incorrect inference. Hence, agent interfaces need to be verified.

An agent interface is a d-sepset if every public node in the interface is a d-sepnode. However, whether a public node x in an interface I is a d-sepnode cannot be determined by the pair of local graphs interfaced with I. It depends on whether there exists a local DAG that contains all parents $\pi(x)$ of x in G. Any local DAG that shares x may potentially contain some parent nodes of x. Some parent nodes of x are public, but others are private. For agent privacy, it is desirable not to disclose parentship. Hence, we cannot send the parents of x in each agent to a single agent for d-sepnode verification. Cooperation among all agents whose subdomains contain x or parents of x is required to verify whether x is a d-sepnode. We refer to the unverified structure of an MAS as a *hypertree DAG union*.

In presenting our method, we will illustrate using examples. Although MAMSBNs are intended for large problem domains, many issues in this paper can be demonstrated using examples of much smaller scale. Hence, we will do so for both comprehensibility as well as space. Readers should keep in mind that these examples do not reflect the scales to which MAMSBNs are applicable. Due to space limit, proofs for some formal results are omitted.

4   Checking Private Parents

A public node x in a hypertree DAG union G may have public or private parents or both. Three cases regarding its private parents are possible: more than one local DAG (Case 1), exact one local DAG (Case 2), or no local DAG (Case 3) contains private parents of x. The following proposition shows that the d-sepset condition is violated in Case (1).

Proposition 4 *Let a public node x in a hypertree DAG union G be a d-sepnode. Then no more than one local DAG of G contains private parent nodes of x.*

Proof: Assume that two or more local DAGs contain private parent nodes of x. Let y be a private parent of x contained in a local DAG G_i and z be a private parent of x contained in $G_j (i \neq j)$. Then there cannot be any one local DAG that contains both y and z. Hence no local DAG contains all parents of x, and x is not a d-sepnode by Def. 2, which is a contradiction. □

Figure 5 shows how this result can be used to detect non-d-sepnodes. We refer to the corresponding operation as **CollectPrivateParentInfo**. To verify if the public node j is a d-sepnode, suppose that agents perform a rooted message passing (shown by arrows in (a)). Agent A_4 sends a count 1 to A_3, signifying that it has private parents of j. A_3 has no private parents of j. It forms its own count 0, adds the count from A_4 to its own, and sends the

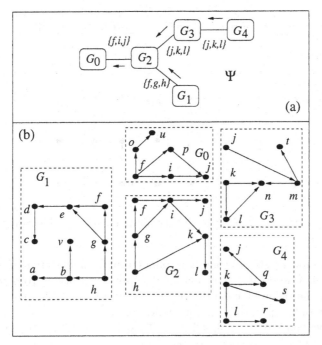

Fig. 5. A hypertree DAG union with the hypertree in (a) and local DAGs in (b).

result 1 to A_2. Because A_1 does not contain j, it does not participate in this operation. Hence, A_2 receives a message only from A_3. Because A_2 has only a public parent i of j, it forms its own count 0, adds the count from A_3 to its own, and sends the result 1 to A_0. Upon receiving the message, A_0 forms its own count 1, for it has a private parent p of j. It adds the count from A_2 to obtain 2 and the message passing halts. The final count signifies that there are two agents which contain private parents of j. Hence, j is a non-d-sepnode and the hypertree DAG union has violated the d-sepset condition.

5 Processing Public Parents

If **CollectPrivateParentInfo** on a public node x results in a final count less than or equal to 1, then no more than one agent contains private parents of x (Cases (2) and (3) above). The hypertree DAG union G, however, may still violate the d-sepset condition. Consider the example in Figure 6. The public nodes are w, x, y, z. No local DAG has any private parent of x or z. Only G_0 has a private parent of y, and only G_2 has a private parent of w. Hence, **CollectPrivateParentInfo** will produce a final count ≤ 1 for each of w, x, y, z. However, no single local DAG contains all parents of x: $\pi(x) = \{w, y\}$. Therefore, x is not a d-sepnode according to Def. 2 and none of the agent interfaces is a d-sepset.

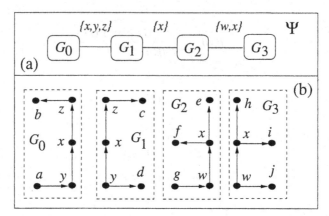

Fig. 6. A hypertree DAG union G with the hypertree in (a) and local DAGs in (b).

The example illustrates that final counts from **CollectPrivateParentInfo** only provide a necessary condition for d-sepset verification. To determine if G satisfies the d-sepset condition conclusively, agents still need to further process the public parents of public nodes.

First, we consider Case 3, where no local DAG contains private parents of x. Case 2 will be considered in Section 6.

5.1   Public parent sequence

We propose the following concept called public parent sequence to describe the distribution of public parents $\pi(x)$ of a public node x on a hyperchain DAG union denoted as $\langle G_0, G_1, ..., G_m \rangle$. We use $X \bowtie Y$ to denote that sets X and Y are incomparable (neither is the subset of the other).

Definition 5 *Let $\langle G_0, G_1, ..., G_m \rangle$ (m $\geq$ 2) be a hyperchain of local DAGs, where x is a public node, each G_i contains either x or some parents of x, and all parents of x are public. Denote the parents of x that G_i (0 < i < m) shares with G_{i-1} and G_{i+1} by $\pi_i^-(x)$ and $\pi_i^+(x)$, respectively. Denote the parents of x that G_m shares with G_{m-1} by $\pi_m^-(x)$. Then the sequence*

$$(\pi_1^-(x), \pi_2^-(x), ..., \pi_m^-(x))$$

*is the **public parent sequence** of x on the hyperchain. The sequence is classified into the following types, where $0 < i < m$:*

Identical *For each i, $\pi_i^-(x) = \pi_i^+(x)$.*
Increasing *For each i, $\pi_i^-(x) \subseteq \pi_i^+(x)$, and there exists i such that $\pi_i^-(x) \subset \pi_i^+(x)$.*
Decreasing *For each i, $\pi_i^-(x) \supseteq \pi_i^+(x)$, and there exists i such that $\pi_i^-(x) \supset \pi_i^+(x)$.*

Concave *One of the following holds:*
1. *For $m \geq 3$, there exists i such that the subsequence $(\pi_1^-(x), ..., \pi_i^-(x))$ is increasing and the subsequence $(\pi_i^-(x), ..., \pi_m^-(x))$ is decreasing.*
2. *There exists i such that $\pi_i^-(x) \bowtie \pi_{i+1}^-(x)$; the preceding subsequence $(\pi_1^-(x), ..., \pi_i^-(x))$ is trivial ($i = 1$), increasing, or identical; and the trailing subsequence $(\pi_{i+1}^-(x), ..., \pi_m^-(x))$ is trivial ($i = m - 1$), decreasing, or identical.*

Wave *One of the following holds:*
1. *There exists i such that $\pi_i^-(x) \supset \pi_i^+(x)$ and $j > i$ such that either $\pi_j^-(x) \subset \pi_j^+(x)$ or $\pi_j^-(x) \bowtie \pi_j^+(x)$.*
2. *There exists i such that $\pi_i^-(x) \bowtie \pi_i^+(x)$ and $j > i$ such that either $\pi_j^-(x) \subset \pi_j^+(x)$ or $\pi_j^-(x) \bowtie \pi_j^+(x)$.*

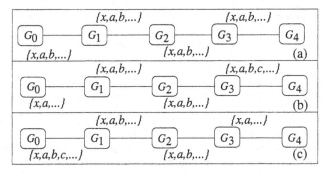

Fig. 7. Public parent sequences. (a) An identical sequence. (b) An increasing sequence. (c) A decreasing sequence.

Figure 7 illustrates the first three sequence types, where only x and its parents are shown explicitly in each agent interface. **Identical** sequence is illustrated in (a). Each G_i contains $\pi(x) = \{a, b\}$, and hence x is a d-sepnode. **Increasing** sequence is exemplified in (b). From $i = 1$ to m, each G_i contains either the identical public parents of x or more. Because G_m contains $\pi(x)$, x is a d-sepnode. **Decreasing** sequence is exemplified in (c). It is symmetric to the increasing sequence; G_0 contains $\pi(x)$ and x is a d-sepnode.

For **Concave** sequence, some parents of x appear in the middle of the hyperchain but not on either end. Figure 8 illustrates two possible cases. In (a), the parent b of x is contained in G_1, G_2, and G_3 but disappears in G_0 and G_4 and c is contained in G_2 and G_3 but disappears in G_0, G_1, and G_4. Two local DAGs (G_2 and G_3) in the middle of the hyperchain contain $\pi(x)$, and hence x is a d-sepnode. In (b), an increasing subsequence ends at $\pi_2^-(x)$, and a decreasing subsequence starts at $\pi_3^-(x)$ with $\pi_2^-(x)$ and $\pi_3^-(x)$ incomparable. Because G_2 contains $\pi(x)$, x is a d-sepnode.

Figure 9 illustrates two possible cases of **Wave** sequence. In (a), a parent d of x appears at one end of the hyperchain, another parent c appears at

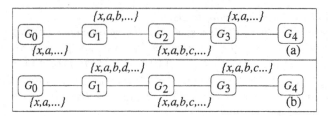

Fig. 8. Concave parent sequences.

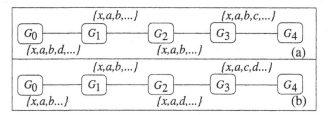

Fig. 9. Wave parent sequences.

the other end, and they disappear in the middle of the hyperchain. In other words, we have $\pi_1^-(x) \supset \pi_1^+(x)$ and $\pi_3^-(x) \subset \pi_3^+(x)$. No local DAG contains all parents of x, and hence x is not a d-sepnode. In (b), we have $\pi_2^-(x)$ and $\pi_2^+(x)$ being incomparable and $\pi_3^-(x) \subset \pi_3^+(x)$.

The following theorem states that the five parent sequences are exhaustive. They are also necessary and sufficient to identify d-sepnode.

Theorem 6 *Let x be a public node in a hyperchain $\langle G_0, G_1, ..., G_m \rangle$ of local DAGs with $\pi(x)$ being the parents of x in all DAGs, where no parent of x is private and each local DAG contains either x or some parents of x.*

1. *There exists one local DAG that contains $\pi(x)$ if and only if the public parent sequence of x on the hyperchain is identical, increasing, decreasing, or concave.*

2. *There exists no local DAG that contains $\pi(x)$ if and only if the public parent sequence of x on the hyperchain is of the wave type.*

Proof:

We prove the statement 1 first:

[Sufficiency] If the sequence type is identical, then every local DAG contains $\pi(x)$. If the type is increasing, then at least G_m contains $\pi(x)$. If the type is decreasing, then at least G_0 contains $\pi(x)$. If the type is concave, for Case (1) (see Definition 5), both G_i and G_{i-1} contain $\pi(x)$. For Case (2), G_i contains $\pi(x)$.

[Necessity] Suppose that there exists a local DAG that contains $\pi(x)$. We show that the parent sequence of x is identical, increasing, decreasing or concave.

If every local DAG contains $\pi(x)$, then $\pi_j^-(x) = \pi_j^+(x)$ for each j and the sequence is identical. Otherwise, if G_0 contains $\pi(x)$, then $\pi_j^-(x) \supseteq \pi_j^+(x)$ for each j and the sequence is decreasing. Otherwise, if G_m contains $\pi(x)$, then $\pi_j^-(x) \subseteq \pi_j^+(x)$ for each j and the sequence is increasing.

Otherwise, if both G_i and G_{i-1} contain $\pi(x)$ for some i ($2 \leq i \leq m-1$), then $\pi_j^-(x) \subseteq \pi_j^+(x)$ for each $j \leq i-1$ and the subsequence $(\pi_1^-(x), ..., \pi_i^-(x))$ is increasing, and $\pi_j^-(x) \supseteq \pi_j^+(x)$ for each $j \geq i$ and the subsequence $(\pi_i^-(x), ..., \pi_m^-(x))$ is decreasing. The entire parent sequence falls under concave type Case (1).

Otherwise, if only one local DAG G_i contains $\pi(x)$, then

$$\pi_i^-(x) \subset \pi(x) \text{ and } \pi_i^+(x) \subset \pi(x).$$

We show $\pi_i^-(x) \bowtie \pi_i^+(x)$ by contradiction. If they are comparable, then

$$\text{either } \pi_j^-(x) \subseteq \pi_j^+(x) \text{ or } \pi_j^-(x) \supset \pi_j^+(x).$$

We have

$$\pi_j^-(x) \subseteq \pi_j^+(x) \subset \pi(x) \text{ or } \pi(x) \supset \pi_j^-(x) \supset \pi_j^+(x),$$

which implies that $\pi(x)$ contains a private parent of x: a contradiction. Furthermore, the subsequence $(\pi_1^-(x), ..., \pi_i^-(x))$ must be trivial, increasing, or identical, and the subsequence $(\pi_{i+1}^-(x), ..., \pi_m^-(x))$ must be trivial, decreasing, or identical. Hence, the entire parent sequence falls under concave type Case (2).

Next, we prove the statement 2:

[Sufficiency] Suppose that the sequence is of the wave type. For wave type Case (1) in Definition 5, we have $\pi_i^-(x) \supset \pi_i^+(x)$. It implies that G_{i-1} and G_i contain a parent, say y, of x that is not contained in G_{i+1}. It cannot be contained in any G_k where $k > i+1$ owing to the hyperchain. If $\pi_j^-(x) \subset \pi_j^+(x)$ holds, then G_{j+1} and G_j contain a parent, say z, of x that is not contained in G_{j-1}. It cannot be contained in any G_k, where $k < j-1$. In summary, only local DAGs $G_0, ..., G_i$ may contain y (not necessarily all of them contain y), and only $G_j, ..., G_m$ may contain z. Because $i < j$, no local DAG contains both y and z.

If $\pi_j^-(x) \bowtie \pi_j^+(x)$, it implies that G_{j+1} and G_j contain a parent, say z, of x that is not contained in G_{j-1}, and G_{j-1} and G_j contains a parent, say w, of x that is not contained in G_{j+1}. Because the same condition as above holds, no local DAG contains both y and z. For wave type Case (2), the same conclusion can be drawn.

[Necessity] Suppose that no local DAG contains $\pi(x)$. Then there exists a pair of local DAGs G_i and G_j ($i < j$) such that the following hold:

1. The DAG G_i contains a parent, say y, of x that is not contained in G_j, and G_i is the closest such local DAG to G_j on the hyperchain.

2. The DAG G_j contains a parent, say z, of x that is not contained in G_i, and G_j is the closest such local DAG to G_i on the hyperchain.
3. No other local DAGs contain both y and z.

Clearly, we have either $\pi_i^-(x) \supset \pi_i^+(x)$ or $\pi_i^-(x) \bowtie \pi_i^+(x)$, and either $\pi_j^-(x) \subset \pi_j^+(x)$ or $\pi_j^-(x) \bowtie \pi_j^+(x)$. Hence, the sequence is of the wave type.   □

5.2 Cooperative verification in hyperchain

To identify the sequence type by cooperation, agents on the hyperchain pass messages from one end to the other, say, from G_m to G_0. Each agent A_i passes a message to A_{i-1} formulated based on the message that A_i receives from A_{i+1} as well as on the result of comparison between $\pi_i^-(x)$ and $\pi_i^+(x)$. Note that A_{i+1} is undefined for A_m.

We partition the five public parent sequence types into three groups and associate each group with a message coded using an integer, as shown in Table 2.

Table 2. Message code according to public parent sequence types

type group	code
decreasing or **identical**	-1
increasing or **concave**	1
wave	0

Agents pass messages according to the algorithm **CollectPublicParentInfoOnChain** as defined below:

Algorithm 1 (CollectPublicParentInfoOnChain)

If A_{i+1} is undefined, agent A_i passes -1 to A_{i-1}. Otherwise, A_i receives a message from A_{i+1}, compares $\pi_i^-(x)$ with $\pi_i^+(x)$, and sends its own message according to one of the following cases:

1. *The message received is -1:*
   If $\pi_i^-(x) \supseteq \pi_i^+(x)$, A_i passes -1 to A_{i-1}.
   Otherwise, A_i passes 1 to A_{i-1}.
2. *The message received is 1:*
   If $\pi_i^-(x) \subseteq \pi_i^+(x)$, A_i passes 1 to A_{i-1}.
   Otherwise, A_i passes 0 to A_{i-1}.
3. *The message received is 0: A_i passes 0 to A_{i-1}.*

We demonstrate how agents cooperate using examples in Figures 7 through 9. In Figure 7 (a), -1 is sent from A_4 to A_3 and is passed along by each agent until A_0 receives it. Interpreting the message code, A_0 concludes that the parent sequence is either **identical** or **decreasing**. Because the actual sequence is **identical**, the conclusion is correct.

In (b), A_3 receives -1 from A_4 and sends 1 to A_2. Afterwards, 1 is passed all the way to A_0, which determines that the sequence is either **increasing** (actual type) or **concave**.

In (c), -1 is sent by each agent. The conclusion drawn by A_0 is to classify the type of sequence as either **identical** or **decreasing** (actual type).

In Figure 8 (a), A_3 receives -1 from A_4 and sends -1 to A_2. Agent A_2 sends 1 to A_1, which passes it to A_0. Agent A_0 then concludes that the sequence type is either **increasing** or **concave**, where **concave** is the actual type. In (b), -1 is sent from A_4 to A_3 and then to A_2. Agent A_2 sends 1 to A_1, which is passed to A_0.

In Figure 9 (a), A_3 receives -1 from A_4 and sends 1 to A_2. Agent A_2 passes 1 to A_1, which in turn sends 0 to A_0. Agent A_0 then interprets the sequence type as a **wave**, which matches the actual type. In (b), A_3 receives -1 from A_4 and sends 1 to A_2. Agent A_2 sends 0 to A_1, which passes 0 to A_0.

In summary, each agent on the hyperchain can pass a code message formulated based on the message it receives and the comparison of the public parents it shares with the adjacent agents. The message passing starts from one end of the hyperchain and the type of the public parent sequence can be determined by the agent in the other end. In this cooperation, no agent needs to disclose its internal structure.

5.3   Cooperative verification in hypertree

We investigate the issue in a general hypertree, and let agents to cooperate in a similar way as in a hyperchain. However, the message passing is directed towards an agent acting as the root of the hypertree.

Consider first the case in which the root agent A_i has exactly two adjacent agents A_1 and A_2. If an agent A_i has a downstream adjacent agent A_k, we denote the parents of x that A_i shares with A_k by $\pi_k(x)$. In Section 5.2, A_i receives message from A_1 and sends message to A_2, so the only information that agent A_i needs to process is the message received from A_1. Here, A_i receives messages from both A_1 and A_2. Thus, A_i has three pieces of information: two messages received from adjacent agents and a comparison between $\pi_1(x)$ and $\pi_2(x)$. The key to determine whether x is a d-sepnode is to detect whether its public parent sequence along any hyperchain, on the hypertree, is the **wave** type. A **wave** sequence can be detected based on one message received by A_i only (when the hyperchain from A_i to a terminal agent is a **wave**), or if not sufficient based on both messages received, or if still not sufficient based in addition on the comparison between $\pi_1(x)$ and $\pi_2(x)$.

The idea can be applied to a general hypertree where A_i has any finite number of adjacent agents. Now A_i must take into account the three pieces of information for each pair of adjacent agents. Consider the hypertree in

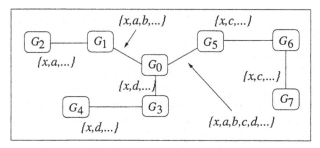

Fig. 10. Parents $\pi(x)$ of a d-sepnode x shared by local DAGs in a hypertree.

Figure 10. If A_0 is the root, then messages will be passed towards A_0 from terminal agents A_2, A_4, and A_7. After agents send messages according to **CollectPublicParentInfoOnChain**, A_0 receives -1 from each of A_1, A_3, and A_5. This implies that the parent sequence type of each hyperchain from A_0 to a terminal agent is either **identical** or **decreasing**. Hence agent A_0 can conclude that itself contains $\pi(x)$ and x is a d-sepnode.

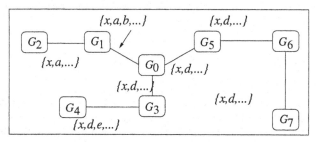

Fig. 11. Parents $\pi(x)$ of a non-d-sepnode x shared by local DAGs in a hyperstar.

In Figure 11, suppose that A_5 is the root. Messages will be passed towards A_5 from terminal agents A_2, A_4, and A_7. Agent A_0 will receives -1 from A_1 and 1 from A_3. It realizes that each hyperchain from A_0 downstream through A_1 is either **identical** or **decreasing** and the hyperchain from A_0 downstream through A_3 is either **increasing** or **concave**. Because the messages are not sufficient to conclude, A_0 compares $\pi_1(x)$ with $\pi_3(x)$. It discovers that they are incomparable. This implies that there exist a hyperchain H_1 from A_0 downstream through A_1 and a hyperchain H_3 from A_0 downstream through A_3 such that when H_1 is joined with H_3 the resultant hyperchain has a **wave** parent sequence. Hence, A_0 will pass the code message 0 to A_5. Based on this message, the root agent A_5 concludes that x is not a d-sepnode. The conclusion is correct because no local DAG contains both a and e.

The following algorithm describes the actions a typical agent A_0 performs.

Algorithm 2 (CollectPublicParentInfo(x))

1. *Receive a message m_i from each downstream adjacent agent A_i.*
2. *(a) If any message is 0, A_0 sends 0 to the upstream agent A_c.*
   (b) Otherwise, if any two messages are 1, A_0 sends 0 to A_c.
   (c) Otherwise, if a message m_i is 1, then A_0 compares $\pi_i(x)$ with $\pi_j(x)$ for each downstream adjacent agent A_j. If j is found such that $\pi_i(x) \not\supseteq \pi_j(x)$, A_0 sends 0. If not found, A_0 sends 1.
   (d) Otherwise, continue.
3. *A_0 compares each $\pi_i(x)$ with the parents $\pi_c(x)$ shared with A_c. If there exists i such that $\pi_c(x) \not\supseteq \pi_i(x)$, then A_0 sends 1 to A_c. Otherwise, A_0 sends -1.*

The following theorem establishes that d-sepnode condition can be verified correctly by agent cooperation through **CollectPublicParentInfo**.

Theorem 7 *Let a hypertree of local DAGs $\{G_i\}$ be populated by a set of agents. Let x be a public node with only public parents in the hypertree. Let agents pass messages according to $CollectPublicParentInfo(x)$.*

Then x is a non-d-sepnode if and only if the root agent returns 0.

6   Cooperative Verification in a General Hypertree

We consider cooperative verification of the d-sepnode condition when both public and private parents of a public nodes are present. Agents who populate such a hypertree can first perform **CollectprivateParentInfo** to find out whether more than one local DAG contains private parents of x. If two or more agents are found to contain private parents of x, then agents can conclude, by Proposition 4, x is a non-d-sepnode. If no agent is found to contain private parents of x, then agents can perform **CollectPublicParentInfo** with any agent being the root to determine if x is a d-sepnode.

On the other hand, if one agent A_0 is found to contain private parents of x, then agents can perform **CollectPublicParentInfo** with A_0 being the root to determine if x is a d-sepnode. Note that it is necessary for A_0 to be the root. For instance, in Figure 10, if A_2 is the only agent that contains the private parents of x, when **CollectPublicParentInfo** is performed with the root A_0, agent A_0 cannot conclude as in Section 5.3. Clearly, although A_0 contains all public parents of x, it does not contain the private parents of x. Hence, it is unknown to A_0 whether there is an agent containing all parents of x. In this case, it depends on whether A_2 is such an agent.

The following algorithm summarizes the method.

Algorithm 3 (VerifyDsepset)

   Let a hypertree DAG union G be populated by multiple agents with one at each hypernode. For each public node x, agents cooperate as follows:

1. *Agents perform **CollectprivateParentInfo**. If more than one agent is found to contain private parents of x, conclude that G violates the d-sepset condition.*

2. *If no agent is found to contain private parents of x, agents perform **CollectPublicParentInfo** with any agent A_0 as the root. If A_0 generates the message 0, conclude that G violates the d-sepset condition. Otherwise, conclude that G satisfies the d-sepset condition.*

3. *If a single agent A_0 is found to contain private parents of x, then agents perform **CollectPublicParentInfo** with A_0 as the root. If A_0 generates the message -1, conclude that G satisfies the d-sepset condition. Otherwise, conclude that G violates the d-sepset condition.*

   It can be proven that **VerifyDsepset** accomplishes the intended task correctly:

Theorem 8 *Let a hypertree DAG union G be populated by multiple agents. After VerifyDsepset is executed in G, it concludes correctly with respect to whether G satisfies the d-sepset condition.*

7   Complexity

We show that multiagent cooperative verification by **VerifyDsepset** is efficient. We denote the maximum cardinality of a node adjacency in a local DAG by t; the maximum number of nodes in an agent interface by k; the maximum number of agents adjacent to any given agent on the hypertree by s; and the total number of agents by n.

   Each agent may call **CollectPrivateParentInfo** $O(k\ s)$ times – one for each shared node. Each call may propagate to $O(n)$ agents. Examination of whether a shared node has private parents in a local DAG takes $O(t)$ time. Hence, the total time complexity for checking private parents is $O(n^2\ k\ s\ t)$.

   Next, we consider processing of public parents after checking private parents succeeds positively. The computation time is dominated by **CollectPublicParentInfo**. Each agent may call **CollectPublicParentInfo** $O(k\ s)$ times. Each call may propagate to $O(n)$ agents. When processing public parent sequence information, an agent may compare $O(s)$ pairs of agent interfaces. Each comparison examines $O(k^2)$ pairs of shared nodes. Hence, the total time complexity for processing public parents is $O(n^2\ k^3\ s^2)$. The overall complexity of **VerifyDsepset** is $O(n^2\ (k^3\ s^2+k\ s\ t))$ and the computation is efficient.

8   Alternative Methods of Verification

Some alternative verification methods to **VerifyDsepset** are worth considering. We analyze alternative methods that deviate from **VerifyDsepset** around two aspects: first, verification by centralizing the parent set information; and second, verification by asynchronous message passing.

According to Definition 2, to determine whether a public node x is a d-sepnode, one needs to know whether a local DAG contains $\pi(x)$. For Case 3 of Section 4, $\pi(x)$ contains only public parents. Hence, it appears that a direct test whether there exists a local DAG containing $\pi(x)$ can be employed.

To put this idea to work, for each such public node x, one needs to centralize the information on $\pi(x)$ somehow. This can be done in at least two ways. The first is to let agents propagate the information on public parents of x through the hypertree. Two passes (inwards and then outwards) are sufficient so that every agent knows about $\pi(x)$. Each can then determine whether $\pi(x)$ is contained locally.

The drawback of this alternative is that information on each element in $\pi(x)$ is disclosed to agents that may not share the element. Note that $\pi(x)$ is *public* only in the sense that each variable in $\pi(x)$ is shared by two or more agents. An agent that shares one variable in $\pi(x)$ may not share another. Hence, this alternative publicizes $\pi(x)$ beyond what is necessary.

An alternative is to collect the information on $\pi(x)$ by a single agent A. Each agent containing x needs to send information on its public parents of x to A. After A collected information on $\pi(x)$, A checks with each agent whether it contains the entire $\pi(x)$.

This method restricts the access to information on $\pi(x)$ to a single agent and is superior than the previous method as far as the privacy issue is concerned. On the other hand, **VerifyDsepset** does not require such a centralized agent at all. One may argue that in the construction of an MAMSBN, an integrator (Section 2) already knows all the public variables and is a suitable candidate for A. Note, however, that the integrator only needs to know what are interface variables. It does not need to know the structural (parent) information on public variables. In summary, although the alternative methods appear to be much simpler, **VerifyDsepset** provides the highest level of privacy for internal structural information of agents.

Next, we consider an alternative method for message passing. **VerifyDsepset** uses a number of *rooted* message propagations. For instance, **CollectPrivateParentInfo** shown in Figure 5 can be performed by first propagating a control message from the root agent A_0 (located at G_0) to the leaf agents A_1 and A_4, and then propagating the private parent information from A_1 and A_4 back to A_0. Alternatively, message passing in a tree structure can be performed in an *asynchronous* fashion such as that used in Shafer-Shenoy belief propagation [13]:

In an asynchronous message passing, each agent on the tree sends one message to each neighbor. It can send a message to a neighbor only after it has received a message from each other neighbor (and, in general, the message sent is dependent on the received messages). Figure 12 illustrates an asynchronous message passing. In (a), agents A_2, A_4 and A_5 are the only

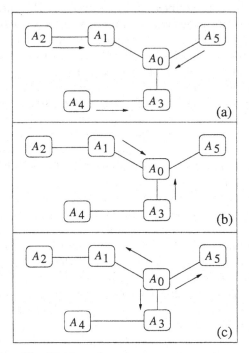

Fig. 12. Asynchronous message passing

ones that are able to send the message (shown by arrows). In (b), A_3 has received message from A_4 and is now ready to send to A_0. Similarly, A_1 is ready to send to A_0. In (c), A_0 has received messages from A_1 and A_5 and is ready to send to A_3. For similar reasons, it is also ready to send to A_1 and A_5. Afterwards, A_1 can send to A_2, and A_3 can send to A_4 (not shown). The asynchronous message passing is then completed.

An asynchronous message passing sends exactly the same set of messages by each agent as in a rooted message passing. Therefore, in principle, **VerifyDsepset** could be performed by asynchronous message passing. However, depending on whether the MAMSBN is open or closed, the rooted message passing can be more advantageous, as analyzed below.

When an MAMSBN is closed (agent membership does not change), interface verification needs to be performed once for all. For rooted message passing, a root agent needs to be selected and agreed by all (otherwise, no agent will act as the root or multiple of them will). For asynchronous message passing, all leaf agents (with exactly one neighbor) must agree on roughly when to start (otherwise, some may start and wait for ever for others). In either case, a value needs to be agreed upon (who is the root or when to start). For how agents can reach such an agreement, see Coulouris et al. [1]. The point is, an equivalent amount of effort is needed in either case. Therefore, there does not seem to be any reason to favor one method of message passing over the other when an MAMSBN is closed.

On the other hand, when an MAMSBN is open, verification may need to be performed again when additional agents join. For asynchronous message passing, all leaf agents must reach another agreement. For rooted message passing, as long as the previous root agent is still a member, it can continue to function as the root. No new agreement is needed. Hence, rooted message passing has an advantage over asynchronous message passing when an MAMSBN is open.

9   Conclusion

We present a method to verify agent interface in an MAS whose knowledge representation is based on MSBNs. To ensure exact, distributed probabilistic inference, agent interfaces must be d-sepsets. Using our verification method, agents only pass concise messages among them without centralized control. A message reveals only partial information about the parenthood of a public node without disclosing additional details on the agent's local DAG. Hence, the method respects agent's privacy, protects agent vendors' know-how, and promotes integration of MAS from independently developed agents.

MSBNs support both modular, exact probabilistic inference in single agent systems and exact, distributed probabilistic inference in MAS. The connection between MSBNs and OOBNs was explored by Koller and Pfeffer [6]. Although OOBNs are intended for single agent systems, the object interfaces also have to satisfy the d-sepset condition. The approach taken was to require all arcs from one network segment to another to follow the same direction. Owing to this requirement, the d-sepset condition is automatically satisfied in a hypertree DAG union. No verification is required. On the other hand, the requirement does restrict the dependency structures to a proper subset of general MSBNs. For instance, in the MAMSBN for monitoring the digital system (Figure 2), arcs may go either way between a pair of adjacent local DAGs. The method presented in this paper allows agent interfaces to be verified efficiently in a general MAMSBN.

Acknowledgements

The funding support from Natural Sciences and Engineering Research Council (NSERC) of Canada to the first author is acknowledged.

References

1. G. Coulouris, J. Dollimore, and T. Kindberg. 2001. *Distributed Systems: Concepts and Design, 3rd Ed.* Addison-Wesley.
2. A.F. Dragoni, P. Giorgini, and L. Serafini. 2001. Updating mental states from communication. In *Intelligent Agents VII: Agent Theories, Architectures and Languages.* Springer-Verlag.
3. P.J. Gmytrasiewicz and C.L. Lisetti. 2000. Using decision theory to formalize emotions for multi-agent systems. In *Second ICMAS-2000 Workshop on Game Theoretic and Decision Theoretic Agents*, Boston.
4. M.N. Huhns and D.M. Bridgeland. 1991. Multiagent truth maintenance. *IEEE Trans. Sys., Man, and Cybernetics*, 21(6):1437–1445.
5. G. Kaminka and M. Tambe. 1999. I'm ok, you're ok, we're ok: experiments in centralized and distributed socially attentive monitoring. In *Proc. Inter. Conference on Automonomous Agents.*
6. D. Koller and A. Pfeffer. 1997. Object-oriented Bayesian networks. In D. Geiger and P.P. Shenoy, editors, *Proc. 13th Conf. on Uncertainty in Artificial Intelligence*, pages 302–313, Providence, Rhode Island.
7. V.R. Lesser and L.D. Erman. 1980. Distributed interpretation: a model and experiment. *IEEE Trans. on Computers*, C-29(12):1144–1163.
8. C.L. Mason and R.R. Johnson. 1989. DATMS: a framework for distributed assumption based reasoning. In L. Gasser and M.N. Huhns, editors, *Distributed Artificial Intelligence II*, pages 293–317. Pitman.
9. P. McBurney and S. Parsons. 2001. Chance discovery using dialectical argumentation. In T. Terano, T. Nishida, A. Namatame, S. Tsumoto, Y. Ohsawa, and T. Washio, editors, *New Frontiers in Artificial Intelligence, Lecture Notes in Artificial Intelligence Vol. 2253*, pages 414–424. Springer-Verlag.
10. H.P. Nii. 1986. Blackboard systems: the blackboard model of problem solving and the evolution of blackboard architectures. *AI Magazine*, 7(2):38–53.
11. J. Pearl. 1988. *Probabilistic Reasoning in Intelligent Systems: Networks of Plausible Inference.* Morgan Kaufmann.
12. A. Rao and M. Georgeff. 1991. Deliberation and its role in the formation of intentions. In B. D'Ambrosio, P. Smets, and P.P. Bonissone, editors, *Proc. 7th Conf. on Uncertainty in Artificial Intelligence*, pages 300–307. Morgan Kaufmann.
13. G. Shafer. 1996. *Probabilistic Expert Systems.* Society for Industrial and Applied Mathematics, Philadelphia.
14. K.P. Sycara. 1998. Multiagent systems. *AI Magazine*, 19(2):79–92.
15. Y. Xiang and V. Lesser. 2000. Justifying multiply sectioned Bayesian networks. In *Proc. 6th Inter. Conf. on Multi-agent Systems*, pages 349–356, Boston.
16. Y. Xiang. 2000. Belief updating in multiply sectioned Bayesian networks without repeated local propagations. *Inter. J. Approximate Reasoning*, 23:1–21.
17. Y. Xiang. 2001. Cooperative triangulation in MSBNs without revealing subnet structures. *Networks*, 37(1):53–65.

Optimal Time–Space Tradeoff
In Probabilistic Inference

David Allen and Adnan Darwiche

University of California, Los Angeles CA 90025, USA

Abstract. Recursive Conditioning, RC, is an any–space algorithm for exact inference in Bayesian networks, which can trade space for time in increments of the size of a floating point number. This smooth tradeoff is possible by varying the algorithm's cache size. When RC is run with a constrained cache size, an important problem arises: Which specific results should be cached in order to minimize the running time of the algorithm? RC is driven by a decomposition structure (a dtree or dgraph). In this research we examine the problem of searching for an optimal caching scheme for a given decomposition structure and present several time–space tradeoff curves for published Bayesian networks. Our results show that the memory requirements of these networks can be significantly reduced with only a minimal cost in time, allowing for exact inference in situations previously impractical. They also show that probabilistic reasoning systems can be efficiently designed to run under varying amounts of memory.

1 Introduction

Recursive Conditioning, RC, was recently proposed as an any–space algorithm for exact inference in Bayesian networks [5]. The algorithm works by using conditioning to decompose a network into smaller subnetworks that are then solved independently and recursively using RC. It turns out that many of the subnetworks generated by this decomposition process need to be solved multiple times redundantly, allowing the results to be stored in a cache after the first computation and then subsequently fetched during further computations. This gives the algorithm its any–space behavior since any number of results may be cached. This also leads to an important question, which is the subject of this research: "Given a limited amount of memory, which results should be cached in order to minimize the running time of the recursive conditioning algorithm?"

We approach this problem by formulating it as a systematic search problem. We then use the developed method to construct time–space tradeoff curves for some real–world Bayesian networks, and put these curves in perspective by comparing them to the memory requirements of state–of–the–art methods based on jointrees [11,15]. The curves produced illustrate that a significant amount of memory can be reduced with only a minimal cost in time. In fact, for much of their domains, the time–space curves we produce appear close to linear, with exponential behavior appearing only near the extreme case of no caching. This dramatic space reduction, without a significant time penalty, allows one to practically reason with Bayesian networks that would otherwise be impractical to handle or in situations where the system memory is constrained.

This chapter is structured as follows. We start in Sect. 2 by providing some background on recursive conditioning and the cache allocation problem. We then formulate this problem in Sect. 3 as a systematic search problem. Time–space tradeoff curves for several published Bayesian networks are then presented in Sect. 4. Finally, in Sect. 5, we provide some concluding remarks.

2   Any–Space Inference

The RC algorithm for exact inference in Bayesian networks works by using conditioning and case analysis to decompose a network into smaller subnetworks that are solved independently and recursively. The algorithm is driven by a structure known as a decomposition tree (dtree), which controls the decomposition process at each level of the recursion. The RC algorithm has also been extended to work on a decomposition graph (dgraph), which has the ability to answer more queries than the dtree version [6,3]. RC can also be augmented to take advantage of determinism in networks by dynamically using logical techniques in the context of conditioning, which in some cases can significantly speedup the inference [2]. We will begin with a review of the dtree structure and then discuss RC.

2.1   Dtrees

Definition 1 *[5] A <u>dtree</u> for a Bayesian network is a full binary tree, the leaves of which correspond to the network conditional probability tables (CPTs). If a leaf node t corresponds to a CPT ϕ, then* vars(t) *is defined as the variables appearing in CPT ϕ.*

Figure 1 depicts a simple dtree. The root node t of the dtree represents the entire network. To decompose this network, the dtree instructs us to condition on variable B, called the cutset of root node t. Conditioning on a set of variables leads to removing edges outgoing from these variables, which for a cutset is guaranteed to disconnect the network into two subnetworks, one corresponding to the left child of node t and another corresponding to the right child of node t; see Fig. 1. This decomposition process continues until a boundary condition is reached, which is a subnetwork that has a single variable.

We will now present some notation needed to define additional concepts with regard to a dtree. The notation t_l and t_r will be used for the left child and right child of node t, and the function vars will be extended to internal nodes t: vars(t) $\stackrel{def}{=}$ vars(t_l) $\cup$ vars(t_r). Each node in a dtree has three more

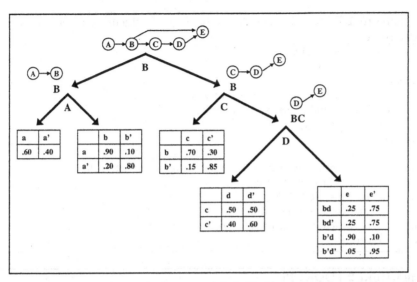

Fig. 1. An example dtree with the cutset labeled below each node and the context next to each node

sets of variables associated with it. The first two of these sets are used by the RC algorithm, while the third set is used to analyze the complexity of the algorithm.

Definition 2 *The* <u>*cutset*</u> *of internal node t in a dtree is:* $\mathsf{cutset}(t) \stackrel{def}{=} \mathsf{vars}(t_l) \cap \mathsf{vars}(t_r) - \mathsf{acutset}(t)$, *where* $\mathsf{acutset}(t)$ *is the union of cutsets associated with ancestors of node t in the dtree.*

Definition 3 *The* <u>*context*</u> *of node t in a dtree is:* $\mathsf{context}(t) \stackrel{def}{=} \mathsf{vars}(t) \cap \mathsf{acutset}(t)$.

Definition 4 *The* <u>*cluster*</u> *of node t in a dtree is:* $\mathsf{cluster}(t) \stackrel{def}{=} \mathsf{cutset}(t) \cup \mathsf{context}(t)$ *if t is a non-leaf, and as* $\mathsf{vars}(t)$ *if t is a leaf.*

The *width* of a dtree is the size of its maximal cluster −1. Figure 1 labels the cutset variables under each dtree node and the context variables beside them.

    The cutset of a dtree node t is used to decompose the network associated with node t into the smaller networks associated with the children of t. That is, by conditioning on variables in $\mathsf{cutset}(t)$, one is guaranteed to disconnect the network associated with node t. The context of dtree node t is used to cache results: Any two computations on the network associated with node t will yield the same result if these computations occur under the same instantiation of variables in $\mathsf{context}(t)$. Hence, a cache is associated with each dtree node t, which stores the results of such computations (probabilities)

Algorithm 1 RC(t): Returns the probability of evidence **e** recorded on the dtree rooted at t

---
 1: **if** t is a leaf node **then**
 2:     return LOOKUP(t)
 3: **else**
 4:     **y** ← recorded instantiation of context(t)
 5:     **if** cache?(t) and cache$_t$[**y**] ≠ nil **then**
 6:         return cache$_t$[**y**]
 7:     **else**
 8:         $p ← 0$
 9:         **for** instantiations **c** of uninstantiated vars in cutset(t) **do**
10:             record instantiation **c**
11:             $p ← p + RC(t_l)RC(t_r)$
12:             un–record instantiation **c**
13:         **when** cache?(t), cache$_t$[**y**] ← p
14:         return p

Algorithm 2 LOOKUP(t)

 ϕ ← CPT of variable X associated with leaf t
 if X is instantiated **then**
     x ← recorded instantiation of X
     u ← recorded instantiation of X's parents
     return $\phi(x|\mathbf{u})$ // $\phi(x|\mathbf{u}) = \Pr(x|\mathbf{u})$
 else
     return 1

---

indexed by instantiations of context(t). This means that the size of a cache associated with dtree node t can grow as large as the number of instantiations of context(t).

For a given Bayesian network, many different dtrees exist and the quality of the dtree significantly affects the resource requirements of RC. The width is one important measure of this, as RC's time complexity is exponential in this value. The construction of dtrees is beyond the scope of this chapter, but in [5,7] it was shown how to create them from elimination orders, jointrees, or directly by using the hMeTiS [12] hypergraph partitioning program. It should also be pointed out that the clusters of a dtree actually form a binary jointree, which was shown to be more efficient than standard jointrees for the Shenoy–Shafer algorithm, at the expense of additional memory [16].

2.2   Recursive Conditioning

Given a Bayesian network and a corresponding dtree with root t, the RC algorithm given in Algorithms 1 and 2 can be used to compute the probability of evidence **e** by first "recording" the instantiation **e** and then calling RC(t), which returns the probability of **e**.

Our main concern here is with Line 5 and Line 13 of the algorithm. On Line 5, the algorithm checks whether it has performed and cached this computation with respect to the subnetwork associated with node t. A computation is characterized by the instantiation of t's context, which also serves as an index into the cache attached to node t. If the computation has been performed and cached before, its result is simply fetched. Otherwise, the computation is performed and its result is possibly cached on Line 13.

When every computation is cached, RC uses $O(n \exp(w))$ space and $O(n \exp(w))$ time, where n is the number of nodes in the network and w is the width of the dtree. This corresponds to the complexity of jointree algorithm, assuming that the dtree is generated from a jointree [5]. When no computations are cached, the memory requirement of RC is reduced to $O(n)$, in which case the time requirement increases to $O(n \exp(w \log n))$. Any amount of memory between these two extremes can also be used in increments of the size of a floating point number, a cache value.

Suppose now that the available memory is limited and we can only cache a subset of the computations performed by RC. The specific subset that we cache can have a dramatic effect on the algorithm's running time. A key question is then to choose that subset which minimizes the running time, which is the main objective of this research. We refer to this as the *secondary optimization problem*, with the first optimization problem being that of constructing an optimal dtree.

Most of our results in this chapter are based on a version of RC which not only computes the probability of evidence **e**, but also posterior marginals over families and, hence, posterior marginals over individual variables. This version of RC uses a *decomposition graph* (dgraph), which is basically a set of dtrees that share structure.

2.3   DGraphs

A dgraph can be constructed from a dtree by orienting the dtree with respect to each of its root nodes [6]. This can be done while maintaining the width, as each of the oriented dtrees will have a width no greater than the original. Figure 2 graphically depicts this process, beginning with the Bayesian network and constructing a dtree, a dtree oriented with respect to leaf $\phi(D|B)$, and finally a complete dgraph. It should be noted that each of the four root nodes in the dgraph corresponds to a valid dtree, so this dgraph actually contains four dtrees which share a significant portion of their structure.

The code in Algorithms 1 and 2 is also used in the dgraph version of RC, where RC(t) is called once on each root t of the dgraph. As a side effect to computing the probability of evidence, RC using a dgraph also computes the posterior marginal of each family in the network [6]. This version of RC uses more memory as it maintains more caches, but it is more meaningful when it comes to comparing our time–space tradeoff curves with the memory

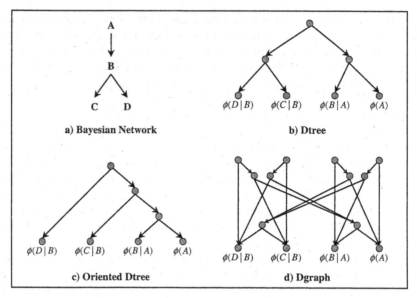

Fig. 2. A Bayesian network and some decomposition structures

requirements of jointree algorithms, as this version of RC is equally powerful to these algorithms.

3   The Cache Allocation Problem

The total number of computations that a dgraph (or dtree) node t needs to cache equals the number of instantiations of context(t). Given a memory constraint, however, one may not be able to cache all these computations, and we need a way to specify which results in particular to cache. A *cache factor cf* for a dgraph is a function which maps each internal node t in the dgraph into a number $cf(t)$ between 0 and 1. Hence, if $cf(t) = .75$, then node t can only cache 75% of these total computations. A *discrete* cache factor is one which maps every internal dgraph node into either 1 or 0: all of the node's computations are cached, or none are cached. The RC code in Algorithms 1 and 2 assumes a discrete cache factor, which is captured by the flag cache?(t), indicating whether caching will take place at dgraph node t.

One can count the number of recursive calls made by RC (and, hence, compute its running time) given any discrete cache factor. Specifically, if t^p denotes a parent of node t in a dgraph, and $S^{\#}$ denotes the number of instantiations of variables S, the number of recursive calls made to node t is [5,6]:

$$calls(t) = \sum_{t^p} cutset(t^p)^{\#}[cf(t^p)context(t^p)^{\#} + (1 - cf(t^p))calls(t^p)]. \quad (1)$$

If the cache factor is not discrete, the above formula gives the average number of recursive calls, since the actual number of calls will depend on the specific computations cached. This equation is significant as it can be used to predict the worst-case expected time requirement of RC under a given caching scheme. RC runs significantly faster as more evidence is set on the network. For example, on Munin1 we have seen instances which would require 12 minutes with no evidence run in just 23 seconds with evidence set on the network.

RC can additionally determine how many times a cache value will be used. This is important because not every cache is useful, as some are never looked up; these are referred to as *dead caches*. Dead caches are those whose context is a superset of its parents context, and can be determined before the cache allocation search. On a dtree, each node only has a single parent and therefore dead caches can be determined by comparing the context of a node with that of its parent. In dgraphs, which are composed of multiple dtrees sharing structure, we will differentiate between *dtree dead caches* and *dgraph dead caches*. Dtree dead caches are those caches which for any individual dtree in the dgraph would not be used. However, when computing all posterior marginals on dgraphs, where nodes may have multiple parents, some of these dead caches would not be useful for any single dtree, but can be useful because one dtree could fill it and another dtree could lookup the stored values. Therefore, a dgraph dead cache is located at a node with only one parent in addition to the context for the node being a superset of its parent.

We focus in this research on searching for an optimal discrete cache factor, given a limited amount of memory, where optimality is with respect to minimizing the number of recursive calls. To this end, we will first define a search problem for finding an optimal discrete cache factor and then develop a depth–first branch–and–bound search algorithm. We will also use the developed algorithm to construct the time–space tradeoff curves for some published Bayesian networks from various domains, and compare these curves to the memory demands and running times of jointree algorithms.

3.1   Cache Allocation as a Search Problem

The cache allocation problem can be phrased as a search problem in which *states* in the search space correspond to partial cache factors that do not violate the given memory constraint, and where an *operator* extends a partial cache factor by making a caching decision on one more dgraph node. The *initial state* in this problem is the empty cache factor, in which no caching decisions have been made for any nodes in the dgraph. The *goal states* correspond to complete cache factors, where a caching decision has been made for every dgraph node, without violating the given memory constraint. Suppose for example that we have a dgraph with three internal nodes t_1, t_2, t_3. This will then lead to the search tree in Fig. 3. In this figure, each node n in the search tree represents a partial cache factor cf. For example, the node in bold corresponds to the partial cache factor $cf(t_1) = 0$, $cf(t_2) = 1$ and $cf(t_3) = ?$. Moreover, if node n is labeled with a dgraph node t_i, then the children of n represent two possible extensions of the cache factor cf: one in which dgraph node t_i will cache all computations (1–child), and another in which dgraph node t_i will cache no computations (0–child).

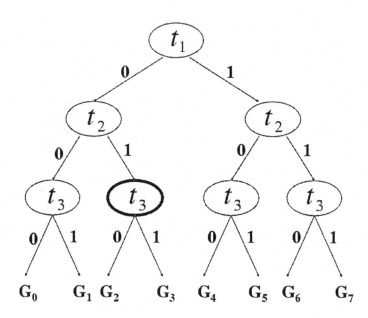

Fig. 3. Search tree for a dgraph with 3 internal nodes

According to the search tree in Fig. 3, one always makes a decision on dgraph node t_1, followed by a decision on dgraph node t_2, and then node t_3. A fixed ordering of dgraph nodes is not necessary, however, as long as the following condition is met: A decision should be made on a dgraph node t_i only after decisions have been made on all its ancestors in the dgraph. We will explain the reason for this constraint later on when we discuss cost functions.

In the search tree depicted in Fig. 3, the leftmost leaf (G_0) represents no caching, while the rightmost leaf (G_7) represents full caching. The search trees for this problem have a maximum depth of d, where d is the number of internal nodes in the dgraph. Given this property, depth–first branch–and–bound search is a good choice given its optimality and linear space complexity [14]. It is also an anytime algorithm, meaning that it can always return its best result so far if interrupted, and if run to completion will return the optimal solution. Hence, we will focus on developing a depth–first branch–and–bound search algorithm.

It should be noted that this search for a cache allocation only needs to be done once, while the user will usually be interested in running multiple probability calculations based on the results.

3.2   Cost Functions

The depth–first branch–and–bound (DFBnB) algorithm requires a cost function f which assigns a cost $f(n)$ to every node n in the search tree. The function $f(n)$ estimates the cost of an optimal solution that passes through n. The key here is that $f(n)$ must not overestimate that cost; otherwise, one loses the optimality guarantee offered by the search algorithm. We will now develop such a cost function $f(n)$ based on the following observations. Since each node n represents a partial cache factor cf, function $f(n)$ must estimate the number of recursive calls made to RC based on an optimal completion of cache factor cf. Consider now the completion cf' of cf in which we decide to cache at each dgraph node that cf did not make a decision on. This cache factor cf' is the best completion of cf from the viewpoint of running time, but it may violate the constraint given on total memory. Yet, we will use it to compute $f(n)$ as it guarantees that $f(n)$ will never overestimate the cost of an optimal completion of cf.

One important observation in this regard is that once the caching decision is made on the ancestors of dgraph node t, we can compute exactly the number of recursive calls that will be made to dgraph node t (see Equation 1). Therefore, when extending a partial cache factor, we will always insist on making a decision regarding a dgraph node t for which decisions have been made on all its ancestors. This improves the quality of the estimate $f(n)$ as n gets deeper in the tree. It also allows us to incrementally compute this estimate based on the estimate of n's parent in the search tree.

3.3   Pruning

As depicted by the search tree in Fig. 3, there is potentially an exponential number of goal nodes in the search tree and the combinatorial explosion of exhaustive search can become unmanageable very quickly. Hence the search algorithm must eliminate portions of the search space while still being able to guarantee an optimal result. One of the key methods of doing this is by pruning parts of the search tree which are known to contain non-optimal results. The DFBnB algorithm does this by pruning search tree nodes when the cost function $f(n)$ is larger than or equal to the current best solution. Hence, more accurate cost functions will allow more pruning. Another major source of pruning is the given constraint on total memory. This is accomplished by pruning a search tree node and all its descendants once it attempts to assign more memory to caches than is permitted by the memory constraint.

3.4   Search Decisions

Now that we have chosen a cost function, we are still left with two important choices in our search algorithm: (1) which child of a search tree node to
expanded first, and (2) in what order to visit dgraph nodes during search.
Expanding the 1–child first is a greedy approach, as it attempts to fully cache
at a dgraph node whenever possible. Results on many different networks have
shown that in many cases, expanding the 1–child before the 0–child appears
to be equal to or better than the opposite [1], and it is this choice that we
adopt in our experiments. The specific order in which we visit dgraph nodes
in the search tree turns out to have an even more dramatic effect on the efficiency of search. Even though we make caching decisions on parent dgraph
nodes before their children, there is still a lot of flexibility. Our experimentation on many networks has shown that choosing the dgraph node t with
the largest context$(t)^{\#}$ is orders of magnitude more efficient than some other
basic ordering heuristics [1]. This choice corresponds to choosing the dgraph
node with the largest cache, and it is the one we use in our search algorithm.

4   Time–Space Tradeoff

The main goal of this section is to present time–space tradeoff curves for a
number of benchmark Bayesian networks, some of which are obtained from
[4] and others are included in the distributions of [9,10]. The main points to
observe with respect to each curve is the slope of the curve, which provides
information on the time penalty one pays when reducing space in probabilistic
inference. The second main point is to compare the produced curves with the
time and space requirement of jointree methods, as the version of RC we are
using provides the same functionality as these algorithms (that is, probability
of evidence and posterior marginals over variables and families). This baseline
comparison is important as it places our results in the context of state–of–
the–art inference systems. The results presented here use the same datasets as
those in [3]. This version however ignores only dgraph dead caches, while the
version in [3] ignored dtree dead caches. Ignoring only dgraph dead caches
adds more memory under full caching, but runs faster under constrained
memory.

   Time–space tradeoff curves. Figures 4 and 5 depict optimal discrete
time–space tradeoff curves for two networks. These curves were generated
as follows. A jointree was first generated for the network using Hugin.[1] The
jointree was then converted into a dtree as described in [5]. The dtree was
finally converted into a dgraph as described in [6]. Two sets of results were
then generated:

---

[1] We used Hugin's default setting: the minimum fill–in weight heuristic in conjunc-
tion with prime component analysis.

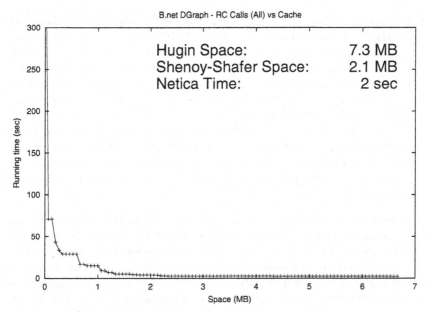

Fig. 4. Time–space tradeoff on B

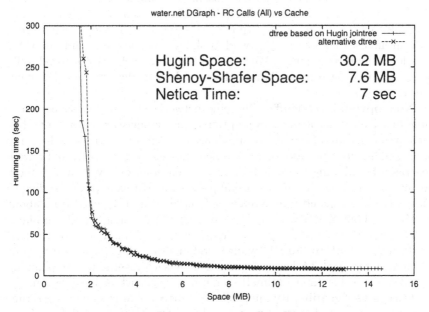

Fig. 5. Time–space tradeoff on Water

- We computed the space requirements for jointree algorithms, using both the Hugin [11] and Shenoy–Shafer [15] architectures (on the non–binary jointree). For the first architecture, we assumed one table for each clique and one table for each separator. For the second, we assumed two tables for each separator (no tables for cliques). We also performed propagation on the jointree using Netica [13], which implements the Hugin algorithm, and recorded the running time.
- We then ran our search algorithm to find an optimal cache factor under different memory constraints, where we generated 100 data points for each curve. For each caching factor that we identified, we computed the number of recursive calls that will be made by RC under that factor and converted the calls to seconds.[2]

A number of observations are in order here. First, RC is using memory efficiently, and would use a similar amount of memory as the Shenoy–Shafer algorithm would if run on the binary jointree determined by the dtree. Second, the curves show that a significant amount of memory can sometimes be reduced from full caching with only a limited increase in the time required; in fact, the exponential growth appears to be occurring only near the lower extreme of no caching. The space requirement for Water (Fig. 5), for example, can be reduced to 30% while only increasing the running time by a factor of 2.9. Moreover, the space requirements for B (Fig. 4) can be reduced to about 3% while increasing the running time by a factor of 19. Finally, we note that each optimal search for the B network took less than a second and for Water took less than three minutes. We stress though that such searches need to be done only once for a network, and their results can then be used for many further queries.

Non–optimal tradeoffs. On some networks, the search space is too large to solve the cache allocation optimally using our search algorithm, but the anytime nature of the algorithm allows us to interrupt the search at any point and ask for the best result obtained thus far. Figures 6, 7, and 8 were generated by allowing the search to run for ten minutes. Even though these curves are not optimal, they are useful practically. For example, according to these curves, the memory requirement of Barley can be reduced from about 54 MB to about 8 MB while only increasing the running time from about 1 to 3 minutes. Moreover, the space requirement of Munin1 can be reduced from about 450 MB to 180 MB, while increasing the running time from about 13 minutes to about 3.5 hours. Encouraged by such results, we are planning to investigate other (non–optimal) search methods, such as local search.

Dtrees vs dgraphs. Running RC on a dtree takes less space than running it on a dgraph, but produces much less information (probability of evidence instead of posterior marginals). To illustrate this difference concretely,

---

[2] Our Java implementation of RC on a Sun Ultra 10, 440 MHz computer with 256 MB of RAM, makes an average number of three million recursive calls per second.

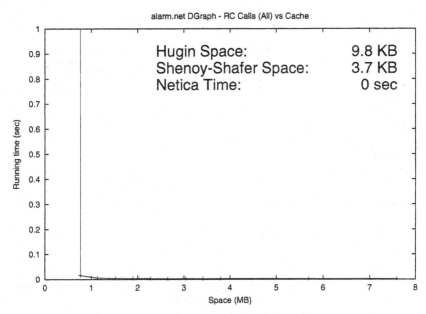

Fig. 6. Time–space tradeoff on Alarm

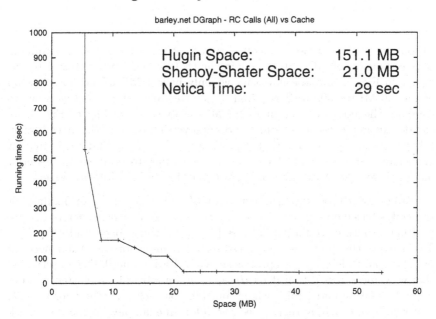

Fig. 7. Time–space tradeoff on Barley

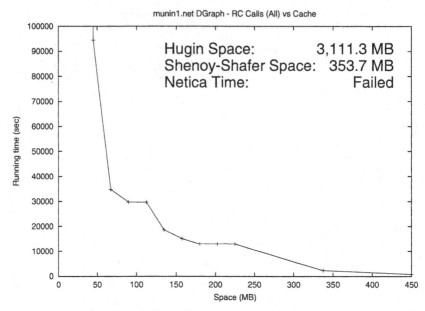

Fig. 8. Time–space tradeoff on Munin1

we present in Fig. 9 two tradeoff curves for the Water network, assuming a dtree version of RC, which require much less memory compared to the curves in Fig. 5. Suppose now that we only have 1 MB of memory, instead of the 7.6 MB or 30.2 MB required by jointree algorithms, and we want to compute the posterior marginals for all variables. According to Fig. 5, we can do this in 693 seconds using the dgraph version of RC. The dtree version takes 16.3 seconds to compute the probability of evidence under this amount of memory, and we would have to run it 85 times to produce all posterior marginals for the Water network (given variable cardinalities in Water).

Effect of dtree/dgraph on tradeoff. Our notion of optimality for tradeoff is based on a given dtree/dgraph; hence, generating different decomposition structures could possibly lead to better time–space tradeoff curves. To illustrate this point, we generated tradeoff curves for the Water network based on multiple dtrees/graphs, as shown in Figs. 5 and 9. One observation that we came across is that dtrees/graphs that are based on jointrees tend to require less time under full caching, but are not necessarily best for tradeoff towards the no caching region; see Fig. 9 for an example. Yet, we used such dtrees/graphs in this chapter in an effort to provide a clear baseline for comparison with jointree methods. If we relax this constraint, however, we can obtain better tradeoff curves than is generally reported here, as illustrated by Figs. 5 and 9. The specific way in which properties of a dtree/dgraph influence the quality of corresponding time–space tradeoff curves is not very

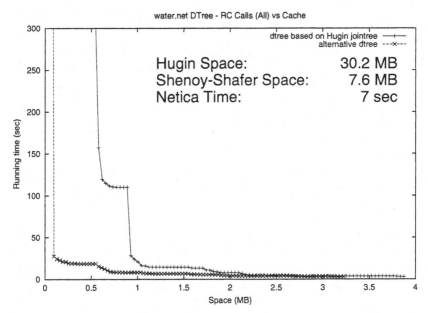

water.net DTree - RC Calls (All) vs Cache

dtree based on Hugin jointree ───+───
alternative dtree ───×───

Hugin Space: 30.2 MB
Shenoy-Shafer Space: 7.6 MB
Netica Time: 7 sec

Fig. 9. Time–space tradeoff on Water for computing probability of evidence

well understood, however, and we hope to shed more light on this in future
work.

Size of search space. It should be noted that the difficulty of obtaining
an optimal time–space tradeoff curve on some networks is not due to a large
space requirement, but is due mostly to the number of nodes in the Bayesian
network as that is what decides the size of search space. To further illustrate
this point, we generated a network randomly with 40 nodes (many of them
non–binary), 86 edges, and a width of 14. This network requires extensive
memory but has a relatively small number of variables. In fact, both Netica
and Hugin were unable to compile the network requiring about 6 GB and 11
GB respectively. We were able, however, to produce an optimal time–space
tradeoff curve for this network. The curve for the dtree version of RC is
shown in Fig. 10. According to this curve, we can compute the probability of
any evidence on this network in about 2 hours using only about 75 MB.

Related work. We close this section by a note on related work for time–
space tradeoff in probabilistic reasoning, which takes a different approach [8].
In this work, large separators in a jointree are removed by combining their
adjacent clusters, which has the effect of reducing the space requirements of
the Shenoy–Shafer architecture (as we now have fewer separators), but also
increasing its running time (as we now have larger clusters). The tradeoffs
permitted by this approach, however, are coarser than those permitted by RC
as discussed in [5]. Furthermore, the secondary optimization problem of which

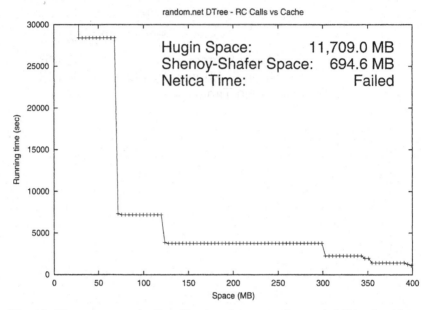

Fig. 10. Time–space tradeoff on Random for computing probability of evidence

separators to remove in order to minimize running time is not addressed in [8] for the proposed approach, as we do in this chapter for the RC approach.

5 Conclusions

The main contribution of this research is a formal framework, and a corresponding working system, for trading space for time when designing probabilistic reasoning systems based on Bayesian networks. The research is based on the algorithm of recursive conditioning, and is accompanied with a set of experimental results showing that a significant amount of memory can sometimes be reduced while only incurring a reasonable penalty in running time. The proposed framework is then beneficial for designing reasoning systems with limited memory, as in embedded systems, and for reasoning with challenging networks on which jointree algorithms can exhaust the system memory.

Recursive conditioning and the described time–space tradeoff system have been implemented in JAVA in the SAMIAM tool, which is available publically [17].

Acknowledgments

This work has been partially supported by NSF grant IIS-9988543 and MURI grant N00014-00-1-0617.

References

1. ALLEN, D., AND DARWICHE, A. Optimal time–space tradeoff in probabilistic inference. In *Proceedings of the First European Workshop on Probabilistic Graphical Models* (2002), pp. 1–8.
2. ALLEN, D., AND DARWICHE, A. New advances in inference by recursive conditioning. To appear in *Uncertainty in Artificial Intelligence: Proceedings of the Nineteenth Conference (UAI)* (2003).
3. ALLEN, D., AND DARWICHE, A. Optimal time–space tradeoff in probabilistic inference. To appear in *Proceedings of the 18th International Joint Conference on Artificial Intelligence (IJCAI)* (2003).
4. BAYESIAN NETWORK REPOSITORY. http://www.cs.huji.ac.il/labs/compbio/Repository/, URL.
5. DARWICHE, A. Recursive conditioning. *Artificial Intelligence 126* (February 2001), 5–41.
6. DARWICHE, A. Decomposition graphs. Tech. Rep. D-134, UCLA, 2002.
7. DARWICHE, A., AND HOPKINS, M. Using recursive decomposition to construct elimination orders, jointrees, and dtrees. In *Proc. 6th European Conf. on Symbolic and Quantitative Approaches to Reasoning and Uncertainty (EC-SQARU'01, Toulouse, France)* (2001), pp. 180–191.
8. DECHTER, R., AND FATTAH, Y. E. Topological parameters for time-space tradeoff. *Artificial Intelligence 125*, 1-2 (2001), 93–118.
9. GeNIe. http://www2.sis.pitt.edu/~genie/, URL.
10. HUGIN EXPERT. http://www.hugin.com/, URL.
11. JENSEN, F. V., LAURITZEN, S. L., AND OLESEN, K. G. Bayesian updating in causal probabilistic networks by local computations. *Computational Statistics Quarterly 4* (1990), 269–282.
12. KARYPIS, G., AND KUMAR, V. Hmetis: A hypergraph partitioning package. http://www.cs.umn.edu/karypis, 1998.
13. NORSYS SOFTWARE CORP. http://www.norsys.com/, URL.
14. PAPADIMITRIOU, C. H., AND STEIGLITZ, K. *Combinatorial Optimization.* Dover Publications, Inc., 1998.
15. SHAFER, G. R., AND SHENOY, P. P. Probability propagation. *Annals of Mathematics and Artificial Intelligence 2* (1990), 327–352.
16. SHENOY, P. P. Binary join trees for computing marginals in the shenoy-shafer architecture. *International Journal of Approximate Reasoning 17* (1997), 239–263.
17. UCLA AUTOMATED REASONING GROUP. SamIam: Sensitivity Analysis, Modeling, Inference And More. http://reasoning.cs.ucla.edu/samiam, URL.

Hierarchical Junction Trees: Conditional Independence Preservation and Forecasting in Dynamic Bayesian Networks with Heterogeneous Evolution

Roberto O. Puch[1], Jim Q. Smith[2], and Concha Bielza[3]

[1] School of Crystallography, Birkbeck College, University of London, London WC1E 7HX, UK
[2] Department of Statistics, University of Warwick, Coventry CV4 7AL, UK
[3] Artificial Intelligence Department, Technical University of Madrid, Madrid 28660, Spain

Abstract. Propagation in decomposable Bayesian networks with junction trees is inferentially efficient: no conditional independence in the Bayesian network is ignored in the junction tree construction and in any propagation task. For non-decomposable Bayesian networks, the junction tree construction uses moralisation and triangulation that ignore some of the conditional independence. The junction tree, therefore, trades inferential efficiency with generality: it can be used to compute the distribution of any set of target nodes given any set of conditioning nodes.

In this chapter inferential efficiency for non-decomposable Bayesian networks is addressed. We present the hierarchical junction tree, a framework that transparently represents the conditional independence in the Bayesian network. We discuss propagation tasks where conditional independence is not ignored in the construction of the hierarchical junction tree and in the propagation tasks. We also discuss their use for efficient exact forecasting in dynamic Bayesian network with heterogeneous evolution.

1 Introduction

Fast probability propagation in a decomposable Bayesian network (BN) is often performed through a secondary structure called the junction tree (JT) [6,3,2]. Conditional independence statements encoded in the BN are used for constructing this structure. A JT provides a general facility for computing probabilities of any set of variables of interest, subsequently called target variables, given the values of any other set of observed variables, subsequently called conditioning variables. However for non-decomposable BNs the JT framework can only provide this facility at the cost of ignoring some of the conditional independence encoded in that BN. Conditional independence statements are ignored both in the moralisation step, where parents of the same child are joined with an undirected edge and then edge direction is dropped, and also in the triangulation step, where undirected edges are added. The addition of these undirected edges, the so-called fill-in's, has the tendency to create large cliques in the JT. The JT propagation algorithm can then become computationally inefficient. This inefficiency can become critical particularly in dynamic settings.

To alleviate these problems, specifically in a non-dynamic setting, Madsen & Jensen [13,14] proposed the Lazy propagation (LP) method. This method records extra conditional independence which had been coded in the BN but not retained in the JT through the potentials of each clique. They preserve each edge direction by recording the potentials as conditional probability distributions, and clique potentials are kept factorised. The LP method uses therefore a hybrid representation, both graphical and algebraic, to store conditional independence statements required for efficient propagation.

It is now timely to stand back and view propagation in a new light. In this article we present a formal framework, the hierarchical junction tree (HJT), for probability propagation that transparently and consistently encodes the edge direction and potential factorisation in the BN. The HJT consistency of graph and potentials representation allows us to prove separation theorems which implies that the HJT is constructed without loss of conditional independence.

For propagation tasks where the conditioning variables are ancestors of the target variables and the ancestral graph of the conditioning variables is decomposable, all conditional independence statements encoded in the BN are transferred to the HJT. Examples of these propagation tasks arise in dynamic models where the primary interest is in forecasting, and so the level of generality that the JT provides is often not required. This setting often occur in the development of decision support systems for emergency planning under uncertainty [16,17]. In decision support systems in the case of a nuclear accident, for example, counter-measures cannot affect the past contamination levels but only future received radiation doses. In addition the information that is usually received is on contamination levels in the past. Although forecasting is a less complex computational task, in this type of problems forecasting is required on-line and as soon as possible so that counter-measures are implemented. In contrast, smoothing is done off-line.

In most dynamic settings, it is natural to construct the BN so that variables appearing earlier in the BN tend to be learned before the variables appearing later in the network. In this setting we also have new descendant variables added with time, and we forecast the future in the light of observing the past. Thus it is typical for the target variables to be descendants of the conditioning variables. Furthermore the use of JT algorithms in settings where new descendant nodes are continuously added to the net has some inherent problems. First, the standard JT algorithms, that is the JT with the Hugin architecture [1], provide no facility to move variables out of the system as they are learned because this is not required for static networks.

So, the number of variables in the system and cliques in the tree steadily increase. Lazy propagation and Nested JTs inherit this feature because they operate on the standard JT. Second, as new variables arrive with time, on-line triangulation is required. This tends to progressively increase the average clique size and further undermines efficiency. Although LP is not highly dependant on the triangulation, on-line triangulation is still required because LP operates on a standard JT.

In this article we also propose the HJT as a efficient framework for exact forecasting in dynamic BNs with heterogeneous evolution. These issues were first addressed in articles for time series with homogeneous time-slices by Kjærulff [10] and Kanazawa *et al.* [7]. However, in the methodology described below we do not assume time homogeneity of structure. The HJT is constructed from the BN without loss of conditional independence when the ancestral graph of the conditioning variables is decomposable [15]. Only this ancestral graph needs to be triangulated when it is not decomposable. In contrast, the whole graph needs to be triangulated in the JT framework. In addition the HJT decreases its size as variables are learned to offset any increase in the descendants, see Section 5.4.

Having developed this framework we can start to formally address and compare issues of efficiency of competing propagation algorithms. Perhaps more importantly we can tailor these algorithms to the predictive needs of the system, often necessary in propagation in dynamic BNs.

In Section 2 BNs are reviewed and the new concept of ancestrality partition is introduced. We review JTs in Section 3 while in Section 4 the using JTs for forecasting in dynamics BNs with heterogeneous evolution is presented. We introduce HJTs in Section 5 and their construction is illustrated through an example. Also in this section we introduced the h-separation criterion for HJTs and a method for propagating probabilities in HJTs. In the last section conclusions are given and future research challenges are discussed.

2 Bayesian Networks

We recall that a *Bayesian network* is a pair (G, P) consisting of a directed acyclic graph $G = (V, E)$ and a probability distribution P. Set V is the index of a set of variables $\{X_v\}_{v \in V}$. Probability distribution P factorises according to G, that is, $P(x_V) = \prod_{v \in V} Pr(X_v = x_v | X_{Pa(v)} = x_{Pa(v)})$ where $Pa(v)$ is the set of parents of node v in the node set V of G.

In the transition from the BN to the HJT, we divide the BN into layers. This division relies on the following concept. A *collider* in a BN is a node having at least two parents which are not joined by an edge. The layers of the BN is then based on the configuration of the colliders. A decomposable BN can be defined as a BN that has no colliders. Some of the nodes of the HJT will be *families* of the nodes of a BN, that is, sets of the form $Fa(v) = \{v\} \cup Pa(v)$.

In fact they will be *maximal* families where family $Fa(v)$ is maximal if it is not contained in another family.

The first step in the transition from the BN to the HJT is the identification of the ancestrality partition of the BN. In the following we will introduce and discuss this new concept.

Let $B = (G, P)$ be a BN where $G = (V, E)$. The *ancestrality partition of* B is the partition of V,
$$\{I(1), H(1), I(2), \ldots, I(m-1), H(m-1), I(m)\}.$$
The elements of this partition are defined as,

$I(1)$ =$\{v : v$ is not a collider and has no collider ancestors$\}$

$H(1)$=$\{v : v$ is a collider and has no collider ancestors$\}$,

$I(i)$ =$\{v : v$ is not a collider, has at least one collider ancestor in $H(i-1)$ and has no collider ancestors in $V \setminus [H(1), \ldots, H(i-1)]\}$.

$H(i)$ =$\{v : v$ is a collider, has at least one collider ancestor in $H(i-1)$ and has no collider ancestors in $V \setminus [H(1), \ldots, H(i-1)]\}$.

$I(i)$ is defined if $H(i-1)$ exists. $H(i)$ is defined if $H(i-1)$ exists and if it is not empty.

The ancestrality partition organises the BN into layers determined by the pattern of colliders. The first layer consists of $I(1)$, the second layer consists of $H(1) \cup I(2)$. In general, the i-th layer consists of $H(i-1) \cup I(i)$, $i = 2, \ldots, m$. These layers help to identify decomposable sections of the original BN from which the JTs of the HJT are constructed, see Section 5.

We can construct the ancestrality partition of a BN given a total order consistent with the partial order induced by the BN, by organising the nodes in layers. Let $(v(1), v(2), \ldots, v(n))$ be a total order and $L(v(i))$ denote the layer where $v(i)$ is. Starting with $v(1)$ the sequential rule is:

1) If $Pa(v(i)) = \emptyset$, $L(v(i)) = 1$

2) If $Pa(v(i)) \neq \emptyset$ and $v(i)$ is not a collider,
   $L(v(i)) = max\{L(u) : u \in Pa(v(i))\}$

3) If $Pa(v(i)) \neq \emptyset$ and $v(i)$ is a collider,
   $L(v(i)) = max\{L(u) : u \in Pa(v(i))\} + 1$.

Then the elements of $I(1)$ are those nodes v with $L(v) = 1$. The elements of $H(i-1)$ are those colliders v with $L(v) = i$ and the elements of $I(i)$ are those non-colliders v with $L(v) = i$.

The ancestrality partition of the BN given in Figure 1 is,
$I(1) = \{a, b, c, d, e, f, g, h, j, l\}$, $H(1) = \{i\}$, $I(2) = \{k\}$, $H(2) = \{m\}$, $I(3) = \emptyset$.

Note that the ancestrality partition is invariant under Markov equivalent classes. Two BNs in the same equivalent class have the same collider structure and therefore the same ancestrality partition.

In the dynamic setting introduced in Section 4, new descendants are added sequentially to a BN as time passes. This sequential addition gives a total order of the nodes that is consistent with the partial order induced by the

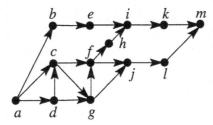

Fig. 1. BN B_2 for the ancestrality partition example

BN, and therefore this order can be used for constructing the ancestrality partition.

3 Junction Trees

In this section we review standard JTs, that is, JTs with the HUGIN architecture [1]. A junction tree for a vector for a potential ϕ_V with *universe* V, is a pair $T = (\mathsf{T}, \phi_V)$. T is a tree of subsets of V such that the so-called JT property holds, that is, the intersection of two sets C_1 and C_2 is contained in all the sets in the unique path between C_1 and C_2. The intersection $C_1 \cap C_2$ is called the separator of edge $\{C_1, C_2\}$. ϕ_V is a potential on X_V, where a potential ϕ_U is a function whose domain is the state space of the variable X_U and whose counter domain is the non-negative real numbers. Potential ϕ_V factorises according to T such that,

$$\phi_V = \frac{\prod_{C \in \mathcal{C}} \phi_C}{\prod_{S \in \mathcal{S}} \phi_S},$$

where $\mathcal{C}$ is the set of nodes of T and $\mathcal{S}$ the set of separators. A directed JT is a JT with directed edges.

We can construct a JT $T = (\mathsf{T}, \phi_V)$ from a given decomposable BN $B = (G, P)$. The nodes of the tree T are cliques of the undirected version of G. The potentials ϕ_C are formed from the product of conditional probabilities $Pr(X_v = x_v | X_{Pa(v)} = x_{Pa(v)})$, seen as a function of $(x_v, x_{Pa(v)})$, where $\{v\} \cup Pa(v) \subseteq C$. The potentials ϕ_S for the separators are initialised to 1. If the BN is not decomposable but its moral graph is chordal we can still construct a JT from the moral graph, but some conditional independence is lost in the moralisation step. If the BN is not decomposable and its moral graph is not chordal, then its moral graph is made chordal through a triangulation step before we construct the JT and this step can be computationally expensive. For a detailed study on triangulation see [8,9] or [12]. Triangulating the moral graph requires us to ignore some of the conditional independence encoded in the BN because we need to add extra edges, the so-called fill-in's. In this case a product of conditional distributions may be assigned to one clique, and therefore losing part of the factorisation encoded in the BN.

Propagation in JTs within the Hugin architecture is based on the operation of a sum-flow from a node to one of its neighbouring nodes. Let C_i and C_j be neighbours with separator $S = C_i \cap C_j$, and let ϕ_{C_i}, ϕ_{C_j} and ϕ_S be their corresponding current potentials. A sum-flow is performed as follows,

1) compute the potential $\phi'_S = \sum_{C_i \setminus S} \phi_{C_i}$;

2) assign $\phi'_{C_j} = \phi_{C_j} * \frac{\phi'_S}{\phi_S}$;

3) assign ϕ'_S to S.

Propagation in the Hugin architecture is done in two steps. Let us suppose that we learn $X_a = x'_a$, or for short, that we learned X_a. First, we choose a clique that contains a, then we multiply the potential of this clique by the finding f_{a,x'_a} for x'_a, where a finding for x'_a is a potential with domain $Sp(X_a)$ that maps x'_a to 1 and 0 for other elements of $Sp(X_a)$. Second, we schedule sum-flows by choosing a clique as a root and then scheduling sum-flows first from the leaves of the tree to the root, the so-called Collect Evidence schedule, and then scheduling sum-flows from the root to the leaves, the so-called Distribute Evidence schedule. After these two schedules have been performed, the JT is sum-consistent, that is, the potentials on the nodes and the separators hold the marginal potentials of the potential P of the JT. For a description of the Hugin architecture see [5,2], and for a more technical description see [3].

4   Forecasting in the Dynamic Setting Using Junction Trees

In this section we describe how JTs can be used for forecasting in a dynamic setting. We first address how a JT can be constructed sequentially and then demonstrate how it can be used for forecasting. The sequential construction is demonstrated for $t = 1$ and the transition from $t = 1$ to $t = 2$. However, the second transition is trivially identical to constructing a JT at time $t + 1$ from a JT valid at time t. Figure 2 displays the BNs B_1 for time $t = 1$ and B_2 for time 2. BN B_1 is surrounded by the inner dashed lines and B_2 is surrounded by the outer dashed lines. Descendant nodes k, l and m were sequentially added into B_1 to obtain B_2.

The construction of the JT T_1 from BN B_1 is done through a standard procedure: the BN is moralised, the moral graph is then triangulated and finally a tree of cliques is constructed from the triangulated graph. In the construction of JT T_1, displayed in Figure 3, the fill-in's are: $\{a, f\}$, $\{a, g\}$, $\{b, f\}$ and $\{e, f\}$.

At time $t = 2$, the nodes k, l and m are added to B_2. If we wish to compute the marginals for X_m on B_2, we need to construct a JT for B_2. One choice would be to start over again and construct a JT for B_2. Another choice would be to use the fill-in's in the triangulation of the moral graph of B_1 as a starting point for the triangulation of the moral graph of B_2. Notice that in both choices on-line triangulation is required. The second choice is

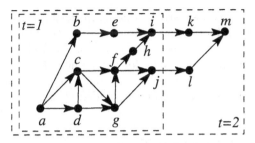

Fig. 2. The BN B_1 at time $t = 1$ is surrounded by the inner dashed box; the BN B_2 at time $t = 2$ is surrounded by the outer dashed box

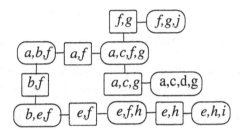

Fig. 3. The JT T_1 constructed from BN B_1 in Figure 2

more computationally efficient because part of the moral graph of B_2 has been already triangulated with the triangulation of the moral graph of B_1. The JT T_2, constructed in this fashion, is displayed in Figure 4. The extra fill-in's are $\{b, g\}$, $\{e, g\}$, $\{f, i\}$, $\{g, i\}$, $\{i, j\}$ and $\{i, l\}$.

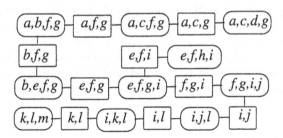

Fig. 4. The JT T_2 constructed from BN B_2 in Figure 2

In addition to the triangulation process being computationally expensive, the sequential on-line triangulation results in a sequential increment of the node/clique sizes. In T_2, for example, node $\{a, b, f\}$ of T_1 is of size 3 and it is included in node $\{a, b, f, g\}$ of T_2 which is of size 4. Note as well that nodes $\{b, e, f\}$ and $\{e, f, g\}$ in T_1 are joined together into node $\{a, e, f, g\}$ in T_2.

Once we have constructed a JT, we can use it for forecasting. The JT presents an inefficiency in that the fill-in's added in the triangulation process may be superfluous once we learn a variable. In the JT T_2, for example, if we learn that $X_a = x'_a$ and we remove a from B_2, then the fill-in's added because of the cycle (a, c, f, h, e, b) in the moral graph of B_2 are no longer needed. A solution would be to construct a JT for the reduced BN, but this process is computationally expensive. We will see in Section 5.4, that the HJT does not have this problem because no fill-in's are added at all.

5   Hierarchical Junction Trees

5.1   Definition

A *hierarchical junction tree* (HJT) $\mathcal{H}$ for a potential ϕ_V with universe V is a directed rooted tree $\mathcal{H} = (T, \mathcal{F})$ where the edge direction is towards the root. The root is a directed JT for ϕ_V. The node set is $T = \{T_1, T_2, \ldots, T_p\}$ where each node T_i, $1 \leq i \leq p$, is a directed JT. The ordered set (T_i, T_j) is in the edge set $\mathcal{F}$ if T_i is a directed JT for ϕ_C where C is a node in the directed JT T_j. The node C is then called a covering node in T_j and a cover of T_i.

The covering nodes of a JT T' in the HJT act as dummy nodes. In fact, the potential associated with a covering node has a JT representation given by the parent T of T' in the HJT. The children of covering nodes are called *upper doors* and the separator of a covering node with an upper door is called an *elevator*. We say that a HJT is sum-consistent if all of its JTs are sum-consistent.

It is important to note that [11] first introduced the idea of JTs with nodes having JTs representations. He proposed the Nested JTs method to alleviate the loss of conditional independence in the construction process of the standard JT, where by standard we mean the JT with the Hugin architecture [1] without the Nested JTs and Lazy propagation methods. The motivation of the HJT is different as it is intended as a propagation framework with preservation and transparent representation of conditional independence.

5.2   Construction

In this section we describe the construction of a HJT from a BN for both the static and dynamic cases. In the first part the construction of the static case is described, which is also a description of the construction of a HJT for the first time step in the dynamic case. In the second part the we show the HJT update given that new nodes have entered the systems. Although the description is from time step 1 to time step 2, the update from time t to time $t + 1$ is analogous. This construction is illustrated with the BNs B_1 and B_2 in Figure 2.

Construction of $\mathcal{H}_1$

In this part we present the general steps for constructing a HJT $\mathcal{H}$ for a BN B. We also illustrate this steps to construct HJT $\mathcal{H}_1$ from BN B_1.

Step 1: Ancestrality partition identification. In general the ancestrality partition is of the form
$$\{I(1), H(1), I(2), \ldots, I(m-1), H(m-1), I(m)\}.$$
We identify this partition using the ancestrality-partition algorithm introduced in Section 2.

In our example, a total order consistent with the partial order induced by BN B_1 is $\{a, b, c, d, e, f, g, h, j, i\}$. The ancestrality partition for B_1 is then: $I_1(1) = \{a, b, c, d, e, f, g, h, j\}$, $H_1(1) = \{i\}$, $I_1(2) = \emptyset$.

Step 2: The construction of the directed JT for Layer 1. This JT is constructed from the BN $B_{I(1)}$ induced in the original BN B by the set $I(1)$ in the ancestrality partition. The nodes in this JT are the maximal families in $B_{I(1)}$. The edges are constructed as follows. A maximal family $Fa(v_1)$ is a parent of maximal family $Fa(v_2)$ if w_1 is a parent of w_2 in $B_{I(1)}$ where $Fa(w_1) \subseteq Fa(v_1)$ and $Fa(w_2) \subseteq Fa(v_2)$, but in most cases v_1 is a parent of v_2. Note that if $B_{I(1)}$ is not a connected BN, a junction forest will be constructed using this algorithm. Each conditional probability distribution $P(X_v|X_{Pa(v)})$ is assigned to the node that contains $Fa(v)$, and separator potentials are initialised to 1.

Let us denote by T_1^1 the JT constructed from the connected BN $B_{I_1(1)}$ displayed in Figure 5. JT T_1^1 is shown in Figure 6. For example, set $\{a, d, c\}$ is maximal family $Fa(c)$ and so, it is a node of T_1^1. Set $Fa(g) = \{d, c, g\}$ is also a node of T_1^1, but $Fa(d) = \{a, d\}$ is not a node because it is a subset of $Fa(c)$. The edge set of T_1^1 is constructed from the edge set of $B_{I_1(1)}$. For example, $Fa(c)$ is a parent of $Fa(g)$.

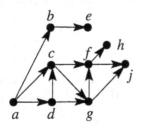

Fig. 5. BN $B_{I_1(1)}$ induced by the set $I_1(1)$ on the BN B_1 in Figure 2

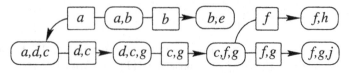

Fig. 6. The directed JT T_1^1 constructed from $B_{I_1(1)}$

Step 3: Construction of the JTs for Layer 2. In general, the JTs in Layer n, $n > 2$ are constructed in three stages. In the first stage JTs are constructed for the BN induced by $Pa(H(n-1) \cup H(n-1) \cup I(n)$. In the second stage covering nodes are identified as the universes of the JTs in Layer $n-1$, and in the third stage covering nodes are connected with directed edges to the JTs constructed in stage 1.

In our example, $Pa(H_1(1) \cup H_1(1) \cup I_1(2) = \{e, h, i\}$ and so, there is one JT consisting of a single node. There is also one covering, $\{a, b, c, d, e, f, g, h, j\}$ which is the universe of the only JT T_1^1 in Layer 1. We joined these two nodes with a directed edge from the covering node to the upper door because their intersection is not empty. we also assign potential $P(X_i|X_{e,h})$ to upper door $\{e, h, i\}$ and initialise the potential of its separator $\{e, h\}$ to 1. The resulting JT is displayed in Figure 7. Notice that covering node $\{a, b, c, d, e, f, g, h, j\}$ is not given a potential because its potential is given by JT T_1^1.

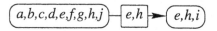

Fig. 7. The directed JT T_1^2 in Layer 2

Step 4: Construction of $\mathcal{H}_1$ edge set. We recall that the nodes of a HJT are JTs. In Steps 2 and 3 the nodes of the HJT were constructed in this step we construct the edge set. The general rule is to join JT T with a directed edge toward T' if T' has a node which is the covering node of T. In our example $\mathcal{H}_1$ has two nodes T_1^1 and T_1^2, and the latter JT contains the covering node of T_1^1. The edge set of $\mathcal{H}_1$ consists of edge (T_1^1, T_1^2). Figure 8 displays HJT $\mathcal{H}_1$.

Construction of $\mathcal{H}_2$

Nodes k, l and m are sequentially added to B_1 to obtain B_2, see Figure 2. In the following steps we update HJT $\mathcal{H}_1$ by accounting for the addition of k, l and m.

Step 1: Ancestrality partition update. The ancestrality partition is sequentially constructed using the algorithm introduced in Section 2 which

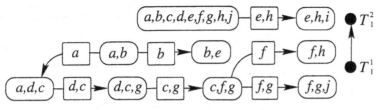

Fig. 8. Hierarchical junction tree $\mathcal{H}_1$

uses a total order of the nodes of the BN. Since the new nodes are descendants of the nodes already in the BN, they can be added at the end of the totally ordered sequence of nodes and apply the ancestrality partition algorithm.

In our example, node k is not a collider and its only parent is $i \in H(1)$, therefore $I_2(2) = \{k\}$. Node l is also not a collider and its only parent is $j \in I_1(1)$ and so, $I_2(1) = \{a, b, c, d, e, g, f, h, j, l\}$. Node m is a collider with parents $k \in I_2(2)$ and $l \in I_2(1)$, thus $H_2(2) = \{m\}$. The updated ancestrality partition is,

$$I_2(1) = \{a, b, c, d, e, g, f, h, j, l\}, \ H_2(1) = \{i\}, \ I_2(2) = \{k\}, \ H_2(2) = \{m\}.$$

Step 2: HJT update. We recall that non-covering nodes in the HJT are maximal families of the BN and that edges are also inherited from this BN. The arrival of a node is therefore translated to adding a new non-covering node in the HJT. If a new node is not a collider only a new non-covering node is added to an existing JT, while if a new node is a collider, a new JT has to be created to accommodate the extra layer added by this collider.

In our example, the addition of node k to B_1 add the non-covering node $Fa(k) = \{i, k\}$ to T_1^2 to obtain T_2^2. Since k is in Layer 2 we add $Fa(k)$ to T_1^2 with an edge from $Fa(i)$ to $Fa(k)$ because i is the parent of k. The addition of l introduces the non-covering node $Fa(l) = \{j, l\}$ to T_1^1 to T_2^1 because l is in $I_2(1)$.

Adding node m requires creating new JT T_2^3 because m is a collider. The upper door for this node is $Fa(m) = \{k, l, m\}$ and intersects JT T_2^2, thus $\{a, b, c, d, e, f, g, h, i, j, k, l\}$ is a covering node in T_2^3. In this JT we also add directed edge from this covering node to upper door $Fa(m)$. Potential $P(X_m | X_k, X_l)$ is assigned to $Fa(m)$ and separator $\{k, l\}$ is initialised to 1.

We have introduced new node T_2^3 and therefore the edge set has to be updated. The new HJT has new directed edge $\{T_2^2, T_2^3\}$. Updated $\mathcal{H}_2$ is displayed in Figure 9.

5.3   Separation

The construction of the HJT preserves the edge direction of the originating BN through the edge direction of the directed JTs and the edge direction

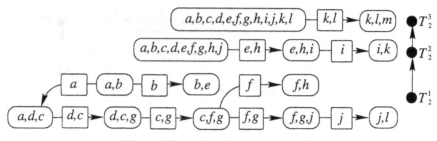

Fig. 9. Hierarchical junction tree $\mathcal{H}_2$

of the HJT joining the JTs. This provides a basis for translating Pearl's d-separation criterion [18] that enables to determine the conditional independence statements implied by the BN to to a criterion that we call *h-separation criterion* that facilitate reading the conditional independence statements implied by the HJT, [15]. In this section we describe this new criterion and and illustrate it using HJT $\mathcal{H}_2$ in Figure 9 constructed from BN B_2 in Figure 1. For a detailed discussion of the h-separation criterion see [15].

The edge direction in the originating BN is mostly preserved through the edge direction of the JTs. For example, edge $(\{a, b\}, \{b, e\})$ in $\mathcal{H}_2$ corresponds to (b, e) in B_2. However, edges from a collider's parents to that collider are not directly visible in the HJT. We called these edges *meta edges*. More specifically, if w is a collider and v is one of its parents, the corresponding meta edge to (v, w) is $(Fa(v), Fa(w))$. For example meta edge $(\{b, e\}, \{e, h, i\})$ corresponds to (e, i). Note that $\{b, e\}$ is in JT T_2^1 while $\{e, h, i\}$ is in T_2^2.

Each edge in a BN corresponds to either an edge in a JT or a meta edge, except for non-maximal families. For a proof see [15] page 121. For example, edge (b, e) corresponds to edge $(\{a, b\}, \{b, e\})$ and edge (e, i) corresponds to $(\{b, e\}, \{e, h, i\})$. $Fa(a)$ is not a maximal family and so, edges (a, b), (a, c) and (a, d) correspond to $(\{a, b\}, \{a, d, c\})$. This exception has no effect on the bijection between the set of conditional independence statements encoded in a BN and the set of conditional independence statements encoded in its corresponding HJT.

The introduction of meta edges gives a natural extension to parents and children of nodes. More formally, for either an edge in a JT of a HJT or a meta edge (C_1, C_2), C_1 is a *meta parent* of C_2 and C_2 is a *meta child* of C_1. For example, $\{b, e\}$ is a meta parent of $\{e, h, i\}$.

Undirected paths in the originating BN can now be mapped to undirected paths in the HJT. We called these paths *undirected meta paths* because meta edges are allowed in them. For example, path (b, e, i, h, f, c) is mapped to meta path $(\{a, b\}, \{b, e\}, \{e, h, i\}, \{f, h\}, \{c, f, g\}, \{d, c, g\}, \{a, d, c\})$. Note that this path contains meta edges $(\{b, e\}, \{e, h, i\})$ and $(\{f, h\}, \{e, h, i\})$. Analogously we can define *directed meta paths* by allowing directed meta edges in directed paths. Path $(\{e, h, i\}, \{i, k\}, \{k, l, m\})$ is a directed meta path.

Meta descendants can now be defined using directed meta paths. A non-covering node C_2 is a *meta descendant* of a node C_1 if there is a directed meta path from C_1 to C_2. Node $\{k, l, m\}$ is a meta descendant of $\{e, h, i\}$.

We need undirected paths that connect elements of the universe of a HJT to read conditional independence from that HJT. More specifically, for two elements v and w of the universe V of a HJT, a $\{v, w\}$ meta path is a meta path from a node that contains v to a node that contains w. Meta path $(\{a, b\}, \{b, e\}, \{e, h, i\}, \{f, h\}, \{c, f, g\}, \{d, c, g\}, \{a, d, c\})$ is actually a $\{b, c\}$ meta path.

In Pearl's d-separation theorem head-to-head nodes in an undirected path play an important role. Undirected meta paths also have head-to-head nodes. In the meta path above, node $\{e, h, i\}$ is a head-to-head because of meta edges $(\{b, e\}, \{e, h, i\})$ and $(\{f, h\}, \{e, h, i\})$ are directed towards this node.

In the construction of a HJT in Secion 5.2 we saw that non-covering nodes are families $Fa(v) = \{v\} \cup Pa(v)$ in the originating BN. If we do not have the BN, we can still write a non-covering node C in this form. In fact, $Pa(v) = \cup_r S_r$ where S_r are separators of C shared with its meta parents in the HJT, and $\{v\} = C \setminus Pa(v)$. For example node $\{b, e\}$ in T_2^1 can be written as $\{e\} \cup Pa(e)$ where $Pa(e) = \{b\}$, because $\{b\}$ is the only separator with its parent $\{a, b\}$. For a no-covering node $C = Fa(v)$, v is called the *head* of C and it is denoted by $\widehat{C}$. For example, the head of $\{b, e\}$ is e. Note that writing non-covering nodes as families gives a basis for reconstructing a BN from a HJT. In fact, there is a bijection between the set of maximal families in a BN and the set of non-covering vertices in its corresponding HJT. For a proof see [15] page 113.

In the d-separation criterion the concept of blocking undirected paths between two nodes is used. Analogously, in the h-separation criterion uses blocking of undirected paths. Formally, a $\{v, w\}$ meta path is *blocked* by a set Z, $v, w \notin Z$, if either

(a) there is a separator in this path contained in Z, or

(b) there is a head-to-head C in this path such that $\widehat{C} \cup \left(\cup_j \widehat{C_j} \right)$ does not intersect Z, where C_j is a meta descendant of C.

For example, path $(\{a, b\}, \{a, d, c\})$ is blocked by $\{a\}$. Meta path $(\{a, b\}, \{b, e\}, \{e, h, i\}, \{f, h\}, \{c, f, g\}, \{d, c, g\}, \{a, d, c\})$ is also blocked by $\{a\}$ because $\{i, k, m\}$ does not intersect $\{a\}$. Set $\{i, k, m\}$ is the set of heads of $\{e, h, i\}$ $\{i, k\}$ and $\{k, l, m\}$, and sets $\{i, k\}$ and $\{k, l, m\}$ are the meta descendants of $\{e, h, i\}$.

The definition of h-separation is based on the concept of blocking paths. For subsets A, B and C of the universe V of a HJT, A and B are *h-separated* by Z if each $\{a, b\}$ meta path, $a \in A$ and $b \in B$ are h-separated by Z. We can now state the h-separation theorem.

H-separation criterion. $X_A \perp\!\!\!\perp X_B | X_C$ if A and B are h-separated by C.

Using this criterion we deduce that $X_a \perp\!\!\!\perp X_c | X_b$ in HJT $\mathcal{H}_2$, because all $\{a, b\}$ paths are blocked by $\{a\}$.

Theorem. The set of conditional independence statements read in a BN using the d-separation criterion is the same as the set of conditional independence statements read in its corresponding HJT using the h-separation criterion. **Proof:** see [15] page 147.

The translation of the d-separation gives a proof that the HJT can be constructed without loss of conditional independence [15]. In particular note that no triangulation was required to construct the HJT.

The Madsen & Jensen's Lazy propagation method [13] also preserves the conditional independence from the originating BN in the construction of their JT. They keep the directionality of the potentials by recording them as conditional distributions, they keep the potentials for the cliques factorised, and they set the separator potentials to 1. The factorisation of the joint distribution in the BN and the JT coincide and so does the conditional independence encoded in these graphical structures. However the JT is not a transparent representation of the conditional independence as it is in the HJT: in a HJT the domain of its potentials is consistent with the nodes. For example, if $\{a, b, c\}$ is a node in a HJT, the domain of its potential would be $Sp(X_a) \times Sp(X_b) \times Sp(X_c)$, where $Sp(X_v)$ denotes the state space of variable X_v. In a JT a node can contain more elements than the domain of its associated potential.

5.4 Propagation

In this section we illustrate how propagation operates in HJTs for a propagation task when we learn a variable in the first layer $I(1)$. If we were to address a propagation task when we learn variables in a higher layer, we would need to first customise the HJT by constructing a JT for the ancestral graph of these conditioning variables. The same propagation principles can then be used for propagation in this customised HJT.

We start with the HJT in Figure 11 constructed from the BN in Figure 10. We compute the propagation task where we learn that $X_a = x'_a$ and we want to compute the distribution of X_h and X_i. At this point the potentials of the non-covering nodes $Fa(v)$ have the conditional distributions $Pr(X_v = x_v | X_{Pa(v)} = x_{Pa(v)})$, except for $Fa(d)$ that has $Pr(X_a = x_a) * P(X_d = x_d | X_a = x_a)$. All separators have been initialised to 1.

Step 1: Propagation within T^1

We enter and propagate that $X_a = x'_a$ using standard JT propagation, described in Section 3. We first choose node $\{a, d\}$ as the root and multiply its potential with the finding for x'_a. We then propagate this information through T^1 by performing a schedule of sum-flows given by a Collect Evidence operation to $\{a, d\}$, followed by a schedule of sum-flows given by a Distribute

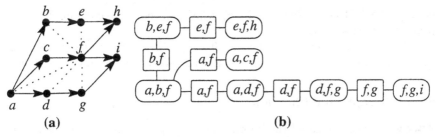

(a)   **(b)**

Fig. 10. (a) The BN in Madsen & Jensen [13]. Edges $\{e, f\}$ and $\{f, g\}$ (*dotted lines*) were added in the moralisation step, and fill-in's $\{a, f\}$, $\{b, f\}$ and $\{d, f\}$ (*dotted lines*) were added in the triangulation step. **(b)** The JT constructed from the BN in (a)

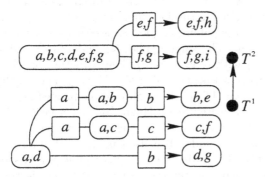

Fig. 11. A HJT for the BN in Figure 10(a)

Evidence operation from $\{a, d\}$. All the sum-flows in the Collect Evidence to $\{a, b\}$ are vacuous because they do not affect any of the potentials. This feature is a consequence of the preservation of the edge direction in the HJT. In the sum-flow from $\{b, e\}$ to $\{a, b\}$, for example, node $\{b, e\}$ has the conditional distribution $Pr(X_e = x_e | X_b = x_b)$ and when we sum-marginalise it, we obtain a potential identical to 1. The potential of the separator $\{b\}$ is already identical to 1. Thus replacing it with the same potential is unnecessary. It is also unnecessary to multiply the potential of node $\{a, b\}$ by 1.

The identification of vacuous sum-flows is also a facility of the LP method [13]. Their method records the edge direction in the potentials which is then used to identify vacuous sum-flows. The LP method is intended for general propagation tasks, that is, the computation of the marginal distribution of any set of nodes conditional of any other set of nodes. This level of generality requires triangulation which does not allow to fully take advantage of the edge direction of the original BN. The sum-flow from node $\{a, c, f\}$ to node $\{a, b, f\}$ is not vacuous because of the fill-in's $\{a, f\}$ and $\{b, f\}$ introduced in the triangulation step in the construction of the JT. In contrast the HJT allows to fully exploit the BN's edge direction and so, none of the sum-flows in the Collect Evidence to $\{a, d\}$ schedule need to be computed for the propagation task at hand.

We only need to compute the sum-flows in the Distribute Evidence from $\{a, d\}$, and after this schedule has been computed JT T^1 is sum-consistent.

Step 2: Graph simplification

After we have propagated the information through T^1, we can then continue propagating this information to the JT in one layer above, but before this we can remove unnecessary zeros encoded in T^1.

The potential of $\{a, d\}$ is a function that maps (x_a, x_d) to 0 if $x_a \neq x'_a$ and to $Pr(X_a = x'_a, X_d = x_d)$ otherwise. The potential of the separator $\{d\}$ maps x_d to $Pr(X_a = x'_a, X_d = x_d)$. The potential for $\{a, d\}$ is in the numerator of the factorisation encoded in the JT, while the potential for separator $\{d\}$ is in the denominator. These potentials quotient is either zero or one and therefore, we can remove them. The same applies for node $\{a, b\}$ and separator $\{b\}$, and for node $\{a, c\}$ with separator $\{c\}$. The potentials of separators $\{a\}$ are constant functions that take the value $Pr(X_a = x'_a)$ and so, they can be also removed. The resulting JT consist of three disconnected nodes: $\{b, e\}$, $\{c, f\}$ and $\{d, e\}$, displayed in Figure 12. Hugin [10] has a facility for compressing the tables that takes many zeros but the graph of the JT is not simplified by removing the learned nodes, because Hugin is a shell for building BNs for static systems. The HJT adjusts its graphs in line with its potentials, and so, it provides a transparent representation of the factorisation of potentials that it encodes.

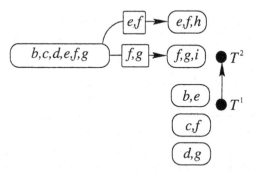

Fig. 12. The HJT after the graph simplification

Step 3: Propagation within T^2

At this point the potentials of the non-covering nodes have the conditional distribution inherited from the original BN, and the separator potentials are set to 1. The sum-flows in the Collect Evidence schedule to the covering node $\{b, c, d, e, f, g\}$ are all vacuous for the same reasons as we discussed in Step 1. The sum-flows in the Distribute Evidence from $\{b, c, d, e, f, g\}$ are computed. Note that for the sum-flow from $\{b, c, d, e, f, g\}$ to $\{e, f, h\}$, we compute the marginals for X_f and X_e from the disconnected nodes and then we assign this product to the separator $\{e, f\}$. If f and e were in the same JT but not in the same node, we would need to use the method of firing variables for computing their joint distribution, see [5]. The nested junction trees method also uses this type of propagation, [11].

Step 4: Graph Simplification

The purpose of the layers in a HJT is to break cycles. Now that JT T^1 has unconnected nodes, we do not need to have the second layer because there is no cycle to break. We can therefore remove the covering node $\{b, c, d, e, f, g\}$ and connect the upper doors $\{e, f, h\}$ and $\{f, g, i\}$ with the nodes that intersect them. The resulting JT is displayed in Figure 13. The HJT is now sum-consistent and so, the potentials of $\{e, f, h\}$ and $\{f, g, i\}$ now encodes the joint distributions of their associated variables. For example the potential of $\{e, f, h\}$ contains the joint distribution of (X_e, X_f, X_h) from where the marginal distribution of X_h can be computed. Analogously, we can compute the marginal distribution of X_i from the potential of $\{f, g, i\}$.

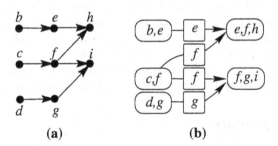

(a)   **(b)**

Fig. 13. (a) The BN obtained from the BN in Figure 10(a) after learning $X_a = x'_a$ and deleting it. (b) The HJT obtained from the HJT in Figure 12 after graph simplification

Notice that if we construct a HJT from the BN in Figure 13(a), we would obtained the HJT in Figure 13(b). This implies that the conditional independence recorded in this BN is the same as the conditional independence in this HJT. This exemplifies the use of incoming conditional independence to simplify the HJT, and so keeping the HJT to the lowest size as possible.

6   Conclusions

We have now defined a formal graphical structure that efficiently codes all conditional independence statements in the BN in a form compatible with the construction of propagation algorithms. It transparently represents the factorisation of the joint distribution of the variables in the system, including features of the factorisation used in the lazy propagation method. This transparency enables us to explore new ways of propagating information for structure inference.

An important building block of the HJT is the ancestrality partition of a BN. This new concept gives a decomposition of a non-decomposable BN into decomposable components. This decomposition facilitated proving the h-separation. We believe that the ancestrality partition may have an important role in providing a framework for proving theorems on BNs.

We have proven with the h-separation theorem that the HJT is constructed with no loss of conditional independence. The HJT construction uses decomposable components of a BN and so, no moralisation and triangulation is required. Propagation tasks only require the moralisation and triangulation of the ancestral graph of the conditioning variables. Total inferential efficiency is achieved when this ancestral graph is decomposable because no conditional independence is lost in this task.

New algorithms can be customised via the HJT to pre-specified inferential tasks and classes of BNs which are not decomposable. In this paper we briefly considered such case. Inferential efficiency gains have important computational cost savings in exact forecasting in dynamic BNs with heterogeneous evolution. We saw that when using JTs for exact forecasting, on-line triangulation is inevitable, even when the lazy propagation method is used because this method operates on a JT. In this dynamic setting, computational savings are achieved with the HJT potentials and graph simplification, which utilises the incoming conditional independence.

We believe that HJTs provide an ideal framework for the development of architectures for probability propagation in highly structured settings.

Acknowledgements

We wish to thank Dr. Lorenz Wernisch for his helpful comments. Part of this research was developed while R. O. Puch was sponsored by Wellcome Trust programme grant GR066790MA.

References

1. SK Andersen, KG Olesen, FV Jensen, and F Jensen. Hugin – a shell for building Bayesian belief universes for expert systems. In *Proceedings of the 11th International Joint Conference on Artificial Intelligence*, pages 1080–5. Morgan Kaufmann, San Mateo, California, NS Shridharan (ed.), 1989.

2. RG Cowell, AP Dawid, SL Lauritzen, and DJ Spiegelhalter. *Probabilistic networks and expert systems.* Springer Verlag, 1999.

3. AP Dawid. Applications of a general propagation algorithm for probabilistic expert systems. *Statistics and Computing,* 2:25–36, 1992.

4. Hugin. A shell for building Bayesian networks. *www.hugin.dk,* 2002.

5. FV Jensen. *An introduction to Bayesian networks.* University College London Press, London, United Kingdom, 1996.

6. FV Jensen, SL Lauritzen, and KG Olesen. Bayesian updating in causal probabilistic networks by local computations. *Computational Statistics Quarterly,* 4:269–282, 1990.

7. K Kanazawa, D Koller, and S Rusell. Stochastic simulation algorithm for dynamic probabilistic networks. pages 346–351. Morgan Kaufmann, San Francisco, California, 1995.

8. U Kjaerulff. Triangulation of graphs – algorithms giving small total state spaces. *Research Report,* Institute of Electronic Systems, Department of Mathematics and Computer Science, Aalborg University, Aalborg, Denmark., 1990.

9. U Kjaerulff. Optimal decomposition of probabilistic networks by simulated annealing. *Statistics and Computing,* 2:7–17, 1992.

10. U Kjaerulff. dHugin: A computational system for dynamic time-sliced Bayesian networks. *International Journal of Forecasting,* Special Issue on Probability Forecasting, 11:89–111, 1995.

11. U Kjærulff. Inference in Bayesian networks using nested junction trees. In *Learning in graphical models, (Ed. M. I. Jordan),* pages 51–74. Kluwer Academic Publishers, Dordrecht, The Netherlands, 1998.

12. P Larrañaga, CMH Kuijpers, M Poza, and RH Murga. Decomposing Bayesian networks: triangulation of the moral graph with genetic algorithms. *Statistics and Computing,* 7:19–34, 1997.

13. AL Madsen and FV Jensen. Lazy propagation in junction trees. In *Proceedings of the 14th Annual Conference on Uncertainty in Artificial Intelligence (Eds. GF Cooper and S Moral),* pages 362–9. Morgan Kaufmann, San Francisco, California, 1998.

14. AL Madsen and FV Jensen. Lazy propagation: a junction tree inference algorithm based on lazy evaluation. *Artificial Intelligence,* 113:203–245, 1999.

15. RO Puch. *Hierarchical junction trees.* PhD thesis, Department of Statistics, University of Warwick, available at "www.warwick.ac.uk/staff/R.Puch-Solis", 2000.

16. JQ Smith, AE Faria, S French, D Ranyard, D Vlesshhouwer, J Bohunova, T Duranova, M Stubna, L Dutton, C Rojas, and A Sohier. Probabilistic data assimilation within RODOS. *Radiation Proteccion Dosimetry,* 73(1-4):57–9, 1997.

17. JQ Smith and KN Papamichail. Fast Bayes and the dynamic junction forest. *Artificial Intelligence,* 107:99–124, 1999.

18. T Verma and J Pearl. Causal networks: Semantics and expressiveness. In *Uncertainty in Artificial Intelligence 4,* pages 69–76, New York, N. Y., 1988. Elsevier Science Publishing Company, Inc.

Algorithms for Approximate Probability Propagation in Bayesian Networks

Andrés Cano[1], Serafín Moral[1], and Antonio Salmerón[2]

[1] Dept. Computer Science and Artificial Intelligence,
University of Granada,
Daniel Saucedo Aranda s/n,
E-18071 Granada, Spain
[2] Dept. Statistics and Applied Mathematics,
University of Almería,
La Cañada de San Urbano s/n,
E-04120 Almería, Spain

Abstract. When a Bayesian network is defined over a very large or complicated domain, computing the posterior probabilities given some evidence may become unfeasible. In fact, probability propagation is known to be an NP-hard problem. Since it is common to find huge domains in practical applications, approximate algorithms have been developed. These algorithms compute estimations of the posterior probabilities with lower requirements, in terms of memory and computing time, than exact algorithms. In this paper we present some of the most recent developments in the area of approximate propagation algorithms.

1 Approximate Probability Propagation

Throughout this chapter, we will consider a Bayesian network defined for a set of variables $\mathbf{X} = \{X_1, \ldots, X_n\}$, where each variable X_i takes values on a finite set Ω_{X_i}. If I is a set of indices, we will write $\mathbf{X}_I$ for the set $\{X_i | i \in I\}$, and $\Omega_{\mathbf{X}_I}$ will denote the Cartesian product $\times_{i \in I} \Omega_{X_i}$. Given $\mathbf{x} \in \Omega_{\mathbf{X}_I}$ and $J \subseteq I$, $\mathbf{x}_J$ is the element of $\Omega_{\mathbf{X}_J}$ obtained from $\mathbf{x}$ by dropping the coordinates not in J.

Definition 1. Let $\mathbf{E} \subset \mathbf{X}$ be a set of observed variables and $\mathbf{ev} \in \Omega_{\mathbf{E}}$ the instantiated value. An algorithm which computes $p(x|\mathbf{ev})$ for each $x \in \Omega_X$, $X \in \mathbf{X} \setminus \mathbf{E}$ is called an *exact propagation algorithm*.

Several exact propagation algorithms that compute the posterior distributions by local computations have been proposed in the last decade [20,25,27,36,37]. Local computation consists of calculating the marginals without actually computing the joint distribution, and it is described in terms of a message passing scheme over a structure called a *join tree*. If the problem is too complicated, however, the application of these schemes may become unfeasible due to a high requirement of resources (computing time and memory).

In order to deal with more complicated problems, approximate propagation algorithms have been developed. These algorithms provide inexact results, but with a lower resource requirement. The formal definition is as follows.

Definition 2. Let $\mathbf{E} \subset \mathbf{X}$ be a set of observed variables and $\mathbf{ev} \in \Omega_{\mathbf{E}}$ the instantiated value. An algorithm which computes $\hat{p}(x|\mathbf{ev})$, an estimation of $p(x|\mathbf{ev})$, for each $x \in \Omega_X$, $X \in \mathbf{X} \setminus \mathbf{E}$ is called an *approximate propagation algorithm*.

Several approximate propagation algorithms have been developed, which can be classified into two groups: *deterministic* and *Monte Carlo* algorithms.

The more recent deterministic algorithms [5–7,12] are based on approximating the operations carried out during the propagation (combination and marginalisation). However, other ideas have been applied before, as replacing by zero low probability values [19], removing weak dependencies in order to reduce the complexity of the network [21] or enumerating the configurations of highest probability [33,34].

On the other hand, Monte Carlo methods consist of transforming the task of probability propagation into an statistical estimation problem: a sample is drawn from the Bayesian network and the posterior probabilities are estimated from it [8,9,17,18,31,32].

2   The Complexity of Probability Propagation

The complexity of probability propagation is determined by the primitive operations over probabilistic potentials in which it is based. A *probabilistic potential* for a set of variables $\mathbf{X}$ with domain $\Omega_{\mathbf{X}}$ is a non-negative real-valued function $f : \Omega_{\mathbf{X}} \rightarrow \mathbb{R}$. Probabilistic potentials represent any probabilistic information [25], as marginal, joint and conditional distributions or intermediate results of operations amongst them. The primitive operations above mentioned are the *marginalisation*, *combination* and *restriction*, which can be defined as follows.

Definition 3. Let f be a potential defined over a set of variables $\mathbf{X}$. Given $\mathbf{Y} \subset \mathbf{X}$ and $\mathbf{Z} = \mathbf{X} \setminus \mathbf{Y}$, the *marginalisation* of f to $\mathbf{Y}$ is defined as

$$f^{\downarrow \mathbf{Y}}(\mathbf{y}) = \sum_{\mathbf{z} \in \Omega_{\mathbf{z}}} f(\mathbf{y}, \mathbf{z}), \quad \forall \mathbf{y} \in \Omega_{\mathbf{Y}} . \tag{1}$$

Definition 4. Given two potentials f_1 and f_2 defined for $\mathbf{X} \cup \mathbf{Y}$ and $\mathbf{Y} \cup \mathbf{Z}$ respectively, with $\mathbf{X} \cap \mathbf{Z} = \emptyset$, its *combination* is a potential f defined for $\mathbf{X} \cup \mathbf{Y} \cup \mathbf{Z}$ as

$$f(\mathbf{x}, \mathbf{y}, \mathbf{z}) = f_1(\mathbf{x}, \mathbf{y}) \cdot f_2(\mathbf{y}, \mathbf{z}) \quad \forall (\mathbf{x}, \mathbf{y}, \mathbf{z}) \in \Omega_{\mathbf{X} \cup \mathbf{Y} \cup \mathbf{Z}} . \tag{2}$$

Definition 5. Let $\mathbf{X}$, $\mathbf{Y}$ and $\mathbf{Z}$ be three disjoint sets of variables, f a potential over variables $\mathbf{X} \cup \mathbf{Y}$, and $(\mathbf{y}, \mathbf{z}) \in \Omega_{\mathbf{Y} \cup \mathbf{Z}}$. The *restriction* of f to $(\mathbf{y}, \mathbf{z})$ is a potential defined for variables $\mathbf{X}$ as

$$f^{R(\mathbf{Y}=\mathbf{y}, \mathbf{Z}=\mathbf{z})}(\mathbf{x}) = f(\mathbf{x}, \mathbf{y}) \quad \forall \mathbf{x} \in \Omega_{\mathbf{X}} . \tag{3}$$

In general, the number of values necessary to specify a potential is an exponential function of the number of variables for which it is defined. Therefore, the complexity of probability propagation arises when the domain of the potential resulting from a combination is too large.

In the next we will consider two exact propagation algorithms, the variable elimination [26,38,11] and the Shenoy-Shafer [37,36] methods. Afterwards, we will show how approximate versions of them can be derived, namely, the mini-bucket scheme [12] and the Penniless propagation [5,7] respectively.

3   The Variable Elimination Algorithm

This method was indepently proposed by Li and D'Ambrosio [26], Zhang and Poole [38] and Dechter [11]. It is designed to compute the posterior distribution over a goal variable $Z \in \mathbf{X}$, where $\mathbf{X}$ is the set of all the variables in the Bayesian network. It is based on sequentially deleting the variables in $\mathbf{Y} = \mathbf{X} \setminus \{Z\}$ from the probabilistic potentials containing them. The deletion of a variable from a set of potentials is carried out by combining the potentials containing it and marginalising it out from the result of the combination. This process is detailed in the next algorithm, which deletes a variable Y from a set of potentials T:

Delete(T**,**Y**)**

1. Let $T_Y = \{f \in T | Y \in \mathrm{dom}(f)\}$.
2. $q_Y := \prod_{f \in T_Y} f$.
3. $r_Y := \sum_{y \in \Omega_Y} q_Y$.
4. $T := (T \setminus T_Y) \cup \{r_Y\}$.
5. Return T.

Using the above procedure, an exact propagation algorithm can be specified, starting with a set T containing the conditional distribution in the Bayesian network (restricted to the observations in the evidence) and the potentials associated with the observations. These *evidence-potentials* are defined as follows.

Definition 6. Let $X \in \mathbf{E}$ be a variable which has been observed to take the value x_0. The *evidence-potential* corresponding to this observation is

$$\delta_X(x; x_0) = \begin{cases} 1 \text{ if } \quad x = x_0 , \\ 0 \text{ otherwise} . \end{cases} \tag{4}$$

The following algorithm computes the posterior distribution for a variable $Z \in \mathbf{X}$ given the observations in $\mathbf{E}$.

Variable_Elimination($\mathbf{X}$,Z,E)

1. Let p_i, $i = 1, \ldots, n$, be the conditional distribution for each variable $X_i \in \mathbf{X}$.
2. $T := \{p_i^{R(\mathbf{E}=\mathbf{ev})}, \; i = 1, \ldots, n\} \cup \{\delta_X | X \in \mathbf{E}\}$.
3. For each $Y \in \mathbf{X} \setminus \{Z\}$,
    $T :=$ **Delete**(T,Y).
4. $p := \prod_{f \in T} f$.
5. Normalise p.
6. Return p.

3.1   Mini-bucket Elimination

The complexity of the variable elimination algorithm is determined by procedure **Delete**(T,Y), since the potential q resulting from the combination in step 2. may be too large. This operation can be simplified if variable Y is marginalised out from the potentials containing it before performing the combination. In general, this is an approximation, but the advantage is that the potentials intervening in the combination are defined over smaller domains. For example, consider two potentials $f_1(x_1, x_2)$ and $f_2(x_2, x_3)$, where X_1, X_2 and X_3 are binary variables. If we want to delete X_2, the exact procedure would be to combine f_1 and f_2, obtaining a potential f defined for X_1, X_2 and X_3, which requires a probability table with $2^3 = 8$ values. However, if we first marginalise out X_2 from f_1 and f_2, the combination requires a probability table with only 4 values.

An intermediate solution could be to combine just some of the potentials before marginalising out. The fundamental of this idea [16] is that there is a case in which combination and marginalisation commute: This is the case in which the potential involved is constant. In general, the more potentials are combined before marginalisation, the more accurate the result is. In relation to this, heuristics can be defined in order to reach a compromise between accuracy of the result and complexity. Some heuristics are proposed in [17].

Taking this into account, procedure **Delete**(T,Y) can be re-formulated, resulting in the so-called *mini-bucket elimination* [12] which can be described as follows.

Mini_Bucket_Elimination(T,Y)

1. Let $T_Y = \{f \in T | Y \in \mathrm{dom}(f)\}$.
2. Let $R = \{R_1, \ldots, R_j\}$ be a partition of T_Y.
3. For $i := 1$ to j
    (a) $q_i := \prod_{f \in R_i} f$.

(b) $q_i := \sum_{y \in \Omega_Y} q_i.$

4. $r := \prod_{i=1}^{j} q_i.$

5. $T := (T \setminus T_Y) \cup \{r\}.$

6. Return T.

4   Shenoy-Shafer Propagation

This algorithm operates over a join tree constructed from the Bayesian network, and the propagation is carried out sending messages between neighbour nodes in the join tree, in order to collect and distribute the information contained in the different parts of the tree. The message from a node V_i to one of its neighbours, V_j, is computed as

$$
\phi_{V_i \to V_j} = \left\{ \phi_{V_i} \cdot \left(\prod_{V_k \in ne(V_i) \setminus V_j} \phi_{V_k \to V_i} \right) \right\}^{\downarrow V_i \cap V_j}, \tag{5}
$$

where ϕ_{V_i} is the initial probability potential on V_i, $\phi_{V_k \to V_i}$ are the messages from V_k to V_i and $ne(V_i)$ are the neighbour nodes of V_i.

The complexity of this scheme is determined by the computation of the messages, which, as it can be seen in Eq. (5), involves the combination of the incoming messages with the potential stored in the node that is sending the message. The result of this operation is a potential of size equal to the product of the number of cases of the variables contained in the node which sends the message. In general, it can be said that the complexity is related to the size of the largest node in the join tree.

A first approach to the reduction of the complexity is to use a flexible data structure that allows the substitution of very similar values by a single one (for instance, the average of them) which is then stored only once. Such data structure can be a probability tree [2,4,5,31].

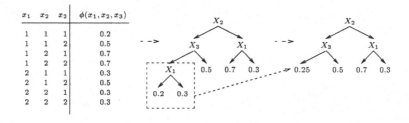

Fig. 1. A potential ϕ, a probability tree representing it and an approximation of it after pruning some branches.

A *probability tree* is a directed labeled tree, where each internal node represents a variable and each leaf node represents a probability value. Each internal node has one outgoing arc for each state of the variable associated with that node. Each leaf contains a non-negative real number. The *size* of a tree $\mathcal{T}$, denoted as $size(\mathcal{T})$, is defined as its number of leaves.

A probability tree $\mathcal{T}$ on variables $\mathbf{X}_I = \{X_i | i \in I\}$ represents a potential $\phi : \Omega_{\mathbf{X}_I} \to \mathbb{R}_0^+$ if for each $\mathbf{x}_I \in \Omega_{\mathbf{X}_I}$ the value $\phi(\mathbf{x}_I)$ is the number stored in the leaf node that is reached by starting from the root node and selecting the child corresponding to coordinate x_i for each internal node labeled with X_i.

A probability tree is usually more a compact representation of a potential than a table. This fact is illustrated in Figure 1, which displays a potential ϕ and its representation using a probability tree. The tree contains the same information as the table, but using five values instead of eight. Furthermore, trees allow to obtain even more compact representations in exchange of loosing accuracy. This is achieved by pruning some leaves and replacing them by the average value, as shown in the second tree in Figure 1.

The basic operations (*combination, marginalisation* and *restriction*) over potentials required for probability propagation can be carried out directly over probability trees, as described in the next algorithms.

The combination is done recursively and basically consists of selecting a starting node and multiply each of its children by the other tree. This operation is illustrated in Figure 2, and the details of the algorithm can be found in [5].

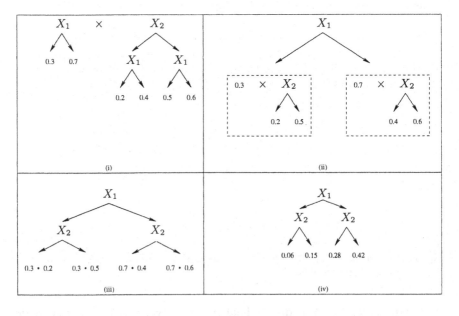

Fig. 2. Combination of two probability trees.

Marginalisation is equivalent to deleting variables. A variable is deleted from a probability tree replacing it by the sum of its children. For example, variable X_3 is marginalised out from the tree in Figure 3 adding its two children, as described in Figure 4. The details can be found in [5].

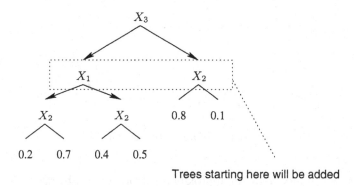

Trees starting here will be added

Fig. 3. Marginalising out variable X_3.

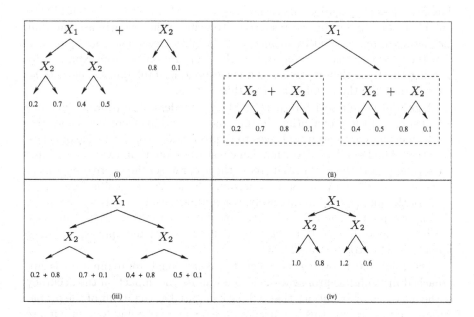

Fig. 4. Addition of two probability trees.

The restriction operation in a tree consists of returning the part of the tree which is consistent with the values of the instantiated variables. It is described in figure 5.

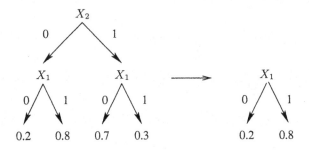

Fig. 5. Restriction of a tree to the value $X_2 = 0$.

Using probability trees, an approximate algorithm can be developed from the Shenoy-Shafer scheme. This algorithm is the so-called *Penniless propagation*.

4.1   Penniless Propagation

The Penniless propagation algorithm is based on Shenoy-Shafer's, but able to provide results under limited resources. To achieve this, we will assume that messages are represented by means of probability trees. The consequence is that the messages that are sent during the propagation are approximated by pruning the trees that represent the corresponding probabilistic potentials. We will describe the pruning later in this section.

Another difference with respect to Shenoy-Shafer's algorithm is the number of stages of the propagation. In the exact algorithm, there are two stages: firstly, messages are sent from leaves to the root and secondly, in the opposite direction. The Penniless algorithm can consist of more than two stages. After the second phase, the goal is to increase the accuracy of the approximate messages at each stage, by taking into account the messages coming from other parts of the join tree. More precisely, when a message is sent throw an edge, it is approximated conditional on the message contained in the same edge but in the opposite direction. The idea behind conditional approximation of potentials was proposed in [22,23].

The criterion according to which the conditional approximation is approached in Penniless propagation is to minimise the impact on the accuracy of the marginal distributions that will be computed as the result of the propagation process. Taking this into account, assume we are going to approximate a message, $\phi_{V_i \to V_j}$, between two nodes V_i and V_j in the join tree, by another potential $\phi'_{V_i \to V_j}$. After the propagation, the marginal probability for variables in $V_i \cap V_j$ will be proportional to $\phi_{V_i \to V_j} \cdot \phi_{V_j \to V_i}$. So $\phi'_{V_i \to V_j}$ should be computed trying to minimise the value of conditional information, given by:

$$D(\phi_{V_i \to V_j}, \phi'_{V_i \to V_j} | \phi_{V_j \to V_i}) = D(\phi_{V_i \to V_j} \cdot \phi_{V_j \to V_i}, \phi'_{V_i \to V_j} \cdot \phi_{V_j \to V_i}) , \quad (6)$$

where $D(\cdot, \cdot)$ is the Kullback-Leibler divergence [24].

The problem is that when a message $\phi_{V_i \to V_j}$ is sent, usually the opposite message, $\phi_{V_j \to V_i}$, is not exact, but an approximation[1] of it, $\phi'_{V_j \to V_i}$, and the distance in Equation (6) is conditioned to this approximation. The first time $\phi_{V_j \to V_i}$ is computed, $\phi'_{V_i \to V_j}$ is not available. Once we have it, we can use it to compute a better approximation of $\phi_{V_j \to V_i}$ by conditioning on it and so on. This process can continue until no further change is achieved in the messages.

If in a given moment some message, say $\phi_{V_i \to V_j}$, is exact[2], then there is no need to try to improve it by iterating as described above. Thus, even though the Penniless algorithm can carry out several iterations, only 2 of them (the first and the last) are necessary for all the messages, the rest will be carried out only through approximate messages. It implies that three different types of stages can be distinguished: the first stage, the intermediate ones and the last stage.

The goal of the first stage is to collect information from the entire join tree in the root node, in order to distribute it to the rest of the graph in a posterior stage. Messages are sent from leaves to the root, and a flag is kept in every message indicating whether its computation was exact or approximate.

In the intermediate stages, the approximate messages are updated according to the information coming from other parts of the tree. To achieve this, messages are sent in both directions: Firstly, from the root towards the leaves and secondly, from the leaves upwards. When a message is going to be sent through an edge, the flag of the message in the opposite direction is checked; if that message is labeled as exact, the messages in the subtree determined by that edge are no further updated. The reason to do this is that sending messages through that edge will not help to better approximate the messages in the opposite directions, since those messages are already exact.

In the last stage the propagation is completed sending messages from the root to the leaves. In this case, messages are sent downwards even if the message stored in the opposite direction is exact. This is done to assure that at the end of the propagation all the cliques in the join tree have received the corresponding messages.

The pseudo-code for the algorithm is as follows.

---

[1] Initially if no message has been sent this message is the potential equal to 1 for all the configurations

[2] A message is exact when all the potentials used in equation 5 to calculate it are exact

ALGORITHM PennilessPropagation($\mathcal{J}$,*stages*)

INPUT: A join tree $\mathcal{J}$ and the number of propagation stages.
OUTPUT: Join tree $\mathcal{J}$ after propagation.

1. Select a root node R.
2. **NavigateUp(R)**
3. *stages* := *stages* − 1
4. WHILE *stages* > 2
        NavigateDownUp(R)
        stages := *stages* − 2
5. IF (*stages* == 1)
        NavigateDown(R)
   ELSE
        NavigateDownUpForcingDown(R)

The first stage is carried out by calling procedure **NavigateUp**. This procedure requests a message from each one of its neighbours, which recursively do the same to all its neighbours downwards until the leaves are reached. Then, messages are sent upwards until the root is reached. This task is implemented by the following two procedures.

NavigateUp(R)

1. FOR each $V \in ne(R)$,
        NavigateUp(R,V)

NavigateUp(S,T)

1. FOR each $V \in ne(T) \setminus \{S\}$,
        NavigateUp(T,V)
2. **SendApprMessage(T,S)**

Once the root node has received messages from all its neighbours, the intermediate stages begin. This is carried out by the next two procedures. Observe that exact messages are not updated.

NavigateDownUp(R)

1. FOR each $V \in ne(R)$
        IF $\phi_{V \to R}$ is not exact
            IF $\phi_{R \to V}$ is not exact
                SendApprMessage(R,V)
            NavigateDownUp(R,V)

NavigateDownUp(S,T)

1. FOR each $V \in ne(T) \setminus \{S\}$
   IF $\phi_{V \to T}$ is not exact
      IF $\phi_{T \to V}$ is not exact
         SendApprMessage(T,V)
      NavigateDownUp(T,V)
2. IF $\phi_{T \to S}$ is not exact
   SendApprMessage(T,S)

Finally, the last step is carried out by calling the procedures described next. After this, the posterior marginal for any variable in the network can be computed by selecting any node containing that variable, calculating the product of the incoming messages and the potentianl in this node, and marginalising this new calculated potential.

Observe that after the intermediate stages, if the total number of stages is odd, we still have to perform two traversals, one downwards and one upwards. The difference with respect to the intermediate steps is that in this case, messages are sent downwards even if they are marked as exact, in order to assure that the posterior marginals can be obtained in any node.

NavigateDownUpForcingDown(R)

1. FOR each $V \in ne(R)$
   IF $\phi_{V \to R}$ is not exact
      IF $\phi_{R \to V}$ is not exact
         SendApprMessage(R,V)
      NavigateDownUpForcingDown(R,V)
   ELSE
      IF $\phi_{R \to V}$ is not exact
         SendApprMessage(R,V)
      NavigateDown(R,V)

NavigateDownUpForcingDown(S,T)

1. FOR each $V \in ne(T) \setminus \{S\}$
   IF $\phi_{V \to T}$ is not exact
      IF $\phi_{T \to V}$ is not exact
         SendApprMessage(T,V)
      NavigateDownUpForcingDown(T,V)
   ELSE
      IF $\phi_{T \to V}$ is not exact
         SendApprMessage(T,V)
      NavigateDown(T,V)
2. **SendApprMessage(T,S)**

However, if the total number of stages is even, in the final stage we just have to send messages downwards, which is implemented by the next procedures.

NavigateDown(R)

1. FOR each $V \in ne(R)$
   SendApprMessage(R,V)
   NavigateDown(R,V)

NavigateDown(S,T)

1. FOR each $V \in ne(T) \setminus \{S\}$
   SendApprMessage(T,V)
   NavigateDown(T,V)

Now we show the details of procedure **SendApprMessage**, used in the algorithms above. A message $\phi_{S \to T}$ is computed by combining all the incoming messages of S except that one coming from T with the potential in S, and then, the resulting potential is approximated according to the message coming from T to S, $\phi_{T \to S}$.

SendApprMessage(S,T)

1. Compute

$$\phi = \left(\prod_{V \in ne(S) \setminus \{T\}} \phi_{V \to S} \right)^{\downarrow S}.$$

2. IF at least one of the messages $\phi_{V \to S}$ is not exact, mark ϕ as approximate.
3. Compute $\phi_{S \to T} = \phi \cdot \phi_S$.
4. IF the size of $\phi_{S \to T}$ is too big
   (a) Approximate $\phi_{S \to T}$ conditional on $\phi_{T \to S}$.
   (b) Mark $\phi_{S \to T}$ as approximate.

5   Monte Carlo Algorithms for Probability Propagation

The approach followed by Monte Carlo algorithms is completely different to the case of deterministic methods. We will try to explain what is the fundamental behind simulation-based propagation algorithms.

A Bayesian network is actually a representation of a joint probability distribution over a set of variables **X**. In fact, a probability distribution can be understood as a means of describing a population. In the case of a Bayesian network, the population described by its associated distribution is composed by all the possible configurations of the variables in **X**. In other words, the distribution indicates the frequency of each individual in the population. If the entire population were available, i.e. we could manipulate the information about all the individuals (configurations of the variables) then the problem of probability propagation could be solved exactly, since the probabilities in Definition 1 could be computed by counting cases in the population.

However usually the size of the population is huge, and counting cases is by no means feasible. Instead of working with the entire population, Monte Carlo methods operate drawing a sample from it using some random mechanism. This sample, whose size is much lower than the population's, is used to estimate the conditional probabilities for each variable of interest.

There are two key issues in a Monte Carlo propagation algorithm:

1. The sampling mechanism.
2. The functions (estimators) which compute the probabilities from the sample.

According to the solution adopted to implement these two features, Monte Carlo algorithms can be divided into two groups: *Markov Chain Monte Carlo* algorithms and *importance sampling* methods.

5.1   Markov Chain Monte Carlo (MCMC)

These methods start off with an initial configuration of the variables in the network consistent with the observations and with positive probability. Afterwards, each new individual in the sample is drawn conditional on the previous one. When a sample is drawn like this, it is said that the individuals are not independent, but they verify the Markov property.

The first MCMC method in Bayesian networks was the so-called *stochastic simulation* proposed by Pearl [29]. The sampling procedure follows the next rules:

1. The variables are not simulated in any special order.
2. Each variable is simulated with a distribution conditional on its *Markov blanket*[3].

For a Bayesian network with variables $\mathbf{X} = \{X_1, \ldots, X_n\}$, the detailed algorithm is as follows.

Stochastic_Simulation

1. Instantiate the variables in the network to a configuration with positive probability.
2. For each $X_i \notin \mathbf{E}$, do $h_i(x_i) = 0$ for all $x_i \in \Omega_{X_i}$.

---

[3] The Markov blanket of a variable X in a Bayesian network is the set of its parents, children and parents of its children except X.

3. From $j = 1$ to m
   (a) For each $X_i \notin \mathbf{E}$,
       i. Compute $p(x_i|w_{X_i})$, where w_{X_i} is the Markov blanket of X_i, as:

$$p(x_i|w_{X_i}) = \alpha p(x_i|pa(x_i)) \prod_{y \in w_{X_i}} p(y|pa(y)) \quad \forall x_i \in \Omega_{X_i} , \quad (7)$$

   where α is a normalisation constant and w_{X_i} is the current configuration of the Markov blanket of X_i.
       ii. Simulate a value $x_i^{(j)} \in \Omega_{X_i}$ for X_i using $p(x_i|w_{X_i})$.
       iii. Update h_i:

$$h_i(x_i^{(j)}) = h_i(x_i^{(j)}) + 1 ,$$

4. Normalise h_i, $i = 1, \ldots, n$. Each h_i is the estimation of the posterior distribution for variable X_i.

where m is the sample size.

This method has two main problems. On the one hand, it may be difficult to find an initial configuration with positive probability. Jensen, Kong and Kjærulff [18] propose to use forward sampling to find such a configuration.

On the other hand, each configuration depends on the preceding one (see Eq. (7)). It is possible then, that when a configuration has appeared, it is generated again many consecutive times, if there are "almost functional" dependencies among the variables, i.e. the distributions generated in formula (7) contain extreme values. The convergence of this method to the exact distribution is guarantied when all the probabilities are strictly positive, even though this convergence can be achieved very slowly, due to the reason explained above. The next example illustrates this fact.

Example 1. Let be a Bayesian network with two variables connected as $X_1 \rightarrow X_2$ such that $p(x_2|x_1) = p(\bar{x}_2|\bar{x}_1) = \delta \simeq 1$. Assume that $p(x_1) = 0.5$ and $X_1 = x_1$, then $p(x_2|w_{x_2}) = p(x_2|x_1) = \delta$. If during the simulation process we obtain $X_2 = x_2$, the distribution that will be used to generate the next value for X_1 is

$$p(x_1|w_{x_1}) = p(x_1|x_2) = \alpha p(x_2|x_1)p(x_1) = \alpha \times 0.5 \times \delta = \delta ,$$

since, by Bayes' rule, $\delta = 1/p(x_2)$, and

$$p(x_2) = p(x_2|x_1)p(x_1) + p(x_2|\bar{x}_1)p(\bar{x}_1) = \delta \times 0.5 + (1 - \delta) \times 0.5 = 0.5 .$$

In this way, we obtain the configuration $(X_1 = x_1, X_2 = x_2)$ with probability very close to 1, and, if a variable ever changed its value, the other one would change as well, thus appearing many times the configuration $(X_1 = \bar{x}_1, X_2 = \bar{x}_2)$. This illustrates how the configuration that is drawn can strongly depend on the previously generated.

Trying to overcome this problem, Jensen, Kong and Kjærulff [18] proposed the so-called *blocking Gibbs sampling*. These authors realised that the problem of stochastic simulation are due to the dependencies in the sample, in the sense that variables are changed one at a time.

Blocking Gibbs sampling is a sophisticated method based on a compromise between dependency in the sample and computational cost, considering the extreme cases:

1. Sampling a single variable given its Markov blanket is computationally easy, but the resulting sample can be strongly dependent.
2. Sampling all the variables simultaneously make the individuals in the sample be independent, but the computational cost can be unaffordable.

The method consists of partitioning the set of variables in the network so that the variables in a group are simulated simultaneously. The bigger a group is, the lower the dependencies among the individuals in the sample will be, but the complexity of obtaining the joint distribution for the variables in the group, required to sample them, will be higher as well.

5.2   Importance Sampling Algorithms

Importance sampling is well known as a variance reduction technique for estimating integrals by means of Monte Carlo methods (see, for instance, [30]). Here we study how to use it to estimate conditional probabilities.

The technique is based on sampling configurations $\mathbf{x} \in \Omega_{\mathbf{X}}$ with a given probability function $p^* : \Omega_{\mathbf{X}} \to [0, 1]$, verifying that $p^*(\mathbf{x}) > 0$ for every point $\mathbf{x}$ such that $p(\mathbf{x}, \mathbf{ev}) > 0$. Then for each configuration, its weight is computed:

$$w(\mathbf{x}) = \frac{p(\mathbf{x}, \mathbf{ev})}{p^*(\mathbf{x})} \ . \tag{8}$$

If we obtain a sample of size m $(\mathbf{x}^1, \ldots, \mathbf{x}^m)$ with its corresponding weights $(w(\mathbf{x}^1), \ldots, w(\mathbf{x}^m))$ we have the following facts:

1. The average of the weights $\sum_{i=1}^m w(\mathbf{x}^i)$ is an unbiased estimator $\hat{p}(\mathbf{ev})$ of the probability of the evidence $p(\mathbf{ev})$.
2. If $p^*(\mathbf{x}) \propto p(\mathbf{x}, \mathbf{ev})$ then the variance of this estimator is 0. In general, this is not possible, but we should select $p^*(\mathbf{x})$ as close as possible to $p(\mathbf{x}, \mathbf{ev})$.
3. For any event A (for example, the event $A \equiv [X_i = x_i]$) the weighted average of the cases in the sample in which the event A is verified is an unbiased estimator $\hat{p}(A|\mathbf{ev})$ of $p(A|\mathbf{ev})$:

$$\hat{p}(A|\mathbf{ev}) = \frac{\sum_{A \text{ is verified in } \mathbf{x}^j} w(\mathbf{x}^j)}{\sum_{j=1}^m w(\mathbf{x}^j)} \ . \tag{9}$$

4. If $p^*(\mathbf{x}) \propto p(\mathbf{x}, \mathbf{ev})$, then the variance of this estimator is $p(A|\mathbf{ev})(1 - p(A|\mathbf{ev}))/m$.

The general structure of an importance sampling algorithm is as follows:

Importance_Sampling

1. For $j := 1$ to m (sample size)
   (a) $T := \{p_i^{R(\mathbf{E}=\mathbf{ev})}, \ i = 1, \ldots, n\} \cup \{\delta_X | X \in \mathbf{E}\}$
   (b) Generate a configuration $\mathbf{x}^{(j)}$ using p^*. Calculate

$$w_j := \frac{\left(\prod_{f \in T} f(\mathbf{x}_f^{(j)})\right)}{p^*(\mathbf{x}^{(j)})} . \tag{10}$$

2. For each $x_i \in \Omega_{X_i}$, estimate $p(x_i, \mathbf{ev})$ using equation (9) where A represents the event $[X_i = x_i]$.
3. Normalise values $p(x_i, \mathbf{ev})$ in order to obtain $p(x_i|\mathbf{ev})$.

In above algorithm $\mathbf{x}_f^{(j)}$ denotes the configuration $\mathbf{x}^{(j)}$ obtained by leaving only the coordinates of the variables for which f is defined.

There are different applications of this general framework that give rise to different importance sampling algorithms. The main differences are relative to the following factors:

- First, we have the computation of the sampling distribution. The following approaches have been proposed:
  - *Sampling distribution based on initial conditional probabilities and observations, and in which variables are simulated in an order compatible with the topological order of the dependence network (from parents to children).*- Logic sampling algorithm [15] is a particular case in which $p^* = \prod_{i=1}^n p_i$, where $p_i(i = 1, \ldots, n)$ are the initial conditional distributions. The likelihood sampling scheme [13,35] is obtained by considering a sampling distribution $p^* = \prod_{i=1}^n t_i$, where $t_i = p_i$ if variable X_i is not observed and equal to δ_{X_i} if $X_i \in \mathbf{E}$.
  - *Sampling distribution based on initial conditional probabilities and observations, with a modified simulation order.*- In this case simulation is also done with (some) initial conditional probabilities and the evidence potentials. Each time a variable X_i is going to be simulated, a potential containing this variable is selected, then it is restricted to the values of variables already simulated, marginalised over X_i, and then a value for the variable is selected according to a distribution proportional to the result. Backward simulation [14] is an example of this approach in which variables are selected starting in observed nodes and working backward (from parents to children). A more sophisticated procedure is proposed in [8], in which variables are selected according to the entropy of the potentials restricted to the values already simulated.

– *Sampling distribution based on precomputation.*- In this case, a sampling distribution is computed from the set of conditional and evidence potentials $T := \{p_i^{R(\mathbf{E}=\mathbf{ev})}, \ i = 1, \ldots, n\} \cup \{\delta_X | X \in \mathbf{E}\}$. An optimal distribution $(p^*(\mathbf{x}) \propto p(\mathbf{x}, \mathbf{ev}))$ can be computed by carrying out a deletion algorithm and then simulating the variables in reverse order of deletion. The obvious question is: Why to carry out an exact algorithm to obtain approximate values? The answer is that, in general, we perform an approximate deletion algorithm which is the basis for the computation of a non-optimal sampling distribution. Different approaches have been applied to compute the sampling distribution. In [17] potentials are represented by probability tables and a mini-bucket elimination is applied. In [31] potentials are represented by means of probability trees and the approximation capabilities of this representation are exploited to reduce the size of the potentials and thus the number of computations in the deletion algorithm.

- *The sampling procedure.*- Different variance reduction techniques can be applied in the sampling generation to reduce the error of the estimators. For example, Bouckaert, Castillo, and Gutirrez [1] have applied a stratified simulation approach in top of a likelihood weighting algorithm and Salmern and Moral [32] have considered the use of antithetic variables on an importance sampling algorithm based on precomputation.

- Most of the algorithms use the same sample to compute all the conditional probabilities $p(x_i|e)$, by considering the weighted average of cases of the sample in which $X_i = x_i$, but there is also the possibility of using a particular sample for each $p(x_i|e)$. This procedure that has been considered by Cano, Hernández and Moral [17], and Cheng and Druzdzel [9] consists of first obtaining an estimation of $p(e)$ by means of $\hat{p}(\mathbf{ev}) = \sum_{i=1}^{m} w(\mathbf{x}^i)$. Then, $\{X_i = x_i\}$ is added to the set of observations, and with a similar procedure we obtain a new sample and an estimation, $\hat{p}(x_i, \mathbf{ev})$ of $p(x_i, e)$. This sample will be obtained with a different sampling distribution that takes into account that $\hat{p}(\mathbf{ev})$ has been added to the set of observations. The estimator of the conditional probability is obtained by computing $\hat{p}(x_i, \mathbf{ev})/\hat{p}(\mathbf{ev})$. In general, in this way more accurate estimators are obtained, but if the number of variables in which we are interested is high, then we have to repeat the sampling for every case of each one of these variables. So the increment of complexity can be significant.

- An important variation that has been recently introduced (Cheng and Druzdzel [9]; Moral and Salmerón [28]) is the possibility that the sampling distribution p^* changes each time a new case case of the sample is obtained. For example, assume that we have drawn a configuration $\mathbf{x}$ and that $p^*(\mathbf{x}, \mathbf{ev}) = 0.0$. In this case, the weight is also 0.0, and this configuration is completely useless in the final estimation. If this happens very often (with zero or close to zero weights), then the quality of the sample will be poor. The dynamic updating of p^* tries to change the sampling distribution so that if we have obtained $p^*(\mathbf{x}, \mathbf{ev}) = 0.0$, then this case is not repeated in future configurations. The main differences between the two existing approaches are the following:

- Cheng and Druzdzel [9] are based on a likelihood sampling algorithm in which the initial sampling potentials are the conditional and evidence potentials. These potentials change in the numerical values $f(\mathbf{x})$, but they are always defined for the same variables.
- Moral and Salmerón [28] use precomputation with potentials represented by means of probability trees for approximation. These trees are dynamically updated. This precomputation detects the variables for which the potential should be defined in order to get the optimal sampling distribution $p^*(\mathbf{x}) \propto p(\mathbf{x}, \mathbf{ev})$. So, if the sample is large enough and we update a sufficient number of values we could arrive to the optimal distribution. This was not the case of Cheng and Druzdzel approach in which potentials are defined for the initial sets of variables appearing in the given conditional probabilities and this is not changed. So the domain does not contain the variables necessary to obtain the distribution $p^*(\mathbf{x}) \propto p(\mathbf{x}, \mathbf{ev})$.

In the following, we describe in some detail the importance sampling algorithm based on precomputation with orobability trees, giving also its extension to the dynamic case.

The algorithm first computes a sampling distribution p^*. This computation is done by means of a variable elimination algorithm in which all the variables are deleted:

Sampling_Distribution($\mathbf{X}$,E)

1. Let p_i, $i = 1, \ldots, n$, be the conditional distribution for each variable $X_i \in \mathbf{X}$.
2. $T := \{p_i^{R(\mathbf{E}=\mathbf{ev})}, \ i = 1, \ldots, n\} \cup \{\delta_X | X \in \mathbf{E}\}$.
3. For each $Y \in \mathbf{X}$,
   $$T :=\textbf{Delete}(T,Y).$$

Potentials are represented by probability trees and when applying procedure **Delete**(T,Y), the result of the operations are always pruned in an exact way (by collapsing leaves with identical numerical values on a single leaf) or by approximation (if the size is too big, close numerical values are represented by its average as in Figure 1).

There are two possibilities for the sampling distribution:

- $p_1^* = \prod_{Y \in \mathbf{X}} q_Y$, where q_Y is the potential calculated when deleting variable Y.
- $p_2^* = \prod_{Y \in \mathbf{X}} \left(\prod_{f \in T_Y} f\right)$, where T_Y is the set computed when deleting variable Y.

If computations are exact, both sampling distributions are the same, as $q_Y = \prod_{f \in T_Y} f$, but there are differences when combinations are approximated: p_1^* is faster and p_2^* is more accurate.

An important fact with a great influence in the quality of the sampling distribution is the order of variables deletion. In general, we can follow any heuristic similar to the ones used for undirected graphs triangulation [3], but two things must be taken into account here: if we approximate a potential and a variable does not appear in the tree representing the potential, then this variable should be removed from the domain of the potential[4]. After this, we have to take into account that non observed descendants have to be deleted first, starting on leaves and going backwards. The reason is the following: imagine that we are deleting a non-observed leaf, Y. The conditional probability distribution of this node given its parents is the only potential in T_Y. When computing r_Y we have to add on the values y, and as this is a conditional distribution, the result is a potential assigning the value 1.0 to all the configurations. This can be represented (in an exact way) by a very simple tree (a single node with the value 1.0) which does not depend on any variable. So all the remaining variables can be removed of the domain of the potential. This is recursively repeated for all the variables with non-observed descendants.

In [31], p_1^* was considered as sampling distribution, but for the dynamic version p_2^* is more appropriate. So, in the following, we will describe how to obtain a configuration with p_2^*.

Sample_Configuration(X)

1. Let $\mathbf{z}$ be an empty configuration for the empty set of variables $\mathbf{Z}$.
2. For every variable $Y \in \mathbf{X}$ (selected in reverse deletion order),
    3. Restrict all the potentials in T_Y to the configuration $\mathbf{z}$:

$$R_Y := \{ f^{R(\mathbf{Z}=\mathbf{z})} | f \in T_Y \} \ .$$

4. $g_Y := \prod_{g \in R_Y} g$.
5. Simulate a value y for variable Y from a probability distribution proportional to g_Y.
6. Add the value y to configuration $\mathbf{z}$ and variable Y to set $\mathbf{Z}$.
7. Return $\mathbf{z}$.

In order to compute the weight of this configuration, we need the value $p_2^*(\mathbf{z})$. This value is calculated at the same time that we are simulating the values for all the variables, and can be done by multiplying the probabilities of obtaining the values of each single variable (the normalised $g_Y(y)$ values).

---

[4] This has consequences in the subsequent selection of the variables to delete. For example, the theoretical maximum size of this potential will be lower after removing the variable.

Dynamic importance sampling is based on updating the potential r_Y (computed when deleting variable Y) if the value of this potential in configuration $\mathbf{z}$ is very different to the normalisation factor of g_Y when simulating a value for Y. To make things more precise, let us call a_Y the value of r_Y in configuration $\mathbf{z}$ and b_Y the addition of $g_Y(y)$ values (the normalisation factor).

r_Y represents the information that is being used to simulate variables in set $\mathbf{Z}$ (simulated before Y) and that has been calculated in this step; a_Y is the value that has been used for this particular configuration $\mathbf{z}$. If all the computations when deleting Y should have been exact, then necessarily $a_Y = b_Y$. So if these values are too different, we have obtained $\mathbf{z}$ using a poor representation of information in r_Y for this configuration. Dynamic importance sampling considers that if the minimum of $\{a_Y/b_Y, b_Y/a_Y\}$ is lower than a given thereshold, then r_Y is updated so that the value of this potential in configuration $\mathbf{z}$ is equal to b_Y.

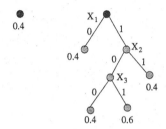

Fig. 6. Example of tree updating.

When updating potentials, the criterion to remove a variable from the domain of a potential has to be changed. If a variable does not appear in a potential and this potential has been approximated, then it is possible that in future updatings we need this variable and can not be removed from the domain. We can only remove a variable from a potential when all the potentials involved in its computation are exact, only exact pruning has been carried out and this variable does not appear in the tree.

Updating a potential f is not a single change of the old value a_Y by the new value b_Y. The problem is that in a probability tree, a number in a leaf represents the value of the potential in several configurations. So, in general, we have to branch the potential by the variables in $\mathbf{Z} \cap dom(f)$ that are not in the path leading to the leaf in which a_Y is stored. In Figure 5.2 we can see an example of updating. Imagine that we have arrived to the leaf in the left with a value of $a_Y = 0.4$ and that the variables in $\mathbf{Z} \cap dom(f)$ that are not in the path to that leaf are X_1, X_2, and X_3, each one of the them taking values in $\{0, 1\}$ and that the values of these variables in the current configuration $\mathbf{z}$ are $1, 0$ and 1 respectively. Assume also that we have to update the value of this configuration in the tree to the new value $b_Y = 0.6$. Then, the result is the tree in the right side of the figure. Details of this procedure can be found in [28].

With this dynamic importance sampling, Moral and Salmerón [28] report very good results even if the approximations in the initial deletion algorithm are very raw.

6 Conclusions

In this chapter, we have reviewed different techniques for approximate inference in Bayesian networks. These techiques have been classified into two groups: deterministic and Monte Carlo algorithms. It is difficult to say which of the two possibilities is the best. Depending on the problem, one class of algorithms can be better than the other. Since approximate propagation for a given precision is an NP-hard problem [10], it is still possible to find problems where the algorithms described here do not provide satisfactory results. It means that approximate propagation is an open research field, where new algorithms can be proposed or the methods reported here improved.

References

1. R.R. Bouckaert, E. Castillo, and J.M. Gutiérrez. A modified simulation scheme for inference in Bayesian networks. *International Journal of Approximate Reasoning*, 14:55–80, 1996.
2. J. Boutilier, N. Friedman, M. Goldszmidt, and D. Koller. Context-specific independence in Bayesian networks. In E. Horvitz and F.V. Jensen, editors, *Proceedings of the 12th Conference on Uncertainty in Artificial Intelligence*, pages 115–123. Morgan & Kaufmann, 1996.
3. A. Cano and S. Moral. Heuristic algorithms for the triangulation of graphs. In B. Bouchon-Meunier, R.R. Yager, and L.A. Zadeh, editors, *Advances in Intelligent Computing*, pages 98–107. Springer Verlag, 1995.
4. A. Cano and S. Moral. Propagación exacta y aproximada con árboles de probabilidad. In *Actas de la VII Conferencia de la Asociación Española para la Inteligencia Artificial*, pages 635–644, 1997.
5. A. Cano, S. Moral, and A. Salmerón. Penniless propagation in join trees. *International Journal of Intelligent Systems*, 15:1027–1059, 2000.
6. A. Cano, S. Moral, and A. Salmerón. Lazy evaluation in Penniless propagation over join trees. *Networks*, 39:175–185, 2002.
7. A. Cano, S. Moral, and A. Salmerón. Novel strategies to approximate probability trees in Penniless propagation. *International Journal of Intelligent Systems*, 18:193–203, 2003.
8. J.E. Cano, L.D. Hernández, and S. Moral. Importance sampling algorithms for the propagation of probabilities in belief networks. *International Journal of Approximate Reasoning*, 15:77–92, 1996.
9. J. Cheng and M.J. Druzdzel. AIS-BN: An adaptive importance sampling algorithm for evidential reasoning in large Bayesian networks. *Journal of Artificial Intelligence Research*, 13:155–188, 2000.

10. P. Dagum and M. Luby. Approximating probabilistic inference in Bayesian belief networks is NP-hard. *Artificial Intelligence*, 60:141–153, 1993.
11. R. Dechter. Bucklet elimination: A unifying framework for probabilistic inference. In E. Horvitz and F.V. Jennsen, editors, *Proceedings of the Twelfth Annual Conference on Uncertainty in Artificial Intelligence (UAI-96)*, pages 211–219, 1996.
12. R. Dechter and I. Rish. A scheme for approximating probabilistic inference. In D. Geiger and P.P. Shenoy, editors, *Proceedings of the 13th Conference on Uncertainty in Artificial Intelligence*, pages 132–141. Morgan & Kaufmann, 1997.
13. R. Fung and K.C. Chang. Weighting and integrating evidence for stochastic simulation in Bayesian networks. In M. Henrion, R.D. Shachter, L.N. Kanal, and J.F. Lemmer, editors, *Uncertainty in Artificial Intelligence*, volume 5, pages 209–220. North Holland (Amsterdam), 1990.
14. R. Fung and B. Del Favero. Backward simulation in Bayesian networks. In R. López de Mántaras and D. Poole, editors, *Proceedings of the 10th Conference on Uncertainty in Artificial Intelligence*, pages 227–234. Morgan & Kaufmann (San Mateo), 1994.
15. M. Henrion. Propagating uncertainty by logic sampling in Bayes' networks. In J.F. Lemmer and L.N. Kanal, editors, *Uncertainty in Artificial Intelligence*, volume 2, pages 317–324. North-Holland (Amsterdam), 1988.
16. L.D. Hernández, S. Moral, and A. Salmerón. Importance sampling algorithms for belief networks based on approximate computation. In *Proceedings of the Sixth International Conference IPMU'96*, volume II, pages 859–864, Granada (Spain), 1996.
17. L.D. Hernández, S. Moral, and A. Salmerón. A Monte Carlo algorithm for probabilistic propagation in belief networks based on importance sampling and stratified simulation techniques. *International Journal of Approximate Reasoning*, 18:53–91, 1998.
18. C.S. Jensen, A. Kong, and U. Kjærulff. Blocking Gibbs sampling in very large probabilistic expert systems. *International Journal of Human-Computer Studies*, 42:647–666, 1995.
19. F. Jensen and S.K. Andersen. Approximations in Bayesian belief universes for knowledge-based systems. In *Proceedings of the 6th Conference on Uncertainty in Artificial Intelligence*, pages 162–169, 1990.
20. F.V. Jensen, S.L. Lauritzen, and K.G. Olesen. Bayesian updating in causal probabilistic networks by local computation. *Computational Statistics Quarterly*, 4:269–282, 1990.
21. U. Kjærulff. Reduction of computational complexity in Bayesian networks through removal of weak dependencies. In *Proceedings of the 10th Conference on Uncertainty in Artificial Intelligence*, pages 374–382. Morgan Kaufmann, San Francisco, 1994.
22. D. Koller, U. Lerner, and D. Anguelov. A general algorithm for approximate inference and its application to hybrid Bayes nets. In K.B. Laskey and H. Prade, editors, *Proceedings of the 15th Conference on Uncertainty in Artificial Intelligence*, pages 324–333. Morgan & Kaufmann, 1999.
23. D. Kozlov and D. Koller. Nonuniform dynamic discretization in hybrid networks. In D. Geiger and P.P. Shenoy, editors, *Proceedings of the 13th Conference on Uncertainty in Artificial Intelligence*, pages 302–313. Morgan & Kaufmann, 1997.

24. S. Kullback and R. Leibler. On information and sufficiency. *Annals of Mathematical Statistics*, 22:76–86, 1951.

25. S.L. Lauritzen and D.J. Spiegelhalter. Local computations with probabilities on graphical structures and their application to expert systems. *Journal of the Royal Statistical Society, Series B*, 50:157–224, 1988.

26. Z. Li and B. D'Ambrosio. Efficient inference in Bayes networks as a combinatorial optimization problem. *International Journal of Approximate Reasoning*, 11:55–81, 1994.

27. A.L. Madsen and F.V. Jensen. Lazy propagation: a junction tree inference algorithm based on lazy evaluation. *Artificial Intelligence*, 113:203–245, 1999.

28. S. Moral and A. Salmerón. Dynamic importance sampling computation in Bayesian networks. In T.D. Nielsen and N.L. Zhang, editors, *Symbolic and Quantitative Approaches to Reasoning with Uncertainty*, volume 2711 of *Lecture Notes in Artificial Intelligence*, pages 137–148, 2003.

29. J. Pearl. Evidential reasoning using stochastic simulation of causal models. *Artificial Intelligence*, 32:247–257, 1987.

30. R.Y. Rubinstein. *Simulation and the Monte Carlo Method*. Wiley (New York), 1981.

31. A. Salmerón, A. Cano, and S. Moral. Importance sampling in Bayesian networks using probability trees. *Computational Statistics and Data Analysis*, 34:387–413, 2000.

32. A. Salmerón and S. Moral. Importance sampling in Bayesian networks using antithetic variables. In S. Benferhat and P. Besnard, editors, *Symbolic and Quantitative Approaches to Reasoning with Uncertainty*, volume 2143 of *Lecture Notes in Artificial Intelligence*, pages 168–179, 2001.

33. E. Santos and S.E. Shimony. Belief updating by enumerating high-probability independence-based assignments. In *Proceedings of the 10th Conference on Uncertainty in Artificial Intelligence*, pages 506–513, 1994.

34. E. Santos, S.E. Shimony, and E. Williams. Hybrid algorithms for approximate belief updating in Bayes nets. *International Journal of Approximate Reasoning*, 17:191–216, 1997.

35. R.D. Shachter and M.A. Peot. Simulation approaches to general probabilistic inference on belief networks. In M. Henrion, R.D. Shachter, L.N. Kanal, and J.F. Lemmer, editors, *Uncertainty in Artificial Intelligence*, volume 5, pages 221–231. North Holland (Amsterdam), 1990.

36. P.P. Shenoy. Binary join trees for computing marginals in the Shenoy-Shafer architecture. *International Journal of Approximate Reasoning*, 17:239–263, 1997.

37. P.P. Shenoy and G. Shafer. Axioms for probability and belief function propagation. In R.D. Shachter, T.S. Levitt, J.F. Lemmer, and L.N. Kanal, editors, *Uncertainty in Artificial Intelligence 4*, pages 169–198. North Holland, Amsterdam, 1990.

38. N.L. Zhang and D. Poole. Exploiting causal independence in Bayesian network inference. *Journal of Artificial Intelligence Research*, 5:301–328, 1996.

Abductive Inference in Bayesian Networks: A Review

José A. Gámez

Departamento de Informática. Universidad de Castilla-La Mancha.
Campus Universitario s/n. Albacete, 02071. Spain.
jgamez@info-ab.uclm.es

Abstract. The goal of this paper is to serve as a survey for the problem of abductive inference (or belief revision) in Bayesian networks. Thus, the problem is introduced in its two variants: total abduction (or MPE) and partial abduction (or MAP). Also, the problem is formulated in its general case, that is, looking for the K best explanations. Then, a (non exhaustive) review of exact and approximate algorithms for dealing with both abductive inference problems is carried out. Finally, we collect the main complexity results appeared in the literature for both problems (MPE and MAP).

Keywords: Abductive inference, belief revision, MPE, MAP, probabilistic reasoning, propagation algorithms, Bayesian networks.

1 Introduction

In the last years artificial intelligence researchers have devoted increasing attention to the development of abductive reasoning methods in a wide range of applications. Probably the most clear application of abductive reasoning is in the field of diagnosis [38,39,42,43], although other applications exist in natural language understanding [6,52], vision [27], legal reasoning [53], plan recognition [4,26], planning [40] and learning [33].

Abduction is defined as the process of generating a plausible explanation for a given set of observations or facts [41]. This kind of reasoning can be represented by the following inference rule:

$$\frac{\psi \to \omega, \omega}{\psi},$$

i.e., if we observe ω and we have the rule $\psi \to \omega$, then we can infer that ψ is a *plausible* hypothesis (or explanation) for the occurrence of ω.

In general, there are several possible abductive hypotheses and it is necessary to choose among them. In order to select the best explanations from the generated set, two kinds of criteria are used: (1) metrics based criteria (probability, weight, ...) and (2) simplicity criteria (the preferred explanation is the simplest available hypothesis).

In the context of probabilistic reasoning, abductive inference corresponds to finding the maximum *a posteriori* probability state of the system variables, given some *evidence* (observed variables). In principle, we can solve this problem by "simply" generating the joint distribution, and then taking it as our starting point, to search for the configuration with maximum probability. However, this way to proceed is intractable even for problems with a small number of variables.

The paper is organized as follows: In the second section we introduce abductive inference in the framework of probabilistic reasoning but focused on Bayesian networks (BNs). In sections three and four we review algorithms for solving both cases of abductive inference problems by using exact and approximate methods. The fifth section is devoted to collect the main complexity results about abduction in BNs. Finally, in section six we conclude.

2   Abductive Inference in Probabilistic Reasoning

In this section we will describe the problem of performing abductive reasoning in the general setting of probabilistic reasoning, although we will assume that the joint probability distribution is represented by the factorization provided by a Bayesian network [37]):

$$P(X_\mathcal{U}) = P(X_1, \dots, X_n) = \prod_{X_i \in X_\mathcal{U}} P(X_i | pa(X_i)), \tag{1}$$

where $pa(X_i)$ is the parent set of X_i.

Before we continue, we give some notation. A lower case subscript indicates a single variable (e.g., X_i). An upper case subscript indicates a set of variables (e.g., X_I). For some particular problems, the propositional variables are denoted by capital letters without subscript $A, B, C, \dots$. The state taken by a variable X_i will be denoted by x_i, and the configuration of states taken by a set of variables X_D will be denoted by x_D. That is, capital letters are reserved for variables and set of variables, and lower case letters are reserved for states and configurations of states.

Given a set of observations[1] x_O for a set of variables X_O, the most common query in probabilistic reasoning is the computation of $P(X_i | x_O)$ for every non observed variable ($X_\mathcal{U} \setminus X_O$). This kind of reasoning is called *probability propagation* or *evidence propagation* or *belief updating*, and a great variety of exact and approximate algorithms have been developed during the last years (see [5] for a description of many of them). However, this paper is not concern with evidence propagation (which can be interpreted as predictive or deductive reasoning), but with abductive or diagnostic reasoning, which consists in looking for the *best* explanation accounting for the observed evidence.

---

[1] The configuration $X_O = x_O$ is known as *evidence*. For the sake of simplicity we will use only x_O in most of the cases.

In the context of probabilistic reasoning and Bayesian networks an explanation for a given set of observations $X_O = x_O$ is a configuration of states for the network variables, $x_{\mathcal{U}}$, such that, $x_{\mathcal{U}}$ is consistent with x_O, that is, $x_{\mathcal{U}}^{\downarrow X_O} = x_O$ (by $x_{\mathcal{U}}^{\downarrow X_O}$ we are denoting the configuration obtained from $x_{\mathcal{U}}$ by removing the literal expressions not in X_O). In fact, the explanation is $x_{\mathcal{U}}^{\downarrow X_{\mathcal{U}} \backslash X_O}$, because the values taken by the variables in X_O are previously known. Given the large number of possible explanations and since we are interested in the best explanation, our goal will be to obtain the *most probable explanation*. Thus, in our setting, abductive inference or *belief revision* [36,37] corresponds to the problem of finding the maximum a posteriori probability state of the network, given the observed variables (the evidence). In a more formal way: if X_O is the set of observed variables and X_U is the set of unobserved variables, then we aim to obtain the configuration x_U^* of X_U such that:

$$x_U^* = \arg\max_{x_U} P(x_U|x_O), \tag{2}$$

where x_O is the observed evidence. Usually, x_U^* is known as the *most probable explanation* (MPE).

In general, this problem cannot be solved by using probability propagation. That is, x_U^* cannot be obtained as

$$x_U^* = (x_1^*, x_2^*, \ldots, x_{|U|}^*), \text{ with } x_i^* = \arg\max_{x_i \in \Omega_{X_i}} P(x_i|x_O). \tag{3}$$

As an example, let us consider the well-known Asia network [28]. If we assume that the observations are (VisitToAsia=yes, PositiveXRay=yes), then figure 1 shows the result of performing different query types over this network by using the Elvira software [7]. In this software we can watch simultaneously the effect of processing different queries. In our example (see detail in figure 1) the first horizontal bar shows the 'a priori' probability for each variable; the second bar shows the 'a posteriori' probability for each variable given the observed evidence; and the third bar shows the state selected for each variable when looking for the most probable explanation (eq. 2). As it can be seen, if we compose the MPE by using probability propagation, we obtain:

(Smoker=yes, Tuberculosis=no, LungCancer=no, Bronchitis=no,
        TuberculosisOrCancer=yes, Dyspnea=yes),

with *unknown* probability. However, when total abductive inference is carried out, we obtain

(Smoker=yes, Tuberculosis=no, LungCancer=yes, Bronchitis=no,
        TuberculosisOrCancer=yes, Dyspnea=yes),

with probability 0.17. As we can see the resulting configuration does not coincide with the previous one, because variable LungCancer takes now the state yes.

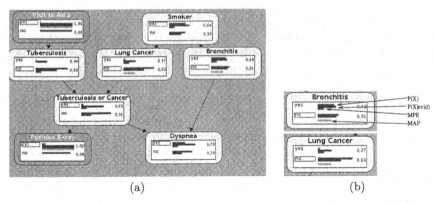

Fig. 1. (a) Different queries performed over Asia network with $X_O = (VisitToAsia = yes, PositiveXRay = yes)$. (b) Detail for nodes *Bronchitis* and *LungCancer*

The previous definition of abductive inference can be generalized by considering a subset of the unobserved variables as the interest target, instead of all of them. These variables are usually termed *explanation set* [30], and the task is known as *Partial* abductive inference (or *maximum a posteriori hypothesis*, MAP) while the previous one which is known as *Total* abductive inference (or MPE). Although this problem seems to be more useful in practical applications (because we can select the relevant variables[2] as the explanation set) than *total* abductive inference it has received much less attention.

Now, if we denote by $X_E \subset X_U$ the explanation set, then we aim to obtain the configuration x_E^* of X_E such that:

$$x_E^* = \arg\max_{x_E} P(x_E|x_O) = \arg\max_{x_E} \sum_{x_R} P(x_E, x_R|x_O), \qquad (4)$$

where $X_R = X_U \setminus X_E$. In general, x_E^* is not equal to the projection of configuration x_U^* onto the variables of X_E. Therefore, we need to obtain x_E^* directly (eq. 4). As an example, let us to retake the example based on figure 1. Now, if we select variables LungCancer and Bronchitis as the explanation set, then by using partial abductive inference we get the configuration:

(LungCancer=no, Bronchitis=no)

as the most probable explanation (see detail in part (b) of the same figure). As we can see, this configuration does not coincide with the one obtained by projecting x_U^* over X_E (LungCancer=yes, Bronchitis=no).

---

[2] those representing diseases in a medical diagnosis problem, those representing critical components (starter, battery, alternator, ...) in a car diagnosis problem, etc ...

To finish this section, just remark that in both queries of abductive inference, the problem is generalized to the one of looking for the K most probable explanations (K MPEs).

3   Solving Total Abduction (MPE) in BNs

As it has been shown in the literature, in total abduction, the MPE can be found by means of probability propagation methods but using maximum as the marginalization operator (fig. 2) instead of summation (due to the distributive property of maximum with respect to multiplication) [12].

$$\psi(A) = \max_B \psi(A, B) = \max_B \begin{pmatrix} & b & \neg b \\ a & 0.35 & 0.2 \\ \neg a & 0.15 & 0.3 \end{pmatrix} = \begin{pmatrix} a & 0.35 \\ \neg a & 0.3 \end{pmatrix}$$

Fig. 2. Marginalizing by maximum (max-marginalization)

Therefore, the process of searching for the most probable explanation (in total abduction) has the same complexity[3] as probabilities propagation. However, in order to look for the K MPEs more work has, in general, to be done. In the next subsections we review some approaches for solving this problem in an exact and approximate way.

3.1   Exact Computation

We review the two main approaches which can deal with networks of unconstrained topology.

Using junction trees. Dawid [12] developed an efficient algorithm to calculate the most probable explanation (MPE) in a junction tree. The algorithm is based on the probabilities propagation algorithm described in [3], but replacing summation (sum-marginalization) by maximum (max-marginalization) in the calculation of messages. When *max-marginalization* is used during the two stages of messages propagation (collect and distribute evidence) the process was termed *max-propagation* by Dawid.

Given an initialized junction tree T with cliques $\{C_0, \ldots, C_t\}$, after max-propagation the following expression holds:

$$\forall C_i \in T \quad \max_{x_U \in \Omega_U} P(x_U, x_O) = \max_{c_i \in \Omega_{C_i}} \psi(c_i) \tag{5}$$

Therefore, if there is only one configuration of maximal probability, then we can identify the MPE x_U^* by inspecting the residual set R_i of each clique

---

[3] See Section 5 for details in this topic.

C_i and picking up the state of maximal probability for the variables in R_i. However, if there are several configurations of maximal probability, then the following algorithm can be used in order to identify one of them.

FindMPE
- input: a rooted junction tree $T = \{C_0, \ldots, C_t\}$ after applying max-prop.
- output: the MPE $x^* = concatenate(c_0^*, r_1^*, \ldots, r_t^*)$

1. For the root C_0: $c_0^* = \arg\max_{C_0} \psi(C_0)$
2. For $j = 1, \ldots, t$ do
   $C_i = parent(C_j)$
   $s_{ij}^* = c_i^{* \downarrow S_{ij}}$
   $r_j^* = \arg\max_{C_j \setminus S_{ij}} \psi(C_j, s_{ij}^*)$

As an example let us consider the figure 3. On the left we have only a configuration of maximal in C_0 and so we can identify it (a, b, c) by inspection. However, in the junction tree on the right there are more than one configuration of maximal probability (highligthed in bold and with an (*) respectively). Thus, we have to apply the previous algorithm in order to identify one of them: (a, b, c) or $(a, \bar{b}, \bar{c})$.

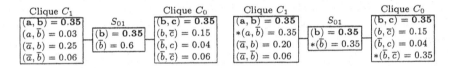

Fig. 3. Identifying the MPE in two different scenarios

Looking for the K MPEs. Looking for the K most probable explanations can be viewed as an extension of the MPE problem. Nilsson [31], has proved that using Dawid's algorithm only the three most probable explanations can directly be identified, but the fourth and subsequent explanations cannot be found directly. Therefore, more complex methods have to be used.

In [47] Seroussi and Goldmard developed a method able to find the K MPEs for every value of K. The method only requires the upward phase of max-propagation (that is, collecting evidence from the root) but modifying the messages that are sent from each clique to its parent. Now, instead of sending a single message/potential from C_i, a vector of K messages is sent, corresponding to the K most probable configurations of the subtree rooted by C_i. The main problem of this algorithm is its high computational cost.

Later, Nilsson [32] has developed a more efficient method for finding the K MPEs for every value of K. As in Seroussi and Goldmard proposal, only the upward phase of max-propagation is necessary. The algorithm for obtaining the K MPEs is based on the combination of Dawid's algorithm with a clever divide and conquer approach. The idea is as follows: after obtaining the best

MPE (x_U^1), the space of possible configurations is partitioned in order to exclude it from the set of possible configurations. As the second MPE (x_U^2) must differ from the best one in at least one residual set, the search space is partitioned into t subspaces by using the value of the residual sets as evidence. The tricky point of Nilsson's algorithm is that no new propagation phases are required in order to consider these new *evidences*, but by inspecting the potentials in the junction tree. After identifying x_U^2 a new partitioning is carried out, now the clique (whose residual changes with respect to x_U^1) is considered as the root for the partition, and so only the cliques below it in the junction tree will be considered. For a complete (and formal) description of the model see [32].

Using variable elimination Variable elimination [48,57] can be directly adapted for dealing with the problem of looking for the best MPE. The main modification is to use max-marginalization instead of sum-marginalization when a variable is being eliminated. Also, we have to store the state of the variable being eliminated which takes the maximum. Finally, the configuration of maximum probability is obtained by assembling the stored states [29,21]. In order to look for the K MPEs more work has to be done. Li and D'ambrosio [29] propose to use the standard variable elimination algorithm to look for the best MPE, and then they give a linear time algorithm which can be invoked (as many times as necessary) in order to get the next MPE.

3.2   Approximate Computation

In this subsection we distinguish between search and non search based methods:

Search methods As abductive inference in BNs can be viewed as a combinatorial optimization problem, several authors have used genetic algorithms to approximate a solution ([24,44,45,58]). In all the cases the idea is to optimize the probability

$$P(x_U|x_O) \propto P(x_U, x_O),$$

so to evaluate the goodness of a given individual x_U the factorization provided by the network is used (eq. 1), requiring only n multiplications (that is, linear in the number of variables).

Below, we describe some relevant points of these algorithms.

- In Gelsema's algorithm [24], a chromosome is a configuration of the unobserved variables, i.e., a vector of integers. In this case, crossover is implemented as the classical one-point operator. It is worth noting that Gelsema uses the '*a priori*' probabilities of the BN and the observed evidence to generate the initial population, so that the search starts in promising regions of the search space.

- In the approach proposed by Rojas and Kramer [44,45], a chromosome of the population is represented as a copy of the graph included in the BN, but in which each variable has been instantiated to one of its possible states. This representation makes it possible to implement the crossover operator as the interchange of a subgraph with the center in the variable X_i, X_i being randomly selected for each crossover.
- Finally, Zhong and Santos [58] propose to group the variables in the network into several clusters attending to the probabilistic correlations among the variables, and to use cluster-based mutation and crossover. The cluster are identified by using reinforcement learning, and the output of the genetic algorithm is used to refine the clustering process in order to get a better grouping.

Recently, Kask and Dechter [25] have studied the applicability of local search to this problem. They compared several stochastic local search algorithms for solving the MPE problem. In their analysis, they found that the greedy approach combined with stochastic simulation outperforms the other analyzed techniques: stochastic simulation (alone), greedy local search (alone) and simulated annealing.

The problem of looking for the K MPEs is solved by these algorithms by returning the K best individuals/chromosomes visited during the search.

Non search methods Non search methods for dealing with the MPE problem are based on the methods discussed in section 3.1:

- The Elvira software [7] allows to solve the MPE problem by using junction tree based approximate propagation. The idea is to use Nilsson's algorithm [32] but dealing with an approximate representation of the potentials associated to cliques and messages. Concretely, approximate probability trees are used instead of probability tables/trees (see [46] for a description on (approximate) probability trees).
- A method based on variable elimination is the so-called mini-bucket approach [22]. The idea of the mini-bucket approach is to move some marginalizations outside of the product, in order to deal with smaller potentials. Of course, the result of this transformation will be, in general, an approximation of the true value. For example, in variable (bucket) elimination we could have the following scenario when eliminating variable X:

$$\max_X \left(f(X, Y_1, Y_2) \times f(X, Y_2, Y_3) \times f(X, Y_4, Y_5) \times f(X, Y_5, Y_6) \right)$$

If each variable has ten different states, then this operation has to build a potential of size 10^7. By using the mini-bucket approach, the previous computation could be organized in the following way:

$$\left\{ \max_X \left(f(X, Y_1, Y_2) \times f(X, Y_2, Y_3) \right), \max_X \left(f(X, Y_4, Y_5) \times f(X, Y_5, Y_6) \right) \right\}$$

which builds two potentials of size 10^4. An extra advantage of the mini-buckets approach is that it gives bounds on the quality of the approximate solution.

4   Solving Partial Abduction (MAP) in BNs

As was stated in Sec. 2 when we are interested in the MPE for a given subset (X_E) of the network's variables, then equation 4 has to be used. Therefore, two types of marginalization operators have to be applied: max-marginalization over the variables in the X_E and sum-marginalization over the rest of variables.

In the next two subsections we describe exact and approximate methods proposed to cope with the problem of partial abductive inference (MAP). In the case of exact computation we focus on junction tree based algorithms, although some remarks (and references) to variable elimination will be provided.

4.1   Exact Computation

Given the algorithms described for total abduction in BNs and the need of using two types of marginalization, it seems easy to solve the problem of partial abductive inference, we can use the following two-stage based algorithm:

1. Marginalize out (by summation) over the variables not in the explanation set. This process will yield a junction tree containing only the variables in the explanation set (X_E).
2. Apply an algorithm of *total* abductive inference.

However, the problem is more complex than it looks like, because due to the non-commutative behaviour of summation and maximum the following constraint has to be considered: *no summation can be carried out over a potential obtained by maximum*. This fact motivates that not all junction trees obtained from the original network are valid. Thus, the previous two-stage algorithm can be only directly applied when X_E is included in a node of the junction tree, or when the variables of X_E constitute a subtree of the complete junction tree. As an example, let us consider the junction tree in figure 4.(a) and the explanation set $X_E = \{A, T, B\}$. Clearly this tree is not valid for the given explanation set, because in clique C_2 we have to sum-marginalize variables $\{E, L\}$ from the potential $\psi(L, E, T)$ whose computation has involved max-marginalizing over B.

Therefore, we have modified the problem to the following one: what happen when the variables of the explanation set are associated with several disconnected subtrees of the junction tree?. Below, we describe several approaches which have been proposed in the literature in the last years.

Adapting a given junction tree. With the proposal of computing marginal values over a set of variables, Xu [55] gives a method for transforming the initial junction tree into another one containing a node/cluster in which the variables of X_E are included. The problem of this approach is that if X_E contains many variables, then the size of the potential associated with that cluster will be to large. Later, Nilsson [32] briefly outlines how to slightly modify Xu's algorithm in order to allow (when possible) that the variables in X_E constitute a sub-tree and not a single cluster. More recently, de Campos et al. [18] have detailed this process, studying heuristics and introducing and intermediate step. In the rest of this section we briefly describe this work.

The goal of obtaining a junction tree $\mathcal{T}_E$ containing only the variables in X_E, from a given junction tree $\mathcal{T}$ can be performed as follows:

1. Identify the smallest subtree $\mathcal{T}'$ of $\mathcal{T}$ that contains the variables of X_E.
2. Pass sum-flows from the rest of the cliques to $\mathcal{T}'$. In this way $\mathcal{T}'$ factorizes over the variables contained in its cliques.
3. While $\mathcal{T}'$ contains variables not in X_E do: Select two neighbors C_i and C_j in $\mathcal{T}'$, and replace them by their fusion into a new node C_{ij} obtained from $C_{ij}^* = C_i \cup C_j$ by deleting the variables that are not necessary to maintain the running intersection property and do not belong to X_E. The potential assigned to C_{ij} is[4]:

$$\psi(C_{ij}) = \sum_{C_{ij}^* \backslash C_{ij}} \frac{\psi(C_i) \cdot \psi(C_j)}{\psi(S_{ij})} \qquad (6)$$

Notice that C_{ij}^* is relevant because though we produce a clique C_{ij}, during the process we have to deal with the potential defined on the whole set C_{ij}^*. Figure 4.(b) shows $\mathcal{T}'$ (wrt $X_E = \{A, T, B\}$) obtained from $\mathcal{T}$ in part (a) of the same figure. Finally part (c) of the same figure shows an example for the Bayesian network Asia taking $X_E = \{A, T, B\}$. Notice that the potential over $\{B, T, E, L\}$ is built during the fusion process.

The following improvements were proposed for this algorithm in [18]:

- Introducing an intermediate step. An optimization can be added to the previous algorithm between steps 2 and 3. The idea is to look for the variables not in X_E that are only included in one clique of $\mathcal{T}'$, then it is clear that these variables can be marginalized out directly. This idea, although it is very simple can improve the efficiency of the fusion process carried out in step 3 and also the quality of the final tree (see [18] for details and examples).
- Defining heuristics for step 3. In [18] the fusion process performed in step 3 is formally defined as a *link* deletion process. It is important to emphasize

---

[4] The division can be omitted if we know that no propagation has been previously carried out over the junction tree, because in such a case S_{ij} contains an unitary potential

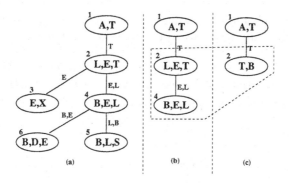

Fig. 4. (a) The initial join tree. (b) $\mathcal{T}'$ for $X_E = \{A, T, B\}$. (c) The junction tree obtained by the fusion process.

that unlike the case of triangulation (deletion of nodes) in this case not all the links have to be removed. Therefore the first think to do is to identify which are the links candidates for deletion.

Definition 1. [18] Given a junction tree $\mathcal{T}'$ and an explanation set X_E, a link (C_i, C_j) with separator $S_{i,j}$ has to be removed if:

i) $S_{i,j} \not\subseteq X_E$, or

ii) $S_{i,j} \subseteq X_E$, but $S_{i,j} = C_i$ or $S_{i,j} = C_j$.

The second condition of the previous definition is necessary to deal with non-maximal nodes, which can be introduced by the intermediate step presented in the previous subsection. To finish with this section, we reproduce here the heuristics which obtained better results in [18] when applied to select the next link to be removed:

– Select the link (C_i, C_j) with yields smallest $s(C_{ij})$.

The goal of this heuristics is the creation of nodes with smallest state space, with the expectation of an earlier deletion of the variables not in X_E.

Looking for an explanation set oriented junction tree. A disadvantage of the previous method is that the quality of the final junction tree, depends on the topology of the initial one, which was obtained for general inference purposes and not thinking about partial abductive inference. An alternative approach could be to search for a specific junction tree given an explanation set X_E [23,18]. This task can be achieved by taking advantage of the available degrees of freedom in the compilation/triangulation process. Concretely, during the triangulation we can constrain the deletion sequence, in such a way that we start to delete the variables of X_E only when all the variables not in X_E have been deleted.

Using this kind of deletion sequences, and adding all the clusters (and not only those which are maximal) to the tree, we can build a junction tree in which a subtree for X_E can be directly identified. If only cliques[5] are added to the tree, it is necessary to apply *maximum cardinality search* (see [28]) beginning with a variable of X_E and breaking ties in favor of the variables in X_E during the numbering of the graph's nodes. Figure 5.a shows a junction tree valid for the Asia network and $X_E = \{A, B, T\}$.

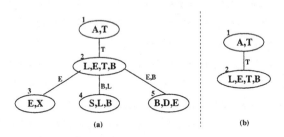

Fig. 5. (a) Specific junction tree for Asia and $X_E = \{A, B, T\}$.(b) Identified $\mathcal{T}'$

Notice that now $\mathcal{T}'$ practically coincides with $\mathcal{T}_E$. In fact the fusion process (step 3) is not necessary because the variables in $\mathcal{T}'$ and not in X_E are in the leaves and will be deleted (by summation) by the intermediate step described above (this is the case in the tree depicted on part (b) of figure 5, where L and E can be directly removed.

The experiments carried out in [18] show that, as expected, the size of the join tree over which partial abductive inference will be performed is smaller when using constrained deletion sequences than when using adapted junction trees. However, the authors also get an unexpected result: in general, the abductive process runs faster over adapted junction trees than over specific ones. De Campos et al. analyze this problem in [20] and conclude that this is due to the fact that when using specific junction trees the abductive inference method deals with very large potentials since the beginning of the process. However, as the initial information, that is the conditional probability tables in the network, is the same in both cases, then there should be lots of regularities in those large potentials. In fact, when using probability trees instead of tables to represent potentials, as regularities and contextual independences are exploited, the efficiency of the method is increased no matter which kind of junction tree (adapted or specific) is used. However, the greater benefit is obtained when using specific ones, to the point that, in general, (as expected initially) the algorithm runs faster over specific junction trees than over adapted ones.

---

[5] Sometimes it is no possible to obtain a valid clique tree [23], but the structure appears valid after applying the intermediate step proposed above.

With respect to variable/bucket elimination, Dechter [21] proposed the algorithm *elim-map* which basically uses a constrained deletion sequence. Firstly, the variables not in the explanation set are eliminated by summation, and then the variables in X_E are eliminated by maximum, storing in each step the state of maximum probability of the variable being eliminated (as in *elim-mpe*). As well as in total abduction, the algorithm of Li and D'Ambrosio [29] could be used to look for the remaining $K - 1$ explanations.

4.2 Approximate Computation

As in the total case, in this subsection we distinguish between search and non search based methods:

Search methods Of course, partial abductive inference can also be defined as a combinatorial optimization problem, being now Ω_{X_E} instead of Ω_{X_U} the search space. Thus, several combinatorial optimization techniques have been used:

- Evolutionary algorithms. As in the total case, genetic algorithms have been used to approach this problem [14,19,15]. Now, eq. 1 cannot be used to evaluate an individual or candidate configuration because we have to remove (by addition) the variables in $X_U \setminus X_E$. That is, if we use eq. 1 we have to invoke it $|\Omega_{X_U \setminus X_E}|$ times, which is impracticable in most of the cases.

  In [14] the fitness $P(x_E, x_O)$ of a configuration x_E is computed by the process described below, where T is a rooted clique tree, being C_0 the root.

  1. Enter the evidence x_O in T,
  2. Enter (as evidence) the configuration x_E in T,
  3. Perform *CollectEvidence* from the root (C_0) (i.e., an upward propagation), and
  4. $P(x_E, x_O)$ is equal to the sum of the potential stored in the root (C_0).

  Therefore, to evaluate a configuration an exact sum-propagation is carried out, or more correctly half propagation, because only the *upward* phase is performed and not the *downward* one. Furthermore, for this propagation we can use a clique tree obtained without constraints and so its size is much smaller than the clique tree used for exact partial abductive inference. In addition, in [14] it is shown how the tree can be pruned in order to avoid the repetition of unnecessary computations when a new chromosome is being evaluated.

  An improvement to this algorithm can be found in [19], consisting in the use of specific genetic operators that allow us to take advantage of the calculations previously carried out when a new individual is being evaluated. In this way the need to perform a whole upward propagation is avoided, although more memory is needed.

In [13] *estimation of distribution algorithms* (EDAs) are used to compare with this two genetic approaches. EDAs are evolutionary algorithms in which genetic has been replaced by estimation and sampling of probability distribution. The same evaluation function is used and the results are really competitive, depending on the type of EDA.

In order to reduce the complexity of evaluating a configuration, in [15] the junction tree is pre-processed by means of explanation set absorption. That is, variables in X_E are considered as evidence, but all its possible states are taken into account. In this way, some cliques store vector of potentials instead of single potentials, and the correct one, which depends on the chromosome being evaluated, is selected in execution time. The advantage of this approach is to deal with smaller junction trees, and the authors show a considerable speed-up in the efficiency of the algorithm. However, the main disadvantage is the need of (much) more memory when many variables are included in the explanation set.

- Simulated annealing. In De Campos et al. [16] a simulated annealing based algorithm is proposed. As in evolutionary algorithms, the evaluation function is based on clique tree propagation, but now neighbor configurations can be evaluated by means of local computations, improving the efficiency with respect to global evaluation. The algorithm maintain a state composed of the current configuration and the current clique C_i, then the neighborhood of that configuration is defined as the set of configurations in which only the variables included in the residual sets of adjacent cliques to C_i can change their value. This assumption allows to organize computations in such a way that to evaluate a neighbor configuration only the current clique and its neighbors in $\mathcal{T}$ are involved. After selecting the new configuration, the algorithm also change its current clique to the neighbor containing (in its residual set) the variable(s) who has change its value (wrt the previous configuration).

- Local search. Park and Darwiche [35] propose to use local search to solve partial abductive inference (or MAP). The algorithm is based in a hill climbing (with restart) strategy in which we move from a configuration x_E to one of its neighbors only if $P(x'_E, x_O) > P(x_E, x_O)$. A neighbor of x_E is defined as the result of changing the value of a single variable, so there are $|X_E|$ possible neighbors. The novelty of this method resides in the way in which configurations are evaluated. The authors propose to use a method based on *differential inference* [11], which allow them to compute the value of all the neighbors in $O(n \exp(w))$, where n is the number of variables and w the width of an (unconstrained) elimination order.

Later, Park [34] extend the approach to larger networks by replacing the evaluation method. In this work, the score for a given configuration is computed by using a method for approximating the probability of evidence based on *loopy belief propagation* [54,56].

Non search methods As wells as in the MPE problem, the mini-bucket approach [22] and the use of approximate probability trees [20] have been used as tools for obtaining approximate solutions to the partial abductive inference problem.

5   Complexity Results

It is well known that exact probabilistic inference can be performed efficiently in networks of restricted topology (trees and polytrees), but in the general case (networks with cycles) propagation is NP-hard [8]. Also, approximate inference in BNs is NP-hard too [9], although the class of solvable problems is wider. Therefore, as abductive inference in BNs is solved by propagation algorithms it is expected to be NP-hard as well. Below we revised some complexity results about abductive inference in BNs.

Looking for the MPE in BNs can be solved efficiently in the same cases as probability propagation, but Shimony [50] proves that in the general case the MPE problem is NP-hard. Besides, with minor modifications to this result, Shimony shows that the problem remains NP-hard even when the topology of the network is restricted to an in-degree (number of parents) of 2 or to an out-degree (number of children) of 2.

With respect to the use of approximations, Abdelbar and Hedetniemi [1] have shown that approximating the MPE problem with a constant ratio bound is also NP-hard. In the same paper the authors also prove that given an explanation, the process of approximating the next one with a constant ratio bound is NP-hard too. Abdelbar and Hedetniemi [1] also gave an additional and interesting result with respect to what they termed *dynamic* abduction: given the MPE (or K MPEs) for a BN and an initial evidence x_O, the problem of finding or approximating the MPE (K MPEs) for a modified evidence x_O' which differs form x_O by the addition or removal of even a single pair (variable,value) is also NP-hard. Finally, as in [50] they extend their results to the case of networks with restricted topology (in-degree=2 or out-degree=2). Later, Abdelbar et al. [2] show that on the contrary to probability propagation [10], finding or approximating the MPE in BNs remains NP-hard even for networks in which probabilities are bounded within the range $[l, u]$ for any $0 \leq l < 0.5 < u \leq 1$.

With respect to partial abduction or MAP, it seems that solving this problem is harder than solving related inference problems as probability propagation or total abductive inference [23,16,34], being really complex even for networks in which the other tasks can be solved efficiently. With respect to the *hardness* of MAP, Park [34] shows that MAP is NP^{PP}-complete (MPE is NP-complete and probability propagation is PP-complete in its decision version). Besides, Park shows that elimination algorithms require exponential resources to perform MAP, even on some polytrees; and that MAP is NP-complete when restricted to polytrees.

Related with MAP, Gámez [23] gives a result which can be used to obtain a lower bound for clique size (or elimination order width) for any given problem. In fact, given a network G and an explanation set X_E, Gámez shows that for all constrained elimination sequence (variables of X_E are eliminated last), and for all variable X_i not in X_E, there is at least one clique containing $X_i \cup acc_E(X_i)$, where

$$acc_E(X_i) = \{X_j \in X_E \mid \exists \text{ an undirected path } (X_i, X_{ij}^1, \ldots, X_{ij}^p, X_j) \text{ in } G,$$
$$\text{such that, } \{X_{ij}^1, \ldots, X_{ij}^p\} \subset X_\mathcal{U} \setminus X_E\}.$$

Therefore, the lower bound for any constrained elimination order for that problem will be:

$$\max_{X_i \text{ not in } X_E} |\{X_i\} \cup acc_E(X_i)|$$

As an example of this result, let us consider the classical Alarm network and all the root nodes (which is very usual) as the explanation set. In this case, there are several variables (not in X_E) such that $acc_E(\cdot) = X_E$, and so there will be at least one clique containing (among others) all the variables in the explanation set.

6 Conclusions

In this paper we have revised the problems of performing total and partial abductive inference in Bayesian networks. A review of exact and approximate algorithm for dealing with both problems and their extension to looking for the K best explanations has been carried out. Of course, we are aware of the fact that not all algorithms have been discussed by our revision, but due to the lack of space we have selected the approaches with wider application (networks of unconstrained topology) and of more recent publication.

There is a very interesting problem related with partial abduction or MAP which has not been covered in this paper. We refer to the problem of selecting the explanation set. In the literature, it is mostly assumed that the explanation set is provided *a priori* by an expert, another algorithm, etc, ... It is an extended practice to select root nodes (which usually represents diseases or disorders) or the evidence ancestors as the explanation set [29]. Shimony [49,51] proposed a method based on irrelevance criteria that does not require a given explanation set, on the contrary the algorithm tries to identify the relevant nodes directly. De Campos et al. [17] propose to start with an explanation set, but to simplify the obtained explanations by using independence-based criteria. However, in our opinion this problem should receive more attention from the BNs research community.

Acknowledgments

This work has been partially supported by the Spanish Ministerio de Ciencia y Tecnología (MCyT), under project TIC2001-2973-C05-05. The author is grateful to Julia Flores for her comments and suggestions on previous versions of this work.

References

1. A.M. Abdelbar and S.M. Hedetniemi. Approximating MAPs for belief networks is NP-hard and other theorems. *Artificial Intelligence*, 102:21–38, 1998.
2. A.M. Abdelbar, S.T. Hedetniemi, and S.M. Hedetniemi. The complexity of approximating MAPs for belief networks with bounded probabilities. *Artificial Intelligence*, 124:283–288, 2000.
3. S.K. Andersen, K.G. Olesen, F.V. Jensen, and F. Jensen. Hugin: a shell for buiding belief universes for expert systems. In *Proceedings of the 11th International Joint Conference on Artificial Intelligence*, Detroit, 1989.
4. D.E. Appelt and M. Pollack. Weighted abduction for plan ascription. Technical report, Artificial Intelligence Center and Center for the Study of Language and Information, SRI International, Menlo Park, California, 1990.
5. E. Castillo, J.M. Gutiérrez, and A.S. Hadi. *Expert Systems and Probabilistic Network Models*. Monographs in Computer Science. Springer-Verlag, New York, 1997.
6. E. Charniak and E. McDermott. *Introduction to Artificial Intelligence*. Addison-Wesley, 1985.
7. Elvira Consortium. Elvira: An environment for creating and using probabilistic graphical models. In *Proceedings of the First European Workshop on Probabilistic Grpahical Models*, pages 220–230, 2002.
8. G.F. Cooper. Probabilistic inference using belief networks is NP-hard. *Artificial Intelligence*, pages 393–405, 1990.
9. P. Dagum and M. Luby. Approximating probabilistic inference in bayesian belief networks is NP-hard. *Artificial Intelligence*, 60:141–153, 1993.
10. P. Dagum and M. Luby. An optimal approximation algorithm for bayesian inference. *Artificial Intelligence*, 93:1–27, 1997.
11. A. Darwiche. A differential approach to inference in bayesian networks. In *Uncertainty in Artificial Intelligence: Proceedings of the Sixteenth Conference (UAI-2000)*, pages 123–132, San Francisco, CA, 2000. Morgan Kaufmann Publishers.
12. A.P. Dawid. Applications of a general propagation algorithm for probabilistic expert systems. *Statistics and Computing*, 2:25–36, 1992.
13. L.M. de Campos, J.A. Gámez, P. Larrañaga, S. Moral, and T. Romero. *Estimation of Distribution Algorithms. A new tool for evolutionary computation*, chapter Partial Abductive Inference in Bayesian Networks: An empirical comparison between Gas and EDAs. Kluwer Academic Publishers, 2001.
14. L.M. de Campos, J.A. Gámez, and S. Moral. Partial Abductive Inference in Bayesian Belief Networks using a Genetic Algorithm. *Pattern Recognition Letters*, 20(11-13):1211–1217, 1999.
15. L.M. de Campos, J.A. Gámez, and S. Moral. Accelerating chromosome evaluation for partial abductive inference in bayesian networks by means of explanation set absorption. *International Journal of Approximate Reasoning*, (27):121–142, 2001.

16. L.M. de Campos, J.A. Gámez, and S. Moral. Partial Abductive Inference in Bayesian Belief Networks by using Simulated Annealing. *International Journal of Approximate Reasoning*, (27):263–283, 2001.

17. L.M. de Campos, J.A. Gámez, and S. Moral. Simplifying explanations in bayesian belief networks. *International Journal of Uncertainty, Fuzziness and Knowledge-based Systems*, (9):461–489, 2001.

18. L.M. de Campos, J.A. Gámez, and S. Moral. On the problem of performing exact partial abductive inference in Bayesian belief networks using junction trees. In B. Bouchon-Meunier, J. Gutierrez-Rios, L. Magdalena, and R.R. Yager, editors, *Technologies for Constructing Intelligent Systems 2: Tools*, pages 289–302, 2002.

19. L.M. de Campos, J.A. Gámez, and S. Moral. Partial Abductive Inference in Bayesian Belief Networks: An Evolutionary Computation Approach by Using Problem Specific Genetic Operators. *IEEE Transaction on Evolutionary Computation*, (6):105–131, 2002.

20. L.M. de Campos, J.A. Gámez, and S. Moral. Partial abductive inference in bayesian networks by using probability trees. In *Proceedings of the 5th International Conference On Enterprise Information Systems (ICEIS'03). Vol 2.*, pages 83–91, Angers, 2003.

21. R. Dechter. Bucket elimination: A unifying framework for probabilistic inference. In *Proceedings of the Twelfth Annual Conference on Uncertainty in Artificial Intelligence (UAI–96)*, pages 211–219, Portland, Oregon, 1996.

22. R. Decther and I. Rish. A scheme for approximating probabilistic inference. In *Proceedings of the Thirteenth Annual Conference on Uncertainty in Artificial Intelligence (UAI–97)*, pages 132–141, San Francisco, CA, 1997. Morgan Kaufmann Publishers.

23. J.A. Gámez. *Abductive inference in Bayesian networks* (in Spanish). PhD thesis, Departamento de Ciencias de la Computación e I.A. Escuela Técnica Superior de Ingeniería Informática. Universidad de Granada, 1998.

24. E.S. Gelsema. Abductive reasoning in Bayesian belief networks using a genetic algorithm. *Pattern Recognition Letters*, 16:865–871, 1995.

25. K. Kask and R. Dechter. Stochastic local search for Bayesian networks. In *Seventh International Workshop on Artificial Intelligence*, pages 469–480. Morgan Kaufmann Publishers, 1999.

26. H. Kautz and J. Allen. Generalized plan recognition. In *Proc. of National Conference on Artificial Intelligence*, pages 32–37. AAAI, 1986.

27. U.P. Kumar and U.B. Desai. Image interpretation using Bayesian networks. *IEEE Transactions on Pattern Analysis and Machine Intelligence*, 18(1):74–78, 1996.

28. S.L. Lauritzen and D.J. Spiegelhalter. Local computations with probabilities on graphical structures and their application to expert systems. *J.R. Statistics Society. Serie B*, 50(2):157–224, 1988.

29. Z. Li and D'Ambrosio B. An efficient approach for finding the MPE in belief networks. In *Proceedings of the 9th Conference on Uncertainty in Artificial Intelligence*, pages 342–349. Morgan and Kaufmann, San Mateo, 1993.

30. R. E. Neapolitan. *Probabilistic Reasoning in Expert Systems. Theory and Algorithms*. Wiley Interscience, New York, 1990.

31. D. Nilsson. An algorithm for finding the M most probable configurations of discrete variables that are specified in probabilistic expert systems. *MSc. Thesis*, University of Copenhagen, 1994.

32. D. Nilsson. An efficient algorithm for finding the M most probable configurations in Bayesian networks. *Statistics and Computing*, 9:159–173, 1998.

33. P. O'Rorke, S. Morris, and D. Schulenberg. Theory formation by abduction: initial results of a case study based on the chemical revolution. Technical Report ICS-TR-89-25, University of California, Irvine, Department of Information and Computer Science, 1989.

34. J.D. Park. MAP complexity results and approximation methods. In *Uncertainty in Artificial Intelligence: Proceedings of the Eighteenth Conference (UAI-2002)*, pages 388–396, San Francisco, CA, 2002. Morgan Kaufmann Publishers.

35. J.D. Park and A. Darwiche. Approximating MAP using local search. In *Uncertainty in Artificial Intelligence: Proceedings of the Seventeenth Conference (UAI-2001)*, pages 403–410, San Francisco, CA, 2001. Morgan Kaufmann Publishers.

36. J. Pearl. Distributed revision of composite beliefs. *Artificial Intelligence*, 33:173–215, 1987.

37. J. Pearl. *Probabilistic Reasoning in Intelligent Systems*. Morgan Kaufmann, San Mateo, 1988.

38. Y. Peng and J.A. Reggia. A probabilistic causal model for diagnostic problem solving. Part One. *IEEE Transactions on Systems, Man, and Cybernetics*, 17(2):146–162, 1987.

39. Y. Peng and J.A. Reggia. A probabilistic causal model for diagnostic problem solving. Part Two. *IEEE Transactions on Systems, Man, and Cybernetics*, 17(3):395–406, 1987.

40. D. Poole and K. Kanazawa. A decision-theoretic abductive basis for planning. In *Proc. of AAAI Spring Symposium on Decision-Theoretic Planning*, pages 232–239. Stanford University, March 1994.

41. H.E. Pople. On the mechanization of abductive logic. In *Proceedings of the 3rd International Joint Conference on Artificial Intelligence*, pages 147–152, 1973.

42. J.A. Reggia. Diagnostic expert systems based on a set covering model. *International Journal of Man-Machine Studies*, 19:437–460, 1983.

43. R. Reiter. A theory of diagnosis from first principles. *Artificial Intelligence*, 32, 1987.

44. C. Rojas-Guzman and M.A. Kramer. Galgo: A genetic algorithm decision support tool for complex uncertain systems modeled with Bayesian belief networks. In *Proceedings of the 9th Conference on Uncertainty in Artificial Intelligence*, pages 368–375. Morgan and Kauffman, San Mateo, 1993.

45. C. Rojas-Guzman and M.A. Kramer. An evolutionary computing approach to probabilistic reasoning in Bayesian networks. *Evolutionary Computation*, 4:57–85, 1996.

46. A. Salmerón, A. Cano and S. Moral. Importance sampling in bayesian networks using probability trees. *Computational Statistics and Data Analysis*, 34:387–413, 2000.

47. B. Seroussi and J.L. Goldmard. An algorithm directly finding the K most probable configurations in Bayesian networks. *International Journal of Approximate Reasoning*, 11:205–233, 1994.

48. R.D. Shachter, B.D. D'Ambrosio, and B.D. Del Favero. Symbolic probabilistic inference in belief networks. In *8th National Conference on Artificial Intelligence*, pages 126–131, Boston, 1990. MIT Press.

49. S.E. Shimony. The role of relevance in explanation I: Irrelevance as statistical independence. *International Journal of Approximate Reasoning*, 8:281–324, 1993.

50. S.E. Shimony. Finding MAPs for belief networks is NP-hard. *Artificial Intelligence*, 68:399–410, 1994.

51. S.E. Shimony. The role of relevance in explanation II: Disjunctive assignments and approximate independence. *International Journal of Approximate Reasoning*, 13:27–60, 1995.

52. M.E. Stickel. A prolog-like inference system for computing minimum-cost abductive explanations in natural language interpretation. Technical Report 451, AI Center, SRI International, 1988.

53. P. Thagard. Explanatory coherence. *Behavioral and Brain Sciences*, 12:435–467, 1989.

54. Y. Weiss. *Approximate inference using belief propagation.* UAI tutorial, 2001.

55. H. Xu. Computing marginals for arbitrary subsets from marginal representation in markov trees. *Artificial Intelligence*, 74:177–189, 1995.

56. J. Yedidia, W. Freeman, and Y. Weiss. Generalized belief propagation. In *NIPS, volume 13*, 2000.

57. N.L. Zhang and D. Poole. Exploiting causal independence in Bayesian network inference. *Journal of Artificial Intelligence Research*, 5:301–328, 1996.

58. X. Zhong and E. Santos Jr. Directing genetic algorithms for probabilistic reasoning through reinforcement learning. *International Journal of Uncertainty, Fuzziness and Knowledge-Based Systems*, 8:167–186, 2000.

Causal Models, Value of Intervention, and Search for Opportunities

Tsai-Ching Lu and Marek J. Druzdzel

Decision Systems Laboratory
School of Information Sciences and Intelligent Systems Program
University of Pittsburgh, Pittsburgh, PA 15260
ching,marek@sis.pitt.edu

Abstract. While algorithms for influence diagrams allow for computing the optimal setting for decision variables, they offer no guidance in generation of decision variables, arguably a critical stage of decision making. A decision maker confronted with a complex system may not know which variables to best manipulate to achieve a desired objective. We introduce the concept of *search for opportunities* which amounts to identifying the set of decision variables and computing their optimal settings, given an objective expressed by a utility function. Search for opportunities is built on value of intervention in causal models.

Keywords: causal models, value of intervention, search for opportunities, causal Bayesian networks, and structural equation models.

1 Introduction

Influence diagrams [4] are popular tools for representing decision problems under uncertainty and identifying optimal strategies. The key problem with using influence diagrams is that we need to specify beforehand all decision alternatives and their consequences explicitly. In complex systems, this may result in a cumbersome, if not a totally unmanageable, modeling process. Ideally, a modeling language should support the prediction of the effects of actions that were not considered in model's construction [1,8,9]. This allows us to search for the best actions to be taken to achieve a set of objectives, a concept that we refer to as *search for opportunities*.

The problem of search for opportunities is related to the problem of *information gathering* [10]. In information gathering, a decision maker tries to decide which information to acquire to reduce the uncertainty over a model, and consequently to improve the quality of the decision at hand. The means of acquiring information is constrained to observations that modify the decision maker's belief over the states of a system. Search for opportunities, in contrast, seeks to apply intervening actions that alter the trajectory of the system toward those outcomes that are preferable to the decision maker.

In decision analysis, the primary tool for information gathering is *value of information* [3]. Value of information is defined as the upper bound on what a decision maker should be willing to pay in employing a clairvoyant to reveal the outcome of a chance variable. Similarly, the concept of *value of control* has been introduced and defined as the upper bound on what a decision maker should be willing to pay a wizard for setting a chance variable into a preferred state. Both value of information and value of control are defined with respect to a decision problem [6]. However, to our knowledge, value of control computation has only been applied to chance nodes with no predecessors in influence diagrams encoded in Howard canonical form [6]. Since influence diagrams may describe probabilistic rather than causal relations, there is no guarantee that converting a chance node with predecessors or a chance node without predecessors in diagrams that are not in Howard canonical form into a decision node will correctly model the effects of control.

In this chapter, we discuss the problem of search for opportunities, where a decision maker seeks for creative decision options in order to achieve a given objective. The basis of the search is a causal model of the system that is subject to the decision. Causal models based on structural equations support causal reasoning and, in particular, prediction of the effects of actions [9,12]. A causal model consists of a self-contained set of simultaneous structural equations, each of which represents a causal mechanism active in the modeled system. Causal models support prediction of the effects of actions by replacing those mechanisms that are impacted by actions with new mechanisms, possibly not contemplated during model's construction, and leaving the rest intact. The problem of search for opportunities, in this formulation, amounts to searching for variables that were not originally contemplated as decision variables and intervening into mechanisms governing these variables in order to affect the outcomes. Therefore, search for opportunities leads to discovery of *novel* actions that help to achieve decision objectives.

To address the problem of search for opportunities, we introduce the concept of *value of intervention*. The value of intervention, related to the value of control, arises from considering jointly the economic factors and effects of actions in causal models. It can be considered as a generalization of the value of control since the intervention operates at the level of mechanism in causal models, but the control operates at the level of variables in influence diagrams. The value of intervention computation is also applicable to influence diagrams in canonical form [2], an extension of Howard Canonical Form that supports causal reasoning, but not to influence diagrams that do not represent causal relations.

The remainder of this chapter is organized as follows. Section 2 gives an overview of causal models. Section 3 introduces augmented models for describing decision problems at hand. Section 4 discusses the concept of value of intervention. Section 5 shows how value of intervention can be used to solve the problem of search for opportunities. Section 6 describes the modeling of non-intervening actions. Finally, we discuss the direction of our future work.

2   Causal Models

A causal model consists of variables and causal relations among them, modeled by a set of simultaneous *structural equations*, where each equation represents a distinct mechanism active in the system.[1] More formally, we denote the set of variables appearing in an equation e as $\mathsf{Vars}(e)$, and in a set of equations $\mathbf{E}$ as $\mathsf{Vars}(\mathbf{E}) \equiv \cup_{e \in \mathbf{E}} \mathsf{Vars}(e)$. A *causal model* $M = \langle \mathbf{X}, \mathbf{E} \rangle$ consists of a *self-contained* set of simultaneous structural equations $\mathbf{E}$ over a set of variables $\mathbf{X} \equiv \mathsf{Vars}(\mathbf{E})$. Each structural equation $e \in \mathbf{E}$ is generally written in its implicit form $e(X_1, X_2, \ldots, X_n) = 0$ where $X_i \in \mathsf{Vars}(e)$. We say that a variable $X_i \in \mathbf{X}$ is *exogenous* to M if its value is determined by factors outside the system, i.e., if there exists a structural equation $e \in \mathbf{E}$, $e(X_i) = 0$, and *endogenous* otherwise. In other words, the set of variables $\mathbf{X}$ consists of two disjoint sets $\mathbf{U}$ and $\mathbf{V}$ of exogenous and endogenous variables respectively. Therefore, a causal model is sometimes denoted as $M = \langle \mathbf{U}, \mathbf{V}, \mathbf{E} \rangle$. Let $\mathsf{D}(X_i)$ be the domain of a variable X_i, and $\mathsf{D}(\mathbf{X}) = \mathsf{D}(X_1) \times \ldots \times \mathsf{D}(X_n)$ be the domain of the set of variables $\mathbf{X} = \{X_1, \ldots, X_n\}$. Given $\mathbf{u} \in \mathsf{D}(\mathbf{U})$, the solutions for endogenous variables $\mathbf{Y} \subseteq \mathbf{V}$, denoted as $\mathbf{Y}_M(\mathbf{u})$ or $\mathbf{Y}(\mathbf{u})$, in a causal model M can always be determined uniquely. The pair $\langle M, \mathbf{u} \rangle$ is called a *causal world*, or simply world. Given a probability distribution $\mathrm{Pr}(\mathbf{u})$ defined over $\mathsf{D}(\mathbf{U})$, the pair $\langle M, \mathrm{Pr}(\mathbf{u}) \rangle$ is called a *probabilistic* causal model where for each $Y \in \mathbf{V}$, $\mathrm{Pr}(Y = y) \triangleq \sum_{\{\mathbf{u} | Y(\mathbf{u}) = y\}} \mathrm{Pr}(\mathbf{u})$. Simon [11] developed an algorithm that explicates the asymmetries among variables in a causal model M and represents them as a *causal graph* $G(M)$. A causal model M is *recursive* if the associated $G(M)$ is a directed acyclic graph, where each node corresponds to a variable, and each family (a node with its parents in $G(M)$) a structural equation [1]. In other words, each structural equation $e(X_1, \ldots, X_n) = 0$ is expressed in its explicit functional form $X_i = f_{X_i}(X_1, \ldots, X_{i-1}, X_{i+1}, \ldots, X_n)$ and is depicted graphically as a family with arcs from nodes representing arguments of f_{X_i} (i.e., $X_1, \ldots, X_{i-1}, X_{i+1}, \ldots, X_n$) to X_i. In the sequel, the term "causal model" refers to a recursive causal model, in which each equation is indexed by the dependent variable of its explicit form.

Example 1. Consider a model for the operational status of a command center (CC). CC depends on the status of communications (C) and radar (R). Radar depends on the antenna structure (A) and the power supplied by the generator (G). Communications rely on the power supplied by the generator. The generator relies on fuel supply (F) to generate power. Each of the variables has state *operational* or *damaged*. We assume that for each of the above relations there is a corresponding exogenous variable, denoted as U_{cc}, U_r, U_c, and U_g, that for each relation summarizes those factors that are outside the model. F and A are themselves exogenous variables. We assume that all exogenous variables $U = \{U_{cc}, U_r, U_c, U_g, F, A\}$ are independent. The set

---

[1] Please see [9] or [12] for general introductions to causal models.

of structural equations, representing the domain of our interest, and its corresponding causal graph are shown in Fig. 1. We have included an explicit graphical representation of variables U_{cc}, U_r, U_c, U_g for the sake of clarity of explanation. In practice, these variables are modeled implicitly by error terms in the corresponding equations and we will omit them in the sequel of this chapter for the sake of clarity.

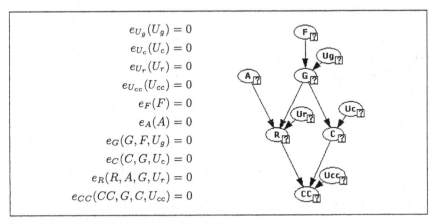

$$e_{U_g}(U_g) = 0$$
$$e_{U_c}(U_c) = 0$$
$$e_{U_r}(U_r) = 0$$
$$e_{U_{cc}}(U_{cc}) = 0$$
$$e_F(F) = 0$$
$$e_A(A) = 0$$
$$e_G(G, F, U_g) = 0$$
$$e_C(C, G, U_c) = 0$$
$$e_R(R, A, G, U_r) = 0$$
$$e_{CC}(CC, G, C, U_{cc}) = 0$$

Fig. 1. Causal model and its corresponding causal graph for modeling the operational status of a command center.

2.1   Actions and Effects

Given a causal model $M = \langle U, V, E \rangle$, an *atomic action*, $do(X = x)$, sets an endogenous variable $X \in V$ to the value $x \in D(X)$ and transforms M to the modified model $M_x = \langle U, V, E_x \rangle$, where $E_x = \{e_Y | Y \in V \setminus \{X\}\} \cup \{X = x\}$. The effect of action $do(X = x)$ is given by M_x. The *action operator* $do(\cdot)$ replaces e_X with its argument, a (probability) function that (probabilistically) assigns X with the value $x \in D(X)$, to derive the modified model M_x and consequently its corresponding effects. In general, the intervention in the argument of $do(\cdot)$ operator may result in non-recursive models. For example, when applying a $do(e'_X(X, Y) = 0)$ to a recursive causal model $M = \langle \{X, Y\}, \{e_X(X) = 0, e_Y(Y, X) = 0\} \rangle$, we have a non-recursive causal model $\langle \{X, Y\}, \{e'_X(X, Y) = 0, e_Y(Y, X) = 0\} \rangle$. However, in this chapter we restrict our analysis to actions that result in recursive models. Pearl and Robins [9] formalized three types of actions in this class: *Conditional action*, denoted as $do(X = x\downarrow_{x=g(z)})$, sets $X \in V$ to the value $x = g(z)$ whenever Z attain values z, where $g : D(Z) \to D(X)$ and Z are non-descendants of X in $G(M)$. *Stochastic action*, denoted as $do(X = x\downarrow_{Pr^*(x)})$, sets $X \in V$ to the value x with probability $Pr^*(x)$ where $Pr^*(x)$ is specified externally.

We may also have *stochastic policy*, denoted as $do(X = x\downarrow_{\Pr^*(x|\mathbf{z})})$, that sets $X = x$ with probability $\Pr^*(x|\mathbf{z})$ whenever $\mathbf{Z}$ attain values $\mathbf{z}$ where $\mathbf{Z}$ are non-descendants of X in $G(M)$ and $\Pr^*(x|\mathbf{z})$ is set externally. In the sequel, we follow Pearl's notation [9] where $\hat{x}$ is the abbreviation of X attaining value $x \in D(X)$ due to an intervention.

Given a world $\langle M, u \rangle$, the *potential response* of $Y \in \mathbf{V}$ to action $do(\cdot)$ on variable $X \in \mathbf{V}$, denoted as $Y_{M_x}(\mathbf{u})$ or $Y_x(\mathbf{u})$, is the solution for Y of the set of equations $\mathbf{E}_x$ of M_x. $Y_x(\mathbf{u})$ can also be interpreted as the *counterfactual* value that Y would obtain had X been x in the counterfactual world brought about by action $do(\cdot)$. Given a probabilistic causal model $\langle M, \Pr(\mathbf{u}) \rangle$, the causal effect on Y of an atomic action $do(X = x)$ is given by $\langle M_x, \Pr(\mathbf{u}) \rangle$ as $\Pr(y|\hat{x}) \equiv \Pr(Y_x = y) \triangleq \sum_{\{\mathbf{u}|Y_x(\mathbf{u})=y\}} \Pr(\mathbf{u})$. The causal effect on Y of a conditional action $do(X = x\downarrow_{x=g(\mathbf{z})})$ is expressed as $\Pr(y|\hat{x})\downarrow_{x=g(\mathbf{z})} \equiv \Pr(Y_x = y)\downarrow_{x=g(\mathbf{z})} \triangleq \sum_{\mathbf{z}} \Pr(y|\hat{x}, \mathbf{z})\downarrow_{x=g(\mathbf{z})} \Pr(\mathbf{z})$. The causal effect on Y of a stochastic action $do(X = x\downarrow_{\Pr^*(x)})$ is expressed as $\Pr(y|\hat{x})\downarrow_{\Pr^*(x)} \equiv \Pr(Y_x = y)\downarrow_{\Pr^*(x)} \triangleq \sum_x \Pr(y|\hat{x}) \Pr^*(x)$. The causal effect on Y of a stochastic policy $do(X = x\downarrow_{\Pr^*(x|\mathbf{z})})$ is expressed as $\Pr(y|\hat{x})\downarrow_{\Pr^*(x|\mathbf{z})} \equiv \Pr(Y_x = y)\downarrow_{\Pr^*(x|\mathbf{z})} \triangleq \sum_x \sum_{\mathbf{z}} \Pr(y|\hat{x}, \mathbf{z}) \Pr^*(x|\mathbf{z}) \Pr(\mathbf{z})$.

Example 2. Suppose the model in Example 1 is an enemy's command center and one objective is to disrupt the enemy's communications. We can act on the communications C by, for example, jamming the signal with noise, and setting C to *damaged*. The modified causal model and its corresponding causal graph are shown in Fig. 2. Please note that the intervention makes the arcs coming into C inactive (arc $G \rightarrow C$).

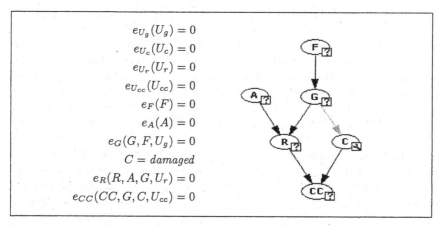

$$e_{U_g}(U_g) = 0$$
$$e_{U_c}(U_c) = 0$$
$$e_{U_r}(U_r) = 0$$
$$e_{U_{cc}}(U_{cc}) = 0$$
$$e_F(F) = 0$$
$$e_A(A) = 0$$
$$e_G(G, F, U_g) = 0$$
$$C = damaged$$
$$e_R(R, A, G, U_r) = 0$$
$$e_{CC}(CC, G, C, U_{cc}) = 0$$

Fig. 2. The modified causal model and its corresponding causal graph after atomic action $do(C = damaged)$.

3   Augmented Models

In order to describe a decision problem at hand, we propose to augment a probabilistic causal model $\langle M, \Pr(\mathbf{u}) \rangle$ by characterizing its variables along three properties: *observability* (observable or unobservable), *manipulability* (manipulable or non-manipulable), and *focus* (focus or non-focus). Let $\mathbf{X}$ denote all variables in M. A variable $X_i \in \mathbf{X}$ is *observable*, denoted as $X_i.o$, if it represents an entity that can be *measured* directly; *unobservable*, denoted as $X_i.\bar{o}$, otherwise. A variable $X_i \in \mathbf{X}$ is *manipulable*, denoted as $X_i.m$, if it represents an entity that can be *manipulated* directly; *non-manipulable*, denoted as $X_i.\overline{m}$, otherwise. We assume that a manipulable variable is always observable, i.e., we assume that we can always observe the effect of our manipulation. A variable $X_i \in \mathbf{X}$ is a *focus* variable, denoted as $X_i.f$, if it represents an objective; *non-focus*, denoted as $X_i.\bar{f}$, otherwise.

The goal of this work is to build a system that suggests decisions. At any stage of working with the system, there may be variables on which users have decided to intervene, whether based on the system's suggestions or the user's prior choices. We represent such decisions by augmenting the model with a set of *decision* variables $\mathbf{D}$ along with their corresponding settings. The domain of each $D_i \in \mathbf{D}$ consists of the choices of setting $X_i.m$, the augmented manipulable variable, to a value $x_i \in \mathsf{D}(X_i)$, denoted as x_i', and a special state *idle* representing the force of nature [9]. Let $\mathbf{PA}_i$ denote the set of parents of X_i in $G(M)$, i.e., $\mathbf{PA}_i = \mathsf{Vars}(e_{X_i}) \setminus \{X_i\}$. We augment the equation $e_{X_i}(X_i, \mathbf{PA}_i) = 0$ to $e'_{X_i}(X_i, \mathbf{PA}_i') = 0$, where $\mathbf{PA}_i' = \mathbf{PA}_i \cup \mathbf{Z} \cup \{D_i\}$ and $\mathbf{Z} \subset \mathbf{X}$, a set of non-descendants of X_i in $G(M)$ brought about by interventions. We define the augmented equation $e'_{X_i}(X_i, \mathbf{PA}_i') = 0$ as

$$e'_{X_i}(X_i, \mathbf{PA}_i') \triangleq \begin{cases} e^*_{X_i}(X_i, \mathbf{PA}_i') = 0 \text{ if } D_i = x_i' \\ e_{X_i}(X_i, \mathbf{PA}_i) = 0 \text{ if } D_i = idle \end{cases}, \tag{1}$$

where the form of $e^*_{X_i}(X_i, \mathbf{PA}_i') = 0$ depends on the type of intervention (see Table 1). To represent *concurrent actions* on X_i and X_j, in addition to D_i and D_j and corresponding augmentations on e_{X_i} and e_{X_j}, we add an additional decision variable, denoted as D_{ij}, to represent the concurrency. The domain of D_{ij} is $\mathsf{D}(D_{ij}) = \mathsf{D}(D_i) \times \mathsf{D}(D_j)$. We add additional *projection* equations $e_{D_i} : \mathsf{D}(D_{ij}) \to \mathsf{D}(D_i)$ and $e_{D_j} : \mathsf{D}(D_{ij}) \to \mathsf{D}(D_j)$ such that $D_i = d_i$ and $D_j = d_j$ for each $d_{ij} \in \mathsf{D}(D_{ij})$.

Finally, to represent our preferences over the given set of objectives and decisions, we augment the model by a set of *utility* variables $\mathbf{UT}$ along with their utility functions $\mathbf{U}$. Each utility function $\mathsf{U}_i \in \mathbf{U}$ can only have focus or decision variables as its arguments. Formally, we can define an augmented model as follows.

Definition 1 (Augmented Model).
An augmented model for a decision problem is
$M_A = \langle \langle M, \Pr(\mathbf{u}) \rangle, C(\mathbf{X}), \langle \mathbf{D}, \mathbf{E}' \rangle, \langle \mathbf{UT}, \mathbf{U} \rangle \rangle$ where:

Table 1. The form of $e^*_{X_i}(X_i, \mathbf{PA}'_i)$ characterized with respect to different types of interventions: atomic action, conditional action, stochastic action, and stochastic policy.

Type	$e^*_{X_i}(X_i, \mathbf{PA}'_i) = 0$
Atomic	$\Pr(x_i \| \mathbf{pa}'_i) = \begin{cases} 1 \text{ if } x'_i = x_i, \\ 0 \text{ otherwise.} \end{cases}$
Conditional	$\Pr(x_i \| \mathbf{pa}'_i) = \begin{cases} 1 \text{ if } x'_i = x_i \text{ and } x_i = g(\mathbf{z}), \\ 0 \text{ otherwise.} \end{cases}$
Stochastic	$\Pr(x_i \| \mathbf{pa}'_i) = \begin{cases} \Pr^*(x_i) \text{ if } x'_i = x_i, \\ 0 \qquad\qquad \text{otherwise.} \end{cases}$
Policy	$\Pr(x_i \| \mathbf{pa}'_i) = \begin{cases} \Pr^*(x_i \| \mathbf{z}) \text{ if } x'_i = x_i \text{ and } x_i = g(\mathbf{z}), \\ 0 \qquad\qquad\quad \text{otherwise.} \end{cases}$

1. $\langle M, \Pr(\mathbf{u}) \rangle$ is a probabilistic causal model.
2. $C(\mathbf{X})$ is a characterization of observability, manipulability, and focus for each $X_i \in \mathbf{X}$.
3. $\langle \mathbf{D}, \mathbf{E}' \rangle$ is a set of decision variables $\mathbf{D}$ and the modified equation $\mathbf{E}'$ with respect to the decisions.
4. $\langle \mathbf{UT}, \mathbf{U} \rangle$ is a set of utility variables $\mathbf{UT}$ and its corresponding utility functions $\mathbf{U}$ over a set of focus variables, characterized by $C(\mathbf{X})$, and a subset of decision variables in $\mathbf{D}$.

Example 3. Suppose that variables F, A, G, R, C, and CC are manipulable and CC is the only focus variable in Example 1. We add a utility node, *Utility*, with utility function $\mathsf{U}(CC)$, to represent our preference over the states of CC. The corresponding causal graph is shown in Fig. 3 (a). Suppose we have a decision option of manipulating the communications C with no direct influence on *Utility*. We then have a causal graph with D_c as a decision variable shown in Fig. 3 (b).

We emphasize that specifying a variable manipulable, which merely acknowledges the possibility of interventions, is not the same as designating a decision variable in an influence diagram, which requires the explicit specifications of a decision variable along with its consequences. Declaring a variable manipulable allows the algorithm of search for opportunities to explore possible interventions that might not be foreseen when the model is constructed. Furthermore, we require neither a manipulable variable being intervened upon, nor an observable variable being observed. It is the task of search for opportunities and information gathering to determine which variable one should intervene or observe and in what order. In other words, we propose to relax not only the assumption of a fixed sequence of intervening actions and observations in influence diagrams, but also the assumption of a fixed operation over a variable. For example, a manipulable variable may be intervened, observed, or unknown, depending on different decision

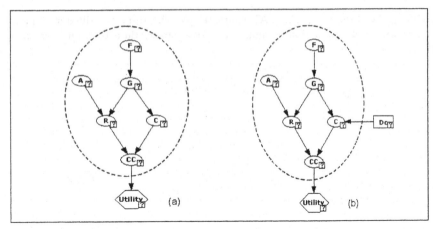

Fig. 3. (a) Example model augmented with utility over variable CC. (b) The model in (a) augmented with the atomic action D_c.

sequences generated by search for opportunities and information gathering. Only when a manipulable variable is augmented by the decision variable and its augmented equation as in the form of Equation 1, we commit ourselves to intervene the manipulable variable with one of the policies specified by the decision variable.

In this chapter, we constrain ourselves in applying augmented models on decision scenarios that contain only intervening actions. We treat an augmented model as an influence diagram and compute the optimal strategy and the maximum expected utility using algorithms for inference diagrams.

4   Value of Intervention

Suppose that we are considering an additional atomic intervention on an unaugmented manipulable variable X_k in an augmented model M_A. We augment M_A by adding a new decision variable D_k and modify e_{X_k} to e'_{X_k} as demonstrated in Equation 1. Let $\mathbf{D}' = \mathbf{D} \cup \{D_k\}$ be the new set of decision variables. We also augment utility function $\mathbf{U}$ to $\mathbf{U}'$ if the intervention directly influences $\mathbf{U}$. Most interventions come at a certain cost and the cost of intervention can be incorporated by augmenting $\mathbf{U}$. Let M'_A denote the newly augmented model, $\pi(M'_A)$ its optimal strategy, and $\mathsf{MEU}(M_A)$ and $\mathsf{MEU}(M'_A)$ the maximum expected utility yielded by the optimal strategies $\pi(M_A)$ and $\pi(M'_A)$ respectively. We define the *value of intervention* on X_k as

$$\mathsf{VOINT}(D_k = d^*_k) \triangleq \mathsf{MEU}(M'_A) - \mathsf{MEU}(M_A), \tag{2}$$

where $d^*_k \in \mathsf{D}(D_k)$ is yielded in $\pi(M'_A)$ by the optimal policy of D_k. Note that value of atomic intervention can account for the concept of the value of control in influence diagrams. Since augmented models support prediction of the effect of actions, we are not constrained to only atomic intervention (control) on those nodes with no predecessors as in the case of value of control in influence diagrams. To compute value of intervention for a conditional, stochastic, or stochastic policy action, we simply substitute the action of interest for the atomic intervention in augmenting model M_A and perform analysis using Equation 2.

Theorem 1. *Let M_A be an augmented model and X_k be an unaugmented manipulable variable in M_A. If M'_A is the augmented model of M_A that considers an intervention on X_k that has no direct impacts on utility functions U, then $\mathsf{MEU}(M'_A) \geq \mathsf{MEU}(M_A)$.*

Proof. When evaluating $\pi(M_A)$, we can decompose the joint probability distribution of M_A according to $G(M_A)$. When considering an additional intervention on X_k that has no direct impacts on U, we augment M_A into M'_A by modifying $\Pr(x_k|\mathbf{pa}_k)$ to $\Pr(x_k|\mathbf{pa}'_k)$. Now, when evaluating $\pi(M'_A)$, we also decompose the joint probability distribution of M'_A according to the $G(M'_A)$. Notice that X_k participates in $\mathsf{MEU}(M_A)$ as

$$\mathsf{MEU}(M_A) = \cdots \sum_{x_k \in \mathsf{D}(X_k)} \Pr(x_k|\mathbf{pa}_k) \cdots \mathsf{U},$$

and in $\mathsf{MEU}(M'_A)$ as

$$\mathsf{MEU}(M'_A) = \cdots \max_{d_k \in \mathsf{D}(D_k)} \sum_{x_k \in \mathsf{D}(X_k)} \Pr(x_k|\mathbf{pa}'_k) \cdots \mathsf{U},$$

where $d_k \in \mathbf{pa}'_k$. Since $\Pr(x_k|\mathbf{pa}_k)$ is also represented in $\Pr(x_k|\mathbf{pa}'_k)$ as $d_k = idle$, $\mathsf{MEU}(M'_A) \geq \mathsf{MEU}(M_A)$ according to the maximization operator.

Consider a simple model M_A which consists of one variable X_k with probability distribution $\Pr(x_k)$, where $x_k \in \mathsf{D}(X_k)$, and a utility variable UT with utility function $\mathsf{U}(x_k)$. We have $\mathsf{MEU}(M_A) = \mathsf{EU}(M_A) = \sum_{x_k \in \mathsf{D}(X_k)} \Pr(x_k) \mathsf{U}(x_k)$. Consider an additional stochastic intervention on X_k that has no direct impact on U. We augment M_A to M'_A with a new decision variable D_k with domain $\mathsf{D}(D_k) = \mathsf{D}(X_k) \cup \{idle\}$, and modify the probability distribution of X_k to

$$\Pr(x_k|d_k) = \begin{cases} \Pr^*(x_k) & \text{if } D_k = x'_k \text{ and } x_k = x'_k, \\ 0 & \text{if } D_k = x'_k \text{ and } x_k \neq x'_k, \\ \Pr(x_k) & \text{if } D_i = idle. \end{cases}$$

We have

$$\mathsf{MEU}(M'_A) = \max_{d_k \in \mathsf{D}(D_k)} \sum_{x_k \in \mathsf{D}(X_k)} \Pr(x_k|d_k) \mathsf{U}(x_k)$$

and the optimal value of setting D_k,

$$d_k^* = \arg\max_{d_k \in D(D_k)} \sum_{x_k \in D(X_k)} \Pr(x_k | d_k) U(x_k).$$

It shows that d_k^* is taken on one of x_k' only if

$$\sum_{x_k \in D(X_k)} \overset{*}{\Pr}(x_k) U(x_k) > \sum_{x_k \in D(X_k)} \Pr(x_k) U(x_k).$$

In other words, it suggests not to act if the stochastic intervention under consideration does no better than the nature. By the same token, d_k^* can always take on the state *idle* for other types of interventions that have no direct impacts on utility functions U, if it will not do better than the nature.

Next, consider an augmented model with variables $D_A \rightarrow A \rightarrow B \rightarrow U$. We have

$$\mathsf{MEU}(M_A) = \max_{d_A \in D(D_A)} \sum_{a \in D(A)} \Pr(a | d_A) \sum_{b \in D(B)} \Pr(b | a) U(b).$$

Consider an additional intervention on B with no direct impacts on U and augment M_A to M_A' correspondingly. We have

$$\mathsf{MEU}(M_A') = \max_{d_A \in D(D_A)} \sum_{a \in D(A)} \Pr(a | d_A) \max_{d_B \in D(D_B)} \sum_{b \in D(B)} \Pr(b | a, d_B) U(b).$$

We see again that D_B will take on the state other than *idle* only if

$$\sum_{b \in D(B)} \Pr(b | a, d_B) U(b) > \sum_{b \in D(B)} \Pr(b | a) U(b).$$

5   Search for Opportunities

Search for opportunities refers to the problem of identifying novel interventions that can improve the outcomes. Ideally, we should consider all possible novel interventions on all unaugmented manipulable variables simultaneously, along with existing decisions, to find the optimal strategy and the maximum expected utility for the model. In a complex system, however, such analysis can easily challenge our modeling and computational capabilities. For example, even if we constrain ourselves to considering only atomic interventions on manipulable variables in the command center example, theoretically, we need to elicit utilities and to evaluate strategies for 3^6 combinations of all possible atomic interventions. In general, if we have n manipulable variables with m states for each, we will have $(m+1)^n$ combinations of utilities and strategies, including one extra dimension for the force of nature.

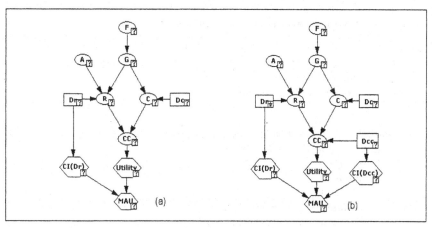

Fig. 4. (a) The model augmented with a possible atomic intervention on R, the cost of intervention CI_{D_R}, and the multi-attribute utility function MAU. (b) The optimal atomic intervention on R is instantiated by setting D_R to d_R^* and the model is augmented with a possible intervention on CC and the cost of intervention $\mathsf{CI}_{D_{CC}}$ in the myopic approach of search for opportunities.

To simplify the problem of modeling, we assume that all novel interventions under consideration have no direct impact on utility functions except by the cost of intervention. We elicit the cost of intervention CI_k for each novel intervention on X_k. Let $\mathbf{CI}$ denote all the costs of interventions under consideration. We assume that our multi-attribute utility function over the existing individual utility functions $\mathbf{U}$ and the costs of interventions is *decomposable*, i.e., there exists a multi-attribute utility function MAU that takes as arguments each $U_i \in \mathbf{U}$ and each $\mathsf{CI}_k \in \mathbf{CI}$ and combine them with a functional form (such as a simple linear or a multiplicative form). In the case of a linearly additive MAU, we need to elicit $n \times (m+1)$ numbers for $\mathbf{CI}$ and at most $n + 1$ numbers for the weights in MAU (if the units of cost are the same, this number can be significantly smaller). Next, we approximate the optimal strategy computation by the *myopic* (greedy) search that considers one intervention at a time and selects the one with the maximum value of intervention to perform. We act according to the selected intervention and perform the myopic search again to select the next intervention to act until there is no intervention that can improve the maximum expected utility over our predefined threshold. Fig. 5 outlines this procedure.

Example 4. Given the model in Fig. 3 (b), suppose that we use the myopic approach to identify that R is the next variable to intervene by an atomic intervention. The model is augmented with an intervention on R as shown in Fig. 4 (a). After intervening on R by setting D_R to d_R^*, suppose that we identify that CC is the next variable to act on by an atomic intervention. We have the model augmented as shown in Fig. 4 (b).

Procedure MyopicSearchForOpportunities

Input: An augmented model M_A, a threshold δ^* for the increase of expected
utility, and Cl_i for each $X_i \in X^m$, the set of unaugmented manipulable
variables in M_A.

Output: A sequence of interventions D on the subset of X^m.

1. $D := \emptyset$; $M^* := M_A$.
2. $Update(M^*)$; $\mu^* := \mathsf{MEU}(M^*)$; $found := $ **false**.
3. **while** $X^m \neq \emptyset$ **and** $found = $ **false**
4. $M := M^*$; $\mu := \mu^*$; $\delta := \delta^*$.
5.   **for each** $X_i \in X^m$
6. $M' := Augment(M^*, do(X_i), Cl_i)$.
7. $Update(M')$; $\mu' := \mathsf{MEU}(M')$.
8. $\triangle\mu := \mu' - \mu^*$. /* $\mathsf{VOINT}(D_i)$ */
9.   **if** $\triangle\mu > \delta$ **then**
10. $\delta := \triangle\mu$; $M := M'$; $\mu := \mu'$;
11. $X := X_i$; $d := \pi_i(M')$;
12. $found := $ **true**.
13.   **end for each**
14.   **if** $found = $ **true then**
15. $M^* := Instantiate(M, X, d)$; $\mu^* := \mu$;
16. $X^m := X^m \setminus \{X\}$; $D := D \cup \{(X, d)\}$;
17. $found := $ **false**,
18.   **else** $found := $ **true**.
19. **end while**.
20. **return** D.

Fig. 5. Myopic approach to search for opportunities. $Update(M)$ computes
the optimal strategy and maximum expected utility for a model M.
$Augment(M, do(X_i), Cl_i)$ denotes the operation of augmenting the model M with
an intervention on X_i with the cost of intervention Cl_i. $Instantiate(M, X, d)$ denotes
the operation of setting the value of D_X to d in M.

The procedure in Fig. 5 can be applied by a robot to find out the next most
effective action. It can also be a useful extension of a modeling environment,
which is how we plan to apply it. As illustrated in Fig. 6, we present users with
a list of ranked values of interventions, generated by the code Lines 5-13 in
Fig. 5.[2] Users may take the suggestion from the myopic search to perform the
intervention at the top of the list, or select any other intervention from the list
to alter the generation of the decision sequences. Once the user has entered the
intervention into the system, the system performs the myopic search again to
update the ranked list of possible interventions. This interactive environment

---

[2] As far as utility and cost of intervention are concerned, we use a simple linearly
additive form of MAU function.

allows the users also to perform "what if" analysis in generating decision sequences.

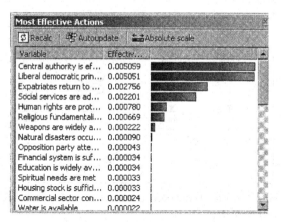

Fig. 6. A ranked list of value of interventions on unaugmented manipulable variables.

6   Non-intervening Action

So far, we have constrained ourselves in applying augmented models on decision sequences containing only intervening actions. In many decision problems, we bring in a new mechanism into the model only when we consider a *non-intervening* action. For example, if we model the relations between heart disease(HD) and blood pressure (BP) as mechanisms $f_{HD}(HD) = 0$ and $f_{BP}(HD, BP) = 0$, we have the causal graph $HD \to BP$. An example of non-intervening action would be the decision of measuring blood pressure (MBP), which brings the variable blood pressure reading (BPR) and the mechanism describing how the blood pressure is measured, $f_{BPR}(BP, BPR, MBP) = 0$, into consideration. We have the causal graph of the augmented model as $HD \to BP \to BPR \leftarrow MBP$. We can also represent the cost of measuring blood pressure (CO_{MBP}) as a value function of MBP, i.e., $U(MBP)$. We extend the causal graph into $HD \to BP \to BPR \leftarrow MBP \to CO_{MBP}$.

Example 5. Give the model in Fig. 4, suppose that we consider a non-intervening action (D_A) that assesses the operational status of antenna (A_S) with the cost of observation CO_{D_A}. The causal graph of the augmented model is shown in Fig. 7.

7   Discussion

We present augmented causal models that support decision making with a flexible set of interventions. We introduce the concept of search for oppor-

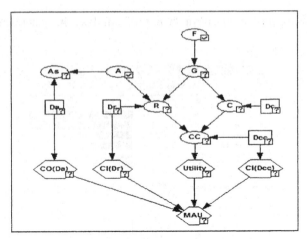

Fig. 7. The model in Fig. 4 augmented with a non-intervening action D_A and the cost of observation CO_{D_A}.

tunities which uses the computation of value of intervention to search for intervening actions on manipulable variables in an augmented model. The result is a greedy search algorithm that suggests the best action to perform.

The concept of value of intervention has also been proposed for causal discovery in active learning [7,13]. Since the focus of that work is on discovering the true model structure, the value of intervention is defined over all possible models. Our approach, on the other hand, assumes the availability of the true model and uses the value of intervention to advise the next intervention to perform to achieve a desired objective.

Jensen and Vomlelová [5] introduced *unconstrained influence diagrams* that address decision problems where the order of decisions and observations is not determined, but partial temporal ordering of decisions and observations is specified. In our framework, we address decision problems where the choice of a variable being observed or intervened is not even determined. We are currently extending our approach to address decision sequences that contain mixed interventions and observations where previous observations can be invalidated by intervening actions.

Acknowledgments

Our research was supported by the Air Force Office of Scientific Research under grant F49620–00–1–0112. The value of intervention computation and search for opportunities are implemented in GeNIe and SMILE, available at http://www2.sis.pitt.edu/~genie.

References

1. Marek J. Druzdzel and Herbert A. Simon. Causality in Bayesian belief networks. In *Proceedings of the Ninth Annual Conference on Uncertainty in Artificial Intelligence (UAI-93)*, pages 3–11, San Francisco, CA, 1993. Morgan Kaufmann Publishers.

2. David Heckerman and R. Shachter. Decision-theoretic foundations for causal reasoning. *Journal of Artificial Intelligence Research*, 3:405–430, 1995.

3. Ronald A. Howard. Information value theory. *IEEE Transactions on Systems Science and Cybernetics*, SSC-2(1):22–26, August 1966.

4. Ronald A. Howard and James E. Matheson. Influence diagrams. In Ronald A. Howard and James E. Matheson, editors, *The Principles and Applications of Decision Analysis*, pages 719–762. Strategic Decisions Group, Menlo Park, CA, 1981.

5. Finn V. Jensen and Marta Vomlelová. Unconstrained influence diagrams. In *Proceedings of the Eighteenth Annual Conference on Uncertainty in Artificial Intelligence (UAI-2002)*, pages 234–241, San Francisco, CA, 2002. Morgan Kaufmann Publishers.

6. James E. Matheson. Using influence diagrams to value information and control. In *Influence Diagrams, Belief Nets and Decision Analysis*, chapter 2, pages 25–48. John Wiley & Sons Ltd., 1990.

7. Kevin Murphy. Active learning of causal Bayes net structure. Technical report, U.C. Berkeley, March 2001.

8. Judea Pearl. *Probabilistic Reasoning in Intelligent Systems: Networks of Plausible Inference*. Morgan Kaufmann Publishers, Inc., San Mateo, CA, 1988.

9. Judea Pearl. *Causality: Models, Reasoning, and Inference*. Cambridge University Press, Cambridge, UK, 2000.

10. Stuart J. Russell and Peter Norvig. *Artificial Intelligence: A Modern Approach*. Prentice Hall, Englewood Cliffs, NJ, 1995.

11. Herbert A. Simon. Causal ordering and identifiability. In William C. Hood and Tjalling C. Koopmans, editors, *Studies in Econometric Method. Cowles Commission for Research in Economics. Monograph No. 14*, chapter III, pages 49–74. John Wiley & Sons, Inc., New York, NY, 1953.

12. Peter Spirtes, Clark Glymour, and Richard Scheines. *Causation, Prediction, and Search*. The MIT Press, Cambridge, MA, second edition, 2000.

13. Simon Tong and Daphne Koller. Active learning for structure in Bayesian networks. In *International Joint Conference on Artificial Intelligence*, pages 863–869, Seattle, Washington, August 2001.

Advances in Decision Graphs

Thomas D. Nielsen and Finn V. Jensen

Aalborg University,
Department of Computer Science,
Research Unit of Decisions Support Systems,
Fredrik Bajers Vej 7E,
9220 Aalborg Ø, Denmark

Abstract. Frameworks for handling decision problems have been subject to many advances in the last years, both w.r.t. representation languages, solution algorithms and methods for analyzing decision problems. In this paper we outline some of the recent advances by taking outset in the influence diagram framework. In particular, we shall focus on advances in representation languages and exact solution algorithms for decision problems with a single decision maker. Moreover, we give a brief outline of recent contributions to methods for performing sensitivity analysis in influence diagrams.

1   Introduction

Decision graphs refer to a general class of models for representing decision problems. These types of models can be characterized by two components: i) a graphical structure, and ii) numerical information in the form of probabilities for representing uncertainty and utilities for representing preferences. The developments in this area were initiated by von Neumann and Morgenstern [43] who proposed the decision tree framework, see also [35]. A decision tree provides a direct representation of a decision problem by modeling all decision scenarios explicitly. Although such an explicit representation has its advantage for some decision problems, it is also one of the main weaknesses: the size of the decision tree grows exponentially in the number of variables. This shortcoming motivated the development of the influence diagram framework [17], which provides a compact representation of symmetric decision problems with a single decision maker. Initially, these models were solved by converting the influence diagram into a corresponding decision tree representation, such that existing decision tree algorithms could be used to solve the decision problem. Unfortunately, this solution technique also suffers from the complexity problem mentioned above. As an alternative, Shachter [38] proposed a solution method that works directly on the influence diagram model, i.e., it does not require a secondary structure. With the introduction of this solution algorithm, the use of influence diagrams found a more widespread interest and since then there have been several advances in the development of new representation languages (which relax some of the assumptions underlying influence diagrams) as well as new methods for solving and analyzing decision problems.

In this paper we outline some of these recent advances. Obviously, we are not able to cover the entire area and, in particular, we shall restrict our attention to frameworks for representing decision problems involving a single decision maker. Note that representation languages and solution methods for decision problems with several decision makers (usually considered in a game setting) have also been proposed, see e.g. [22]. Moreover, we only deal with frameworks for handling discrete variables although influence diagrams with mixed variables have also been considered, see e.g. [26]. Additionally, we will focus on exact solution algorithms for these models; for an overview of approximate solution methods, the interested reader is referred to [7] and the references within.

Finally, we outline a few advances in methods for analyzing decision problems. More specifically, we shall consider methods for performing sensitivity analysis in influence diagrams although the general task of model analysis also covers other areas such as value of information analysis, see e.g. [12,37].

2   Influence Diagrams

The *influence diagram* (ID) framework [17] serves as an efficient modeling tool for symmetric decision problems with several decisions and a single decision maker. An influence diagram can be seen as a Bayesian network (BN) augmented with *decision nodes* and *value nodes*, where value nodes have no descendants. Thus, an influence diagram is a directed acyclic graph $G = (\mathcal{U}, \mathcal{E})$, where the nodes $\mathcal{U}$ can be partitioned into three disjoint subsets; *chance nodes* $\mathcal{U}_C$, decision nodes $\mathcal{U}_D$ and value nodes $\mathcal{U}_V$. In the remainder of this paper we will use the concept of node and variable interchangeably if this does not introduce any inconsistency. We will also assume that no *barren nodes* are specified by the influence diagram since they have no impact on the decisions [38]; a chance node or a decision node is said to be barren if it has no children, or if all its descendants are barren. Furthermore, in an influence diagram we have a total ordering of the decision nodes indicating the order in which the decisions are made (the ordering of the decision nodes is traditionally represented by a directed path which includes all decision nodes).

With each chance variable and decision variable X we associate a finite *state space* $sp(X)$, which denotes the set of possible outcomes/decision options for X. For a set $\mathcal{U}'$ of chance variables and decision variables we define the state space as $sp(\mathcal{U}') = \times \{sp(X) | X \in \mathcal{U}'\}$, where $A \times B$ denotes the Cartesian product of A and B. The uncertainty associated with each chance variable C is represented by a *conditional probability potential* $P(C|pa(C)) : sp(\{C\} \cup pa(C)) \rightarrow [0; 1]$, where $pa(C)$ denotes the parents of C in the influence diagram. The domain of a conditional probability potential $\phi_C = P(C|pa(C))$ is denoted $\text{dom}(\phi_C) = \{C\} \cup pa(C)$.

The decision maker's preferences is described by a multi-attribute utility potential, and in the remainder of this paper we shall assume that this utility potential is linearly-additive with equal weights, see e.g. [42]; the set of value nodes $\mathcal{U}_V$ defines the set of *utility potentials* which appear as additive components in the multi-attribute utility potential.[1] Each utility potential indicates the local utility for a given configuration of the variables in its domain. The domain of a utility potential ψ_V is denoted $\text{dom}(\psi_V) = pa(V)$, where V is the value node associated with ψ_V. Analogously to the concepts of variable and node we shall sometimes use the terms value node and utility potential interchangeably.

A *realization* of an influence diagram I is an attachment of potentials to the appropriate variables in I, i.e., a realization is a set $\{P(C|pa(C))|C \in \mathcal{U}_C\} \cup \{\psi_V(pa(V))|V \in \mathcal{U}_V\}$. So, a realization specifies the quantitative part of the model whereas the influence diagram constitutes the qualitative part.

The arcs in an influence diagram can be partitioned into three disjoint subsets, corresponding to the type of node they go into. Arcs into value nodes represent functional dependencies by indicating the domain of the associated utility potential. Arcs into chance nodes, termed *dependency arcs*, represent probabilistic dependencies, whereas arcs into decision nodes, termed *informational arcs*, imply information precedence; if there is an arc from a node X to a decision node D, then the state of X is known when decision D is made.

Let $\mathcal{U}_C$ be the set of chance variables and let $\mathcal{U}_D = \{D_1, D_2, \ldots, D_n\}$ be the set of decision variables. Assuming that the decision variables are ordered by index, the set of informational arcs induces a partitioning of $\mathcal{U}_C$ into a collection of disjoint subsets $C_0, C_1, \ldots, C_n$. The set C_j denotes the chance variables observed between decision D_j and D_{j+1}. Thus the variables in C_j occur as immediate predecessors of D_{j+1}. This induces a *partial order* $\prec$ on $\mathcal{U}_C \cup \mathcal{U}_D$, i.e., $C_0 \prec D_1 \prec C_1 \prec \cdots \prec D_n \prec C_n$.

The set of variables known to the decision maker when deciding on D_j is called the *informational predecessors* of D_j and is denoted $\text{pred}(D_j)$. By assuming that the decision maker remembers all previous observations and decisions, we have $\text{pred}(D_i) \subseteq \text{pred}(D_j)$ (for $D_i \prec D_j$) and in particular, $\text{pred}(D_j)$ is the variables that occur before D_j under $\prec$. This property is known as *no-forgetting* and from this we can assume that an influence diagram does not contain any no-forgetting arcs, i.e., $pa(D_i) \cap pa(D_j) = \emptyset$ if $D_i \neq D_j$.

Example 1 (The reactor problem). An electric utility firm is considering building a reactor, and must decide (B) whether to build an advanced reactor (a), a conventional reactor (c) or no reactor at all (n). If an advanced reactor (A) is built, the profit is larger than for a conventional reactor (C) assuming that

---

[1] Note that Tatman and Shachter [42] also considers the case where the utility is defined as the product of a set of utility potentials; such a utility potential can be transformed to decompose additively by taking the logarithm of the utilities.

no accidents occur. However, past experience indicates that an advanced reactor is more probable of having accidents than a conventional reactor. If the
firm builds a conventional reactor, the profits are \$8B if it is a success (cs)
or -\$4B if there is a failure (cf). On the other hand, the profits of building
an advanced reactor are \$12B if it is a success (as), -\$6B if there is a limited
accident (al) and -\$10B if there is a major accident (am).

Before deciding on what reactor to build, the firm has the option of having
a test (T) performed on the components of the advanced reactor. The results
(R) of the test ($T = t$) can be classified as either bad (b), good (g) or excellent
(e); the cost of performing the test is \$1B. If the test results are bad, the
nuclear regulatory commission will not allow an advanced reactor to be built.

An influence diagram representation of the reactor problem can be seen
in Fig. 1. Note that neither the state spaces nor the realization have been
specified.

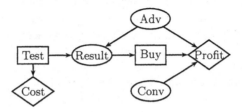

Fig. 1. An influence diagram representation of the reactor problem.

When evaluating an influence diagram we identify a strategy for the decisions involved; a strategy can be seen as a prescription of responses to earlier
observations and decisions. The evaluation is usually performed according to
the *maximum expected utility principle*, which states that we should always
choose an alternative that maximizes the expected utility.

Definition 1. Let I be an influence diagram and let $\mathcal{U}_D$ denote the decision
variables in I. A *strategy* is a set of functions $\Delta = \{\delta_D | D \in \mathcal{U}_D\}$, where δ_D
is a *policy* given by:
$$\delta_D : sp(\mathrm{pred}(D)) \rightarrow sp(D).$$

A strategy that maximizes the expected utility is termed an *optimal strategy*,
and each policy in an optimal strategy is termed an *optimal policy*.

In general, the optimal policy for a decision variable D_k is given by:[2]
$$\delta_{D_k}(\mathcal{C}_0, D_1, \ldots, D_{k-1}, \mathcal{C}_{k-1}) = \tag{1}$$
$$\arg\max_{D_k} \sum_{\mathcal{C}_k} P(\mathcal{C}_k | \mathcal{C}_0, D_1, \ldots, \mathcal{C}_{k-1}, D_k) \rho_{D_{k+1}},$$

---

[2] For the sake of simplifying notation we shall assume that for all decision variables
D_i there is always exactly one element in $\arg\max_{D_i}(\cdot)$.

where $\rho_{D_{k+1}} = \sum_{V \in \mathcal{U}_V} \psi_V$ if $k = n$; otherwise $\rho_{D_{k+1}}$ is the maximum expected utility potential for decision D_{k+1}:

$$\rho_{D_{k+1}}(\mathcal{C}_0, D_1, \ldots, D_k, \mathcal{C}_k) =$$
$$\max_{D_{k+1}} \sum_{\mathcal{C}_{k+1}} P(\mathcal{C}_{k+1} | \mathcal{C}_0, D_1, \ldots, \mathcal{C}_k, D_{k+1}) \rho_{D_{k+2}}.$$

As the domain of a policy function grows exponentially with the number of variables in the past, it is important to weed out variables irrelevant for the decision.

Definition 2. Let I be an influence diagram and let D be a decision variable in I. The variable $X \in \mathrm{pred}(D)$ is said to be *required* for D if there exists a realization of I, a configuration $\bar{y}$ over $\mathrm{dom}(\delta_D) \backslash \{X\}$, and two states x_1 and x_2 of X s.t. $\delta_D(x_1, \bar{y}) \neq \delta_D(x_2, \bar{y})$.

A way of determining the variables required for D would be to analyze δ_D. However, then we would not have avoided the computational problem. Instead, methods for structural analysis of relevance have been constructed, see e.g. [37,29,24]. Common for these methods is that they start off by determining the required past for the last decision D. When this is done, D is replaced by a chance variable with D's required past as parents, and the methods then recursively work their way backwards in the temporal order.

To analyze relevance for the last decision D, let U be a utility node which is a descendant of D. A variable $X \in \mathrm{pred}(D)$ is then required for D, if X is not d-separated from U given $\mathrm{pred}(D) \backslash \{X\}$.[3] This is illustrated in Fig. 2a, which is used for analyzing the required past for D_2; the analysis for D_1 is performed on the network in Fig. 2b.

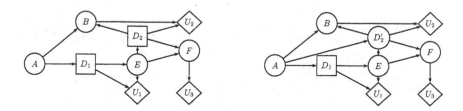

Fig. 2. Figure a: D_1 is not required for D_2 because D_1 is d-separated from U_2 and U_3 given $\mathrm{pred}(D_2) \backslash \{D_1\}$. Figure b: The influence diagram used for analyzing the required past for D_1 (A is required for D_1 due to U_2 and U_3).

Finally, observe that Equation 1 conveys that in order to determine an optimal policy for a decision variable, we have to perform a series of alternating

---

[3] Different algorithms for testing for d-separation have been proposed by Lauritzen et al. [23], Geiger et al. [16], and Shachter [36].

max-marginalizations and sum-marginalizations to eliminate the variables. The order in which the variables are eliminated must respect the partial order induced by the influence diagram. Thus we define a *legal elimination ordering* as a bijection $\alpha : \mathcal{U}_C \cup \mathcal{U}_D \leftrightarrow \{1, 2, \ldots, |\mathcal{U}_C \cup \mathcal{U}_D|\}$, where $X \prec Y$ implies $\alpha(X) < \alpha(Y)$. Note that a legal elimination ordering is not necessarily unique, since the chance variables in the sets $\mathcal{C}_j$ can be commuted. Even so, any two legal elimination orderings result in the same optimal strategy since the decision variables are totally ordered and sum-operations commute; the total ordering of the decision variables ensures that the relative elimination order for any pair of variables of opposite type is invariant under the legal elimination orderings (this is needed since a max-operation and a sum-operation do not commute in general).

3   Modeling Decision Problems

3.1   Non-sequential Decision Scenarios

Consider the classical situation: we would like to buy a used car, but only if the car is in a satisfactory state. We cannot observe the state C directly but we can get information I by just looking at the car and we may - or may not - put an effort into additional tests T_A and T_B before we decide to buy the car (we assume that the tests may be performed in any order). Although this decision problem can be represented as an influence diagram, the representation is awkward (see Fig. 3a). Instead, we may wish to represent the decision problem more directly. This is done in Fig. 3b which is also called a *partial influence diagram* (PID).

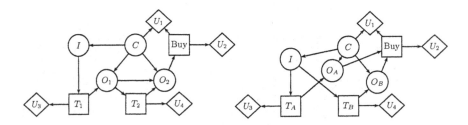

Fig. 3. Figure (a) shows an influence diagram representing two tests and a buy option. The test nodes have three options, t_A, t_B and $no - test$. The O nodes have five states, $pos_A, pos_B, neg_A, neg_B, no - test$. The arc $O_1 \rightarrow O_2$ indicates that repeating a test will give identical results. Figure (b) gives a more direct representation of the test and buy example by not specifying a temporal ordering of T_A and T_B.

In general, a partial influence diagram is like an influence diagram, but without the requirement of a linear ordering of the decisions (and observations). For instance, the partial order of observations and decisions induced by the PID in Fig. 3b is given by $I \prec T_A \prec O_A \prec$ Buy and $I \prec T_B \prec O_B \prec$ Buy.

As the temporal order is only partially specified, the question is whether it matters, i.e., whether the PID is well-defined. That is, will the expected utility of an optimal strategy be affected by the order in which the observations and decisions are taken. This means that we would in principle need to investigate all linear orders extending the partial order (such linear orders are called *admissible*). As there is nothing gained by delaying a (cost free) observation, we introduce the convention that an observation is performed whenever it can be done. We say that an observation is *free* when all its preceding decisions have been made, and we say that the last of these decisions *releases* the observation. In Fig. 3b, the decision T_A releases O_A, and T_B releases O_B.

Definition 3. Let $<$ be an admissible order for the PID I, and let O be a chance variable which may be observed before the last decision is taken. Let D be the decision node immediately preceding O in $<$. O is *well-placed* in $<$ if D releases O. An observable node which is not well-placed is *misplaced*. An admissible order without misplaced variables is called *strictly admissible*. An influence diagram extending the partial order of I to a strictly admissible order is called *admissible* over I.

Definition 4. A PID I is *well-defined* if for any realization of I it holds that any admissible influence diagram over I yields the same expected utility for an optimal strategy.

Theorem 1. *(Nielsen and Jensen [29]) Let I be a PID. Then i) and ii) are equivalent.*

i) *Let $<_1$ and $<_2$ be any two strictly admissible orders, and let D be any decision variable. Then the set of chance variables in the required past of D in $<_1$ is identical to the set of chance variables in the required past of D in $<_2$.*
ii) *I is well-defined.*

To investigate whether a PID is well-defined it is sufficient to investigate all admissible influence diagrams. Furthermore, as two neighboring decisions (and observations) can be permuted without affecting the expected utility, we need not care about such permutations in the ordering.

For example, to investigate the PID in Fig. 3b we should investigate the two strictly admissible orders illustrated in Fig. 4. As the test nodes have different required pasts in the two orders, the PID is not well-defined.

If a PID is not well-defined, the system may return it to the modeler requesting further specification. This request may be followed by indications of how the partial order may be extended, for instance according to the number of fill-ins making the PID well-defined, see [28].

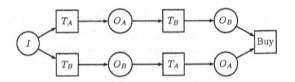

Fig. 4. A directed graph representing the possible admissible orders of the graph in Fig. 3b.

3.2   Unconstrained Influence Diagrams

An ill-defined PID may also be considered as an optimization problem: what strategy with respect to the order of decisions and observations should be followed in order to maximize the expected utility? Notice that in this case, the answer is not a single strictly admissible order.

For example, consider the following story. The beautiful princess in the kingdom Lovania has a wooer. It is rather convenient for the king as he considers retirement. Furthermore, in case he starts a war with the neighbor king, he needs a good general. As customary, the king shall confront the wooer with three tasks. One of the tasks (T_1) shall be either to kill a unicorn or a dragon. Another task (T_2) will be to spend a night in the royal tomb or in the haunted castle tower. The third type of task (T_3) is to swim across the river or to climb the highest mountain in the kingdom.

The king can decide to retire (Rt) or to start a war (Wr) at any time. However, he cannot start a war after retirement, and he cannot give his daughter to the wooer before he has been confronted with all three tasks.

To represent this type of decision problem we extend the language of PIDs so that chance nodes which may be observed are specified directly. We call them *observables* (depicted by double circles), and the language is called *unconstrained influence diagrams* (UIDs). A UID for the kings problem is given in Fig. 5.

As the next step in a strategy (decision or observation variable) may be dependent on the past, a strategy for a UID is not one unique strictly admissible order together with policies for the decision variables. It is rather a DAG over decision nodes and observables. The set of strictly admissible orders is organized in a so-called normal form S-DAG: a directed acyclic graph where each path from source to sink is a strictly admissible order, and where all strictly admissible orders are represented.

A normal form S-DAG may represent any optimal strategy for the corresponding S-DAG. Jensen and Vomlelova [19] give a precise definition of normal form S-DAGs together with an algorithm for constructing them. A normal form S-DAG for the kings problem can be seen in Fig. 6, and Fig. 4 shows a normal form S-DAG for a UID corresponding to the PID in Fig. 3b.

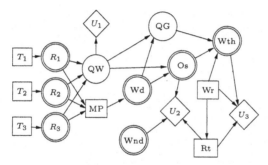

Fig. 5. An unconstrained influence diagram for the kings problem. Abbreviations used: R_i is the result of task T_i, Wnd (wooer's noble descent), QW (quality of wooer), Qg (wooer's quality as a general), MP (marry princess), Wd (wedding), Os (offspring) and Wth (wealth).

3.3   Limited Memory Influence Diagrams

The frameworks presented in the previous sections all rely on the no-forgetting assumption, i.e., at any point in time the decision maker remembers all previous observations and decisions. Unfortunately, this assumption may not always be valid, and, more importantly, it may make the computational complexity impractical as the required past for a decision may become intractably large. A way to restrict the size of the required past is to use *information blocking*. That is, by introducing variables which, when observed, d-separates some of the past from the present decision [18]. Alternatively, we can explicitly pinpoint which variables are remembered when taking a particular decision, thereby dropping the no-forgetting assumption. The latter approach is pursued in the *limited memory influence diagrams* (LIMIDs) by Lauritzen and Nilsson[24], where informational arcs are used to indicated which variables are known when taking a certain decision.

Example 2 (The North sea fishing problem). Every year, the European Union undertakes very delicate political and biological negotiations to determine a volume of fishing for most kinds of fish in the North sea. Oversimplified, you can say that each year we have a test for the volume of fish, and based on this test the volume of allowable catch is decided; this decision also has an impact on the volume of fish for next year. A LIMID representation for a three year strategy is illustrated in Fig. 7, where we only remember the last decision and observation (these two variables are therefore also the only variables in the policy for a given decision variable).

Alternatively, by representing the North sea fishing problem using an influence diagram with the no-forgetting assumption, we find that the required past for decision FV_i constitute the entire past for that decision. Hence, the policies grows exponentially large; this analysis can be performed using the method described in Section 2.

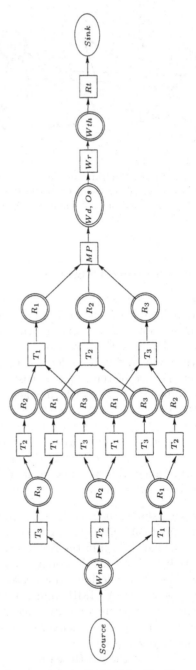

Fig. 6. A normal form S-DAG for the kings problem.

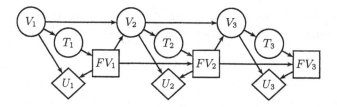

Fig. 7. A LIMID representation of the North sea fishing problem, where only the last decision and observation are remembered.

3.4 Asymmetric Decision Problems

The frameworks described above have mainly been developed for representing symmetric decision problems with a single decision maker. However, another important type of decision problem is the class of so-called asymmetric decision problems; these decision problems cannot be represented efficiently using e.g. influence diagrams or *valuation networks* [39]. There is currently no complete consensus about the definition of an asymmetric decision problem although most authors agree that a decision problem is asymmetric if the number of scenarios is less than the cardinality of the Cartesian product of the state spaces of all chance and decision variables, see e.g. [34,2,40,30]. For example, the decision problem described in Example 1 is asymmetric as can be seen by unfolding the influence diagram into a decision tree.

Various frameworks have been proposed as alternatives to the influence diagram when dealing with asymmetric decision problems. Covaliu and Oliver [9] extend the influence diagram with another diagram, called a *sequential decision diagram*, see Fig. 8. The sequential decision diagram is used for describing the asymmetric structure of the problem, as complementary to the influence diagram which is used for specifying the probability model; the functional and numerical information from these two diagrams are combined in a so-called *formulation table* similar to that of Kirkwood [21].

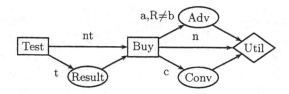

Fig. 8. A sequential decision diagram representation of the reactor problem.

Smith et al. [41] introduce the notion of *distribution trees* within the framework of influence diagrams, see Fig. 9. The use of distribution trees allows the possible outcomes of an observation to be specified, as well as the legitimate decision options for a decision variable, see also Qi et al. [34].

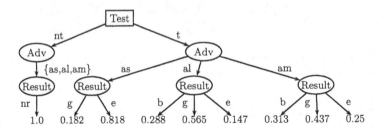

Fig. 9. The distribution tree for Result.

However, as the distribution trees are not part of the influence diagram, the structure of the decision problem cannot be deduced directly from the graphical structure. Moreover, the sequence of decisions and observations is predetermined, i.e., previous observations and decisions cannot influence the temporal order of future observations and decisions. Finally, distribution trees have a tendency of creating large conditionals during the evaluation, since they encode both numeric information and information about asymmetry. To overcome this problem, Shenoy [40] presents the *asymmetric valuation network* as an extension of the valuation network for modeling symmetric decision problems. The asymmetric valuation network uses so-called *indicator functions* to encode asymmetry, thereby separating it from the numeric information (see Fig. 10). However, asymmetry is still not represented directly in the model and, as in Smith et al. [41], the sequence of observations and decisions is predetermined. Further details and comparisons of these methods can be found in [2].

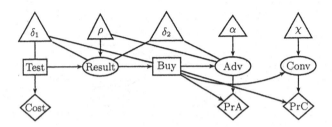

Fig. 10. An asymmetric valuation network representation of the reactor problem. The triangles ρ, α and χ are probability valuation, whereas δ_1 and δ_2 are indicator valuations. Observe that information precedence is encoded by the directed paths involving chance and decision nodes.

Nielsen and Jensen [30] propose another framework (called *asymmetric influence diagrams*) which extends the influence diagram representation by introducing *guards* at the graphical level: guards on arcs and nodes specify conditions for when the associated arcs and nodes are part of the decision problem. Moreover, *restrictive functions* (associated with decision variables) are introduced to allow the state space of a decision variable to depend on variables in its past; the domain of a restrictive function is indicated by dashed arcs into the associated decision node (see Fig. 11). Having guards associated with informational arcs also supports the specification of decision problems where the temporal order of a set of variables depends on previous observations and decisions.

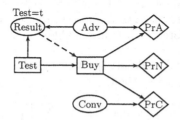

Fig. 11. An asymmetric influence diagram representation of the reactor problem. The domain of the restrictive function associated with Buy is specified by the dashed arc.

The use of guards has also been pursued by Demirer and Shenoy [11] who propose a framework, called *sequential valuation networks*, which is a combination of the sequential decision diagram and the asymmetric valuation network. Basically, this framework takes outset in the sequential decision diagram and augments this representation with valuation functions as found in the asymmetric valuation network.

Although several frameworks have been proposed for modeling asymmetric decision problems, there is currently no complete specification and comparison of their strengths and weaknesses.[4] Additionally, the expressive power of the frameworks is not clear, i.e., what types of decision problems can be modeled in the various frameworks without introducing redundant information such as artificial states or duplicated variables.

---

[4] Note that a comparison of sequential decision diagrams, asymmetric valuation networks and extended influence diagrams can be found in [2].

4   Evaluating Decision Problems

4.1   Lazy Evaluation of (Partial) Influence Diagrams

An influence diagram is solved through dynamic programming using Equation 1. That is, we start by determining an optimal policy for the last decision, and then move backwards in the temporal order to determine a policy for the other decisions. When a policy for the last decision is found, the utility functions are substituted by a utility function representing the expected utility of an optimal choice for that decision.

This procedure has been translated to a variable elimination procedure, where variables are eliminated in reverse admissible order Shenoy et al. [39,20,10,27]. Taking advantage of lazy propagation, the elimination procedure is described as follows [25]: The method keeps two sets of potentials: Φ, a set of probability potentials; Ψ, a set of utility potentials. When a variable X is eliminated, the potential sets are modified in the following way:[5]

1. Set $\Phi_X := \{\phi \in \Phi \mid X \in dom(\phi)\}$ and $\Psi_X := \{\psi \in \Psi \mid X \in dom(\psi)\}$.
2. If X is a chance variable, then

$$\phi_X := \sum_X \prod \Phi_X \text{ and } \psi_X := \sum_X \prod \Phi_X (\sum \Psi_X).$$

If X is a decision variable, then

$$\phi_X := \max_X \prod \Phi_X \text{ and } \psi_X := \max_X \prod \Phi_X (\sum \Psi_X).$$

3. Let $\Phi := (\Phi \setminus \Phi_X) \cup \{\phi_X\}$ and $\Psi := (\Psi \setminus \Psi_X) \cup \{\frac{\psi_X}{\phi_X}\}$.

Similar to probability propagation for Bayesian networks, the solution method for influence diagrams can be expressed in terms of junction trees. However, the partial order of the variables adds some constraints to the corresponding triangulation, and the structure used is called a strong junction tree Jensen et al. [20].

4.2   Evaluation of Unconstrained Influence Diagrams

When a normal form S-DAG for a UID has been established, it is solved in almost the same manner as influence diagrams. That is, variables are eliminated in reverse temporal order.

When a branching point is met, the elimination is branched out, and you work with particular potential sets for each branch. When paths meet they have to meet in a chance variable, and the variable immediately after

---

[5] Note that the operation $\max_X \prod \Phi_X$ simply corresponds to removing X from the domain of $\prod \Phi_X$ since $\prod \Phi_X$ is a constant function over X.

must be a decision variable in all branches. So, assume that two branches meet in A from D_1 and D_2, respectively. Now, the sets of probability potentials are the same (they represent sum-marginalizations of the same variables in various orders), but the utility potentials may be different. The unified utility potentials are determined through maximization. For simplicity, let $U_1(B, A)$ and $U_2(B, A)$ be the potentials. Then the potential for A is $\max(U_1(B, A), U_2(B, A))$.

4.3   Evaluation of LIMIDs

The evaluation of a LIMID is based on an iterative improvement of the policies for the decision variables, and is closely connected to the method of policy iteration for Markov decision processes.

As a starting point, the decision variables are assigned random policies, i.e., policies which specify probability distributions over the state spaces of the associated decision variables given their parents; recall that the variables known to the decision maker when deciding on a decision D are the parents of the corresponding node. If a random policy specifies a unique alternative for each configuration of $pa(D)$, then the policy is called a *pure policy*; a pure policy corresponds to the traditional notion of a policy if $pa(D)$ contains the variables being required for D in the corresponding ID representation.

Given the initial policies of the decision variables, the LIMID is converted into a junction tree. However, as opposed to the construction of a strong junction tree, informational arcs are not removed before moralization and the triangulation need not respect any partial or total ordering of the nodes. The junction tree is then initialized by associating each probability potential and utility potential to a clique which can accommodate it. Afterwards all potentials assigned to a clique are combined thereby following the approach of Jensen et al. [20]. Note that in the "lazy" version of the algorithm the latter step should not have been performed.

Based on the initialized junction tree structure, the evaluation algorithm proceeds by iteratively improving the policy for each decision variable. This is performed by making a collect propagation (in the usual fashion) to the clique containing the decision variable in question. The current policy is then substituted by another policy which maximizes the expected utility for that decision; the algorithm converges after having updated the policy for each decision variable. Obviously, this method only provides an approximate solution, and for lower and upper bounds of the expected utility of the approximation the reader is referred to [32].

4.4   Evaluation of Asymmetric Decision Problems

Existing methods for solving asymmetric decision problems can roughly be characterized based on how asymmetry is represented in the associated frameworks.

For those frameworks which use secondary structures to represent asymmetry, such as distribution trees and indicator functions, the solution algorithms incorporate the information about asymmetry into the solution algorithm. That is, information about asymmetry is directly combined with the numerical information during the solution phase, see e.g. [41] and [40]. For example, the algorithm for solving asymmetric valuation networks, known as the *fusion algorithm*, works simultaneously on all three sets of valuations defined in the model: utility valuations $\mathcal{V}$, probability valuations Γ and indicator valuations Υ. Similarly to standard evaluation algorithms for influence diagrams, the fusion algorithm marginalizes out the variables in reverse temporal order by combining all relevant valuations. However, the fusion algorithm only works on the *effective state spaces* of the variables, i.e., the legitimate state configurations as defined by the indicator valuations. For instance, the combination of two probability valuations ρ_1 and ρ_2 is defined as:

$$(\rho_1 \otimes \rho_2)(\bar{x}) = (\rho_2 \otimes \rho_1)(\bar{x}) = \rho_1\left(\bar{x}^{\downarrow dom(\rho_1)}\right)\rho_2\left(\bar{x}^{\downarrow dom(\rho_2)}\right),$$

where $\bar{x}$ is in the effective state space of the variables in the domain of both in ρ_1 and ρ_2.

To specify the fusion algorithm w.r.t. a variable X we denote by $\mathcal{V}_X$, Γ_X and Υ_X the utility, probability and indicator valuations including X in their domain. The fusion operation $\mathrm{Fus}_X(\mathcal{V} \cup \Gamma \cup \Upsilon)$ w.r.t. the valuations $\mathcal{V} \cup \Gamma \cup \Upsilon$ is then defined differently depending on the type of variable X. If X is a chance variable, then:[6]

$$\mathrm{Fus}_X(\mathcal{V} \cup \Gamma \cup \Upsilon) =$$

$$\left\{ \left[v \otimes \left(\frac{\rho}{\rho^{\downarrow(dom(\Gamma_X \cup \Upsilon_X)\backslash\{X\})}} \right) \right]^{\downarrow(dom(\mathcal{V}_X \cup \Gamma_X \cup \Upsilon_X)\backslash\{X\})} \right\} \tag{2}$$

$$\cup \left\{ \rho^{\downarrow(dom(\Gamma_X \cup \Upsilon_X)\backslash\{X\})} \right\} \cup \{\mathcal{V}\backslash\mathcal{V}_X\} \cup \{\Upsilon\backslash\Upsilon_X\} \cup \{\Gamma\backslash\Gamma_X\},$$

where

$$\rho = (\otimes\{\iota | \iota \in \Upsilon_X\}) \otimes (\otimes\{\rho | \rho \in \Gamma_X\}) \text{ and } v = \otimes\{v | v \in \mathcal{V}_X\}.$$

If X is a decision variable, then:

$$\mathrm{Fus}_X(\mathcal{V} \cup \Gamma \cup \Upsilon) = \{v^{\downarrow(dom(\mathcal{V}_X)\backslash\{X\})}\} \cup$$

$$\{(\rho \otimes \zeta_X)^{\downarrow(dom(\mathcal{V}_X \cup \Gamma_X \cup \Upsilon_X)\backslash\{X\})}\} \tag{3}$$

$$\cup\{\mathcal{V}\backslash\mathcal{V}_X\} \cup \{\Upsilon\backslash\Upsilon_X\} \cup \{\Gamma\backslash\Gamma_X\},$$

---

[6] Note that the projection notation, i.e. $\downarrow$, refers to the standard marginalization of a variable except that we only work with the effective state space.

where ζ_X is the indicator valuation representing the optimal policy for X.

Note that Shenoy [40] also derives cases where the fusion algorithm can be optimized based on the specific valuations involved.

Alternatively, in the frameworks where asymmetry is represented explicitly in the graphical structure, the solution algorithms use this information to construct a secondary structure for solving the model. The secondary structure is used to decompose (or unfold) the original asymmetric decision problem into a collection of symmetric subproblems that can be solved independently; the actual decomposition is performed by traversing in the temporal order and iteratively instantiating the variables which "produce" asymmetry. For instance, in the asymmetric influence diagram we instantiate the variables which appear in the domain of either a guard or a restrictive function. The resulting subproblems can then be organized in a tree structure (see Fig. 12) having the property that a solution to the original decision problem can be found be combining the solutions from the "smaller" symmetric decision problems.

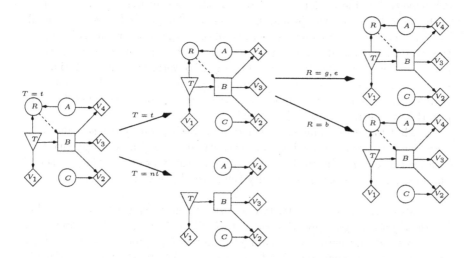

Fig. 12. The figure illustrates the decomposition tree for the reactor problem depicted in Figure 11. Note that decomposing w.r.t. R produces subproblems which are structurally identical but differ in the state space of the decision variable B.

Obviously, both types of frameworks have their advantages and disadvantages. For example, the asymmetric valuation network relies on artificial states to represent asymmetry. From a computational perspective this has the undesirable effect that we in general work with larger valuations during the solution phase, as compared to e.g. the sequential valuation networks or asymmetric influence diagram. On the other hand, when e.g. evaluating an asymmetric influence diagram, the calculations are performed according to its decomposition tree. This implies that the same calculations may be required in different subproblems, i.e., we may perform redundant calculations. Obviously, this problem does not occur when evaluating an asymmetric valuation network using the fusion algorithm. One way around this problem could be to maintain a cache of previous calculations e.g. a hash table indexed with the calculated potentials, see also Cano et al. [3].

5   Model Analysis

5.1   Sensitivity analysis in influence diagrams

When solving an influence diagram the aim is to determine an optimal policy for each decision involved. These policies (as well as the maximum expected utility of the influence diagram) are sensitive to variations in both the utilities and the probabilities. Unfortunately, it is usually the case that these values are difficult to elicit and subject to second-order uncertainty.[7] This makes it desirable to be able to determine how robust the solution is to variations in both the utilities and the probabilities, see also [33]. Based on such an analysis, the modeler may focus his/her attention on parameter values for which the solution is particularly sensitive (a situation known as *one-way* sensitivity analysis). Another important type of analysis concerns the joint variation of a set of parameters (*n-way* sensitivity analysis). This type of analysis yields little information about the individual parameters, but provides general insight into the overall robustness of the model.

In what follows we make a distinction between *value sensitivity* and *decision sensitivity* when referring to sensitivity analysis. Value sensitivity concerns variations in the maximum expected utility when changing a set of parameters, and decision sensitivity refers to changes in the optimal strategy.

Analysis of decision sensitivity can furthermore be characterized as either *threshold proximity* or *probabilistic sensitivity analysis*. Threshold proximity uses a distance measure to determine the values a parameter (or a set of parameters) can be assigned without changing the optimal strategy found w.r.t. the initial values of the parameters. Probabilistic sensitivity analysis assigns a probability distribution to each parameter under investigation. Hence, the analysis is based on the probability of obtaining a different optimal strategy rather than determining the admissible domain in which the parameters can be varied, see e.g. Doubilet et al. [13].

A method based on value sensitivity has been proposed in [14]. This method uses the expected value of perfect information as a sensitivity indicator, and similar to probabilistic sensitivity analysis it requires that a

---

[7] Note that methods exist for learning the probabilities from a database, and methods for learning the utilities have also recently been proposed, see e.g. Chajewska et al. [4–6].

probability distribution is assigned to each parameter under investigation. More precisely, let $\bar{t}$ be the uncertain parameters, and let Δ^0 be the optimal strategy found w.r.t. the initial values of the parameters. If ρ_Δ and ρ_{Δ^0} denote the expected utilities of the influence diagram under the strategies Δ and Δ^0, respectively, then the expected value of perfect information (denoted by $EVPI(\bar{t})$) is given by:

$$EVPI(\bar{t}) = E_{\bar{t}}[\max_\Delta [\rho_\Delta(\bar{t})] - \rho_{\Delta^0}(\bar{t})].$$

This expectation can be very difficult to evaluate. Hence simulations are usually applied. A detailed description and comparison of the methods outlined above, can be found in [14] and [15].

A drawback of employing sensitivity analysis based on probability distributions over the uncertain parameters is the elicitation of these distributions. Given that the decision maker has a (partial) preference ordering over the outcomes, the uncertain utility parameters are not independent. I.e., we work with joint probability distributions over the utility parameters.[8] This makes it difficult to elicit these distributions, in particular, when the utilities have no straightforward interpretation. Parameter dependence also has an impact on the computational complexity. Even when the parameters are assumed to be independent, the calculations can be rather cumbersome, and Monte Carlo methods are usually applied to sample values for the parameters. For instance, when applying the method based on the expected value of perfect information we need to compute $\text{diff}_i = [\max_\Delta [\rho_\Delta(\bar{t}^i)] - \rho_{\Delta^0}(\bar{t}^i)]$, for each generated sample $\bar{t}^i$; the expectation is then approximated by $EVPI(\bar{t}) \approx \frac{1}{N} \sum_{i=1}^N \text{diff}_i$, where N is the number of generated samples, see e.g. [1]. However, when the parameters are not independent we will in general require a larger set of samples in order to get a representative sample from the joint probability distribution over all the parameters. Thus, dependent parameters may induce a serious computational overhead since the decision problem is evaluated for each generated sample.

With this outset, Nielsen and Jensen [31] propose two methods for performing decision sensitivity analysis based on threshold proximity. The methods mainly focus on uncertain utility parameters, and works by visiting the decision variables in reverse temporal order. For each decision variable, a set of linear constraints is computed which in turn forms the basis for the analysis; as all constraints are linear, n-way sensitivity analysis comes free of charge after one-way sensitivity analysis has been performed. Moreover, when working with uncertain utility parameters the time complexity of the algorithms is roughly linear in the number of uncertain parameters (the basic unit of computation is one propagation); unfortunately, in case of n-way sensitivity analysis w.r.t. probability parameters the time complexity is exponential in the number of uncertain parameters.

---

[8] Analogously to the standard assumptions within the learning community, see e.g. [8], it is usually assumed that the probability parameters are independent.

Finally, it should be noted that decision sensitivity has recently been subject to criticism on the ground that it tends to overestimate the sensitivity of a model (different strategies may yield almost the same expected utility), see e.g. [15]. On the other hand, one may argue that if a parameter induces significant changes in the expected utility without influencing the optimal strategy, then the model is also insensitive to the parameter. To overcome the problem of overestimating model sensitivity, one may perform the decision analysis iteratively. That is, after decision sensitivity has been analyzed we can calculate the expected utility of the optimal strategy, Δ', induced by changing the parameters. If the expected utility is not significantly different from the expected utility of the initial strategy, then the analysis is repeated and the strategy is allowed to change accordingly (i.e., Δ' is disregarded). Thus, by analyzing decision sensitivity iteratively, we may avoid the problem of overestimating the sensitivity of the model.

References

1. Concha Bielza, Sixto Ríos-Insua, Manuel Gómez, and Juan A. Fernández del Pozo. Sensitivity analysis in IctNeo. In *Lecture Notes in Statistics*, volume 152 of *Robust Bayesian Analysis*, pages 317–334. Springer, 2000.
2. Concha Bielza and Prakash P. Shenoy. A comparison of graphical techniques for asymmetric decision problems. *Management Science*, 45(11):1552–1569, 1999.
3. Andrés Cano, Serafín Moral, and Antonio Salmerón. Lazy evaluation in penniless propagation over join trees. *Networks*, 39:175–185, 2002.
4. Urzula Chajewska, Lise Getoor, joseph Norman, and Yuval Shahar. Utility elicitation as a classification problem. In Gregory F. Cooper and Serafin Moral, editors, *Proceedings of the Fourteenth Conference on Uncertainty in Artificial Intelligence*, pages 79–88, 1998.
5. Urzula Chajewska and Daphne Koller. Utilities as random variables: Density estimation and structure discovery. In Craig Boutilier and Moisés Goldszmidt, editors, *Proceedings of the Sixteenth Conference on Uncertainty in Artificial Intelligence*, pages 63–71, 2000.
6. Urzula Chajewska, Daphne Koller, and Dirk Ormoneit. Learning an Agent's Utility Function by Observing Behavior. In Carla Brodley and Andrea Danyluk, editors, *Proceedings of the 18th International Conference on Machine Learning (ICML '01)*, pages 35–42, 2001.
7. John C. Charnes and Prakash P. Shenoy. A multi-stage Monte Carlo method for solving influence diagrams using local computations, March 2002. Working paper no. 273.
8. Gregory F. Cooper and Edward Herskovits. A Bayesian method for constructing Bayesian belief networks from databases. In Bruce D. D'Ambrosio, Philippe Smets, and Piero P. Bonissone, editors, *Proceedings of the Seventh Conference on Uncertainty in Artificial Intelligence*, pages 86–94, 1991.
9. Zvi Covaliu and Robert M. Oliver. Representation and solution of decision problems using sequential decision diagrams. *Management Science*, 41(12):1860–1881, 1995.

10. Robert G. Cowell. Decision networks: a new formulation for multistage decision problems. Research report 132, Department of Statistical Science, University College London, London, 1994.

11. Riza Demirer and Prakash P. Shenoy. Sequential valuation networks: A new graphical technique for asymmetric decision problems. In *Lecture Notes in Artificial Intelligence*, pages 252–265. Springer, 2001.

12. Søren L. Dittmer and Finn V. Jensen. Myopic value of information in influence diagrams. In Dan Geiger and Prakash Pundali Shenoy, editors, *Proceedings of the Thirteenth Conference on Uncertainty in Artificial Intelligence*. Morgan Kaufmann Publishers, 1997.

13. P. Doubilet, C. B. Begg, M. C. Weinstein, P. Braun, and B. J. McNeil. Probabilistic sensitivity analysis using Monte Carlo simulation. *Medical Decision Making*, 5(2):157–177, 1985.

14. James C. Felli and Gordon B. Hazen. Sensitivity analysis and the expected value of perfect information. *Medical Decision Making*, 18:95–109, 1998.

15. James C. Felli and Gordon B. Hazen. Do sensitivity analysis really capture problem sensitivity? An empirical analysis based on information value. *Risk, Decision and Policy*, 4(2):79–98, 1999.

16. Dan Geiger, Thomas Verma, and Judea Pearl. d-separation: From theorems to algorithms. In Max Henrion, Ross D. Shachter, Laveen N. Kanal, and John F. Lemmer, editors, *Uncertainty in Artificial Intelligence 5*, pages 139–148, 1990.

17. Ronald A. Howard and James E. Matheson. Influence diagrams. In Ronald A. Howard and James E. Matheson, editors, *The Principles and Applications of Decision Analysis*, volume 2, chapter 37, pages 721–762. Strategic Decision Group, 1981.

18. Finn V. Jensen. *Bayesian Networks and Decision Graphs*. Springer-Verlag New York, 2001. ISBN: 0-387-95259-4.

19. Finn V. Jensen and Marta Vomlelová. Unconstrained influence diagrams. In Adnan Darwiche and Nir Friedman, editors, *Proceedings of the Eighteenth Conference on Uncertainty in Artificial Intelligence*, pages 234–241. Morgan Kaufmann Publishers, 2002.

20. Frank Jensen, Finn V. Jensen, and Søren L. Dittmer. From influence diagrams to junction trees. In Ramon Lopez de Mantaras and David Poole, editors, *Proceedings of the Tenth Conference on Uncertainty in Artificial Intelligence*, pages 367–373. Morgan Kaufmann Publishers, 1994.

21. Craig W. Kirkwood. An algebraic approach to formulating and solving large models for sequential decision under uncertainty. *Management Science*, 39(7):900–913, July 1993.

22. Daphne Koller and Brian Milch. Multi-agent influence diagrams for representing and solving games. In *Proceedings of the International Joint Conference on Artificial Intelligence (IJCAI)*, pages 1027–1034, 2001.

23. Steffen L. Lauritzen, A. Philip Dawid, Birgitte N. Larsen, and Hans-Georg Leimer. Independence properties of directed Markov fields. *Networks*, 20(5):491–505, August 1990.

24. Steffen L. Lauritzen and Dennis Nilsson. Representing and Solving Decision Problems with Limited Information. *Management Science*, 47(9):1235–1251, 2001.

25. Anders L. Madsen and Finn V. Jensen. Lazy evaluation of symmetric Bayesian decision problems. In Kathryn B. Laskey and Henri Prade, editors, *Proceedings*

of the Fifthteenth Conference on Uncertainty in Artificial Intelligence, pages 382–390. Morgan Kaufmann Publishers, 1999.

26. Anders L. Madsen and Frank Jensen. Mixed influence diagrams. In Thomas D. Nielsen and Nevin L. Zhang, editors, Proceedings of the Seventh European Conference on Symbolic and Quantitative Approaches to Reasoning with Uncertainty. Springer, 2003. To appear.

27. Pierre C. Ndilikilikesha. Potential influence diagrams. International Journal of Approximate Reasoning, 10:251–285, 1994.

28. Thomas D. Nielsen. Graphical Models for Partially Sequential Decision Problems. PhD thesis, Aalborg University, Department of Computer Science, Fredrik Bajers vej 7E, 9220 Aalborg Ø, 2002.

29. Thomas D. Nielsen and Finn V. Jensen. Welldefined decision scenarios. In Kathryn B. Laskey and Henri Prade, editors, Proceedings of the Fifthteenth Conference on Uncertainty in Artificial Intelligence, pages 502–511. Morgan Kaufmann Publishers, 1999.

30. Thomas D. Nielsen and Finn V. Jensen. Representing and solving asymmetric Bayesian decision problems. In Craig Boutilier and Moisés Goldszmidt, editors, Proceedings of the Sixteenth Conference on Uncertainty in Artificial Intelligence. Morgan Kaufmann Publishers, 2000.

31. *Thomas D. Nielsen and Finn V. Jensen. Sensitivity Analysis in Influence Diagrams. Accepted for publication in IEEE Transactions on Systems, Man and Cybernetics Part B, 2003.

32. Dennis Nilsson and Michael Höhle. Computing bounds on expected utilities for optimal policies based on limited information. Research report 94, Danish Informatics Network in the Agricultural Sciences, October 2001.

33. Kim Leng Poh and Eric J. Horvitz. Reasoning about the value of decision-model refinement: Methods and applications. In David Heckerman and Abe Mamdani, editors, Proceedings of the Ninth Conference on Uncertainty in Artificial Intelligence. Morgan Kaufmann Publishers, 1993.

34. Runping Qi, Nevin Lianwen Zhang, and David Poole. Solving asymmetric decision problems with influence diagrams. In Proceedings of the Tenth Conference on Uncertainty in Artificial Intelligence, pages 491–497. Morgan Kaufmann Publishers, 1994.

35. Howard Raiffa. Decision Analysis, Introductory Lectures on Choices under Uncertainty. Addison-Wesley, Massachusetts, 1968.

36. Ross D. Shachter. Bayes ball: The rational pastime (for determining irrelevance and requisite information in belief networks and influence diagrams). In Gregory F. Cooper and Serafin Moral, editors, Proceedings of the Fourteenth Conference on Uncertainty in Artificial Intelligence, pages 480–487. Morgan Kaufmann Publishers, 1998.

37. Ross D. Shachter. Efficient value of information computation. In Kathryn B. Laskey and Henri Prade, editors, Proceedings of the Fifthteenth Conference on Uncertainty in Artificial Intelligence, pages 594–601. Morgan Kaufmann Publishers, 1999.

38. Ross D. Shachter. Evaluating influence diagrams. Operations Research Society of America, 34(6):79–90, February 1986.

39. Prakash P. Shenoy. Valuation-based systems for Bayesian decision analysis. Operations Research, 40(3):463–484, 1992.

40. Prakash P. Shenoy. Valuation network representation and solution of asymmetric decision problems. *European Journal of Operations Research*, 121(3):579–608, 2000.

41. J. E. Smith, S. Holtzman, and J. E. Matheson. Structuring conditional relationships in influence diagrams. *Operations research*, 41(2):280–297, March/April 1993.

42. Joseph. A. Tatman and Ross. D. Shachter. Dynamic programming and influence diagrams. *IEEE Transactions on Systems, Man and Cybernetics*, 20(2):365–379, March/April 1990.

43. J. von Neumann and O. Morgenstern. *Theory of Games and Economic Behavior*. John Wiley & Sons, New York, first edition, 1944.

Real-World Applications of Influence Diagrams

Manuel Gómez

Dpt. of Computer Science and Artificial Intelligence
University of Granada
c/ Periodista Daniel Saucedo Aranda, s/n
Granada 18071, Spain

Abstract. It is very well known the difficulties involved in making decisions between different alternatives. A substantial amount of empirical evidence demonstrates that this is something inherent in human beings. That is the reason why there is a major focus on methods and techniques for aiding to overcome the deficiencies of human judgment and decision making. Decision theory is the mathematical foundation for rational decision making, combining probability theory and utility theory. Decision analysis studies the application of decision theory to decision problems. This research have provided several results, one of them being Influence Diagrams (IDs). An ID is a graphical model with two important features: a) gives a powerful tool to capture all the decision problem elements and b) can be evaluated providing optimal policies for decision makers. These reasons explain the extensive body of work devoted to IDs and the use of this tool in real-world applications.

1 Decision Making Under Uncertainty

All of us are familiar with the difficulties involved when having to decide between several alternatives. It seems to be something inherent to human beings, and not a particular feature of certain individuals. This has been proven through several experiments, showing the difficulties related to making decisions, specially in critical situations and under pressure. Hence, making decisions on complex systems (industrial operations planning, medical treatments, economical investments, control of nuclear plants, etc.) seems to exceed our cognitive capabilities.

So, it is clear the need for methods and tools for helping people to deal with these situations. Several subjects, like Statistics, Economy and Operations Research, among others, have developed methods with that objective. Recently, these methods, based on techniques coming from psychology, artificial intelligence and computer science, have been used to implement computer systems called Decision Support Systems (DSS).

One of the most important reason for the difficulties related to making decisions lies in the uncertainty about the variables which have a bearing on the problem, about their mutual influence and the future consequences of the decisions made as well. This is due to the decisor does not have a complete knowledge about the problem. Moreover, the construction of a model for a decision problem is itself a source of uncertainty, just because any representation will be only a simplification of the real world [1].

One subject where the uncertainty plays a central role, for example, is medical reasoning [2]. All the factors implied in medical decision making add its own load of uncertainty: patients can not exactly describe what happen to them, medical personnel can not manage all their observations, laboratories give results with a certain degree of error, medical research is able neither to determine the way some illnesses act nor the reaction of the human body, etc. The development of new treatments and the huge number of medical publications make very hard to be updated [3], and paradoxically incorporate a new source of uncertainty. Rector [4] states that many mistakes in medical practice are caused by the inability to process all the available information. All these reasons justify the need for methods, tools, systems and techniques to improve medical decision making. The existence of the *Society for Medical Decision Making* proves that this subject is ideal for the use of DSS [5].

2   Artificial Intelligence, Expert Systems and Decision Support Systems

Early work on expert systems for diagnosis used Bayesian and decision theoretic approaches. However, soon many researches lost interest in decision theory. This disenchantment arose, in part, for a perception that it was hopelessly intractable and inadequate for expressing the rich structure of human knowledge [6]. The main problems related with the use of probability and decision analysis are related with obtaining the probabilistic and preference information and with the need for efficient methods for reasoning with it. So, AI investigators considered heuristic techniques to get their systems tractable. However, theoretical work demonstrated clear parallelisms between the shortcoming of several heuristic calculi and the probabilistic approach [7].

Throughout the 1980s, increasing attention was given to the integration of probabilistic and decision-theoretic reasoning methods with AI approaches. Some of the reasons for this approximation are outlined below:

- To recognize that a major drawback of the expert systems approaches was the lacking of general representations for the preferences or beliefs of decision makers [8].
- To perceive that rule-based methodologies are epistemologically inadequated for capturing uncertain knowledge [7].
- The development of efficient and expressive representations of probabilistic knowledge, called belief networks or Bayesian networks (BNs) and IDs [9]. These tools allow people to express qualitative and quantitative knowledge about beliefs, preferences and decisions. The key feature of BNs is the fact that they provide a method for decomposing a probability distribution into a set of local distributions. This has three important consequences: a) reduces the number of values required to specify the quantitative knowledge; b) allows to get efficient algorithms to make inference transmitting information between the local distributions rather

than working with the full joint distribution; and c) separates qualitative and quantitative information, what supposes a significant advantage for knowledge engineering [10]. In building a BN model, one can first focus on specifying the qualitative structure of the domain and then quantifying the influences.

- The advances in representation and computation with these graphical modeling formalisms. Because the fixed models have proven to be inadequately expressive when a broad range of situations must be handled, many researchers have combined the strengths of flexible knowledge representation with the normative status and well understood properties of decision modeling formalisms and algorithms [11].

- The growing capabilities of computers. Today, a simple desktop computer can perform, in few seconds, a complex computation, only affordable in 1980's with very expensive computers.

The most common normative approach to decision making is Bayesian decision theory, whose central elements are: a) a set of available acts; b) a set of possible outcomes of acts; c) a conditional probability distribution specifying the probability of each outcome given each available act; and d) a preference order ranking the possible outcome distributions according to its desirability. The most straightforward way to apply this theory in a DSS is to construct a representation of the decision problem directly in terms of these basic elements. This can be done with IDs, being possible too to evaluate the model, calculating the expected utility of each alternative.

3   Techniques to Build Complex DSSs

As long as DSSs have been applied to more and more complex decision problems the techniques to built them have evolved as well. The first systems were built to deal with fixed problems but soon this was considered insufficient. The evolution was directed toward DSSs with the ability to build decision models for several problems, related to the questions of the users. At the same time, there have been several modifications on IDs definition and capabilities. These modifications are directed to augment its expressiveness and to make easier the modeling of large and complex systems.

3.1   Approaches to the Construction of DSSs

An ideal computational decision system would possess general and broad knowledge of a domain, but would also have the ability to reason about the particular circumstances of any given decision problem within the domain. One obvious approach, usually called Knowledge-Based Model Construction (KBMC) (see Breese *et al.* [12]), is to generate decision models dynamically, at run-time, based on the problem description and the information gathered.

Model construction consists of selection, instantiation and assembly of causal and associational relationships from a board knowledge base of general relationships among the domain concepts. In a first approach, the systems only allow particular specifications for quantitative knowledge.

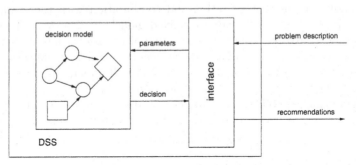

Fig. 1. A DSS with parameterized model

This approach has been adopted by the developers of Intellipath [13], who have constructed a large probabilistic network for the diagnosis in several subspecialties of surgical pathology. However, there are situations where the customization of the model itself is required, and not only of quantitative parameters. In these cases this is needed to express the domain facts and relationships in a general purpose knowledge representation language. When the DSS is faced with a particular situation, it operates over the knowledge base to dynamically construct a decision model customized for the problem at hand (see Wellman *et al.* [11]).

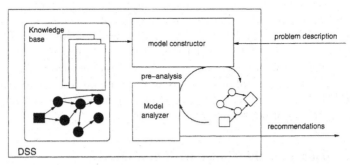

Fig. 2. A DSS with knowledge based model construction

Following these ideas, Sonnenberg *at al.* [14] have developed a system that contains two separate knowledge bases. One includes frames encoding knowledge of the medical domain: the evaluation of pulmonary disease in patients infected with HIV. The other adds rules of correct decision model construction that guide the selection of items from the domain knowledge base and their insertion into a decision model. This approach has a central problem: the need to integrate basic science and clinical knowledge, just because the ontological differences between these kind of knowledges [15].

Other direction outlined by Horvitz and Seiver [16] is related with the identification of common features between different problems. For example, these authors developed an general ID for a class of time critical decision problem (*pathological process problems*). This model is applied to trauma-care transportation and to propulsion system control and management. Although these problems belong to different domains, they are typically associated with well characterized sequences of states or processes that occur over time. The same ideas can be found in the following works:

- Heathfield and Wyatt [17] think that a major challenge for the development of DSSs for medical problems is to discover significant classes of common features and to define modeling techniques appropriated to them.
- Egar and Musen [18] explain that models in domains, such as medicine, exhibit certain prototypical patterns that can guide the modeling process. Medical concepts can be classified according to semantic types that have characteristic positions and typical roles in an ID model. The authors developed a graph-grammar production system that uses such inherent interrelationships among medical terms to facilitate the modeling of medical decision problems. The graphical representation they use is called *Qualitative Contingent IDs* (QCIDs).
- Mahoney and Laskey [19] propose that network engineering follows a rapid prototyping approach to network construction. They describe methods for identifying network modules, with the aim of reusing them. They propose an object oriented representation for BNs, to capture the semantic as well as representational knowledge embedded in the variables, their values and their parameters.

Perhaps the main problem with KBMC approach is to decide whether to include or not a given factor. For example, in medicine it seems that any event can be related to any other by some conceivable path of associations. Because reasoning with decision models can not begin until the model is completed, an exhaustive constraint delays the production of any results from the DSS. This problem is afforded in the work of Poh and Horvitz [20]. These authors investigate the value of extending the completeness of a decision model along different dimensions of refinement: quantitative, conceptual and structural. Goldman and Breese [21] try to superate the separation between construction and evaluation in most of DSSs. They develop an approach to combine incremental construction and evaluation, trying to get a trade-off between exactness and feasibility.

3.2   Modified Definitions of IDs to Deal with Real Problems

Although IDs are well suited to model and evaluate decision problems, its use involves limitations when dealing with complex problems. In order to overcome the difficulties related to modeling real problems (inherent to this task itself and not particularly when IDs are used) several modifications have been done to wide the expressive capabilities of IDs. In this subsection some of them will be analyzed.

- Several attempts have been done to work with large networks. One method exists for easy representation of replicated processes: the so-called frames, where a rectangular box around a part of the network means that this part is going to be replicated (see Gilks *at al.* [22]).
- Nilsson and Lauritzen [23] present *Limited Memory IDs* (LIMIDs), obtained from IDs but with a simpler structure. This representation is based on the identification of information not required for computing the optimal policies. This result allows an efficient computation for decision problems without loosing the expressiveness of IDs.
- Poh and Wu [24] propose a flexible approach to knowledge-based model construction. To achieve it the authors extend the standard ID representation to a multilevel structure in order to facilitate decision model construction at multiple levels of abstraction and define the formal relations between the IDs at different levels of abstraction, but representing all of them the same decision problem.
- Provan [25] has introduced the concept of Temporal IDs (TIDs), which adequately capture the temporal evolution of a dynamic system without prohibitive data and computational requirements. The author looks for procedures able to get a trade-off between exactness and tractability. Using these ideas Provan and Clarke [26] have developed an ID model for the diagnosis and treatment of acute abdominal pain.
- Ramoni *et al.* [27] perceive that one important drawback of IDs is the large amount of information needed to build a model, this information being not always easy to obtain. To overcome this limitation they develop a class of IDs, called *Ignorant IDs* (IIDs), able to reason on the basis of incomplete information and to improve the accuracy of the decisions as a monotonically increasing function of the available information. They explore the properties of IIDs using them in a clinical decision problem: the assessment of the optimal dosage of a drug to prevent the development of Graft Versus Host Disease (GVHD) after allogeneic bone marrow transplantation [28].
- Renooij and van der Gaag [29] try to avoid the obstacle related with the large number of probabilities required to model complex decision problems. For that purpose they propose the use of *qualitative abstractions*. They present *Qualitative IDs* (QIDs) and an efficient algorithm for qualitative decision making.

- Xiang and Poh [30],[31] presents a formalism called *Time-Critical Dynamic IDs* (TCDIDs) that provide the capability for temporal and space abstractions. In TCDIDs, time and temporal oriented relations are explicitly represented and modeled through the use of temporal arcs and time sequences.

4   Real World Applications

In this section real-world problems will be considered, being them afforded with DSS where IDs play a central role. Some of them are real applications while others are only prototypes to analyze the capabilities of IDs. The works are grouped according to the field of application.

4.1   Agriculture and Livestock Management

- Höhle *at al.* [32] show the way to apply LIMIDs [23] for modeling pig production scenarios.
- Jensen and Jensen [33] present MIDAS, a DSS for the management of mildew disease in winter wheat.
- Jorgensen [34] analyzes the potential use of IDs for herd management.
- Sahin and Bay [35] propose a biological intelligent agent model to solve a herding problem, designed by combining BNs and IDs.
- Xuan *at al.* [36] discuss the design of the Islay Land Use DSS (ILUDSS), a decision support system for strategic land-use planning in rural areas.

4.2   Economy

- Ahn and Ezawa [37] present a DSS for real-time telemarketing operations.
- The work of Freund and Dittus [38] outlines some of the basic principles of pharmacoeconomic analysis. They consider the use of different frameworks to make this analysis, one of them being IDs.
- Gmytrasiewicz and Tseng [39] propose the use of IDs for stock portfolio selection. They use an algorithm that applies mutual information as the metric to guide the refinement of the model.
- Poh [40] presents an intelligent DSS that combines decision analysis and traditional investment evaluation and analysis.
- Regan and Holtzman have developed *R & D Decision Advisor* [41], a commercial intelligent decision system for evaluating corporate research and development projects and portfolios.

4.3   Environmental Impact and Natural Resources Management

Due to the escalating pressure that mankind puts on natural resources and the environment, there is a growing need to develop management schemes and approaches that assist policy makers for better actions.

- Apostolakis and Bell [42] describe a structured approach for including multiple stakeholders in risk-based decision making in environmental restoration. This methodology is demonstrated with the modeling of a groundwater contamination problem.
- Casman *et al.* [43] consider the effects of climate change on drinking-waterborne infections.
- Heger and White [44] demonstrates the use of IDs in data worth analysis, applying to a monitor-and-treat problem often encountered in environmental decision problems. IDs are used to assess the worth of any set of additional data required for the decision problem.
- Kuikka *et al.* [45] analyzed the Baltic cod fisheries management with an ID model.
- Kuikka [46] discusses the use of BNs (with and without decision variables and utility function) to fisheries management.
- Rahman and de Castro [47] present an ID-based technique for modeling ans studying the effects of uncertain environmental factors in the electric utility planning process.
- Siddhaye and Sheng [48] uses IDs for determining the environmental sensitivity related to the design of electronic products.
- Sistonen [49] made a comprehensive survey of the priorities of the stakeholders for the rehabilitation process of Kyrnjoki river (Finland) and structured the various components of the problem using IDs.
- Varis [50] deals with cost-effective upgrading of wastewater treatment plants in a watershed on the basis of river water quality.
- Varis *at al.* [51] made an ID-based DSS to manage the water quality in the Kermana river, Finland.
- Varis *et al.* [52] built an ID to analyze restoration measures for Lake Tuusulanjrvi (Finland), harmed by eutrophication. Varis, in a previous work [53] analyzed the capabilities of IDs and decision analysis for dealing with environmental problems.
- Varis and Kuikka [54] used an ID to find a sustainable level of fishing effort for Lake Kariba, situated on the Zambezi River.

4.4   Industrial Applications

- Ang *et al.* [55] present a model for evaluating electricity distributed generation investment under uncertainty.
- Attoh-Okine [56] uses IDs to assist highway managers in making consistent decisions related to maintenance and rehabilitation of pavements.
- Buede [57] relates the design process of systems engineering and decision analysis with the aim of enhancing decision making by systems engineers.
- Cox *et al.* [58] used IDs to represent and analyze the knowledge of experts and users about chemical hazards.
- Kim [59] uses an ID knowledge base to represent the relationships between symptoms and process failures. The output of the system is a set

of corrective actions in response to sensor readings. This framework is applied to a milling machine process diagnosis.

- Lehtilä *et al.* [60] faced the problem of forecasting the demand and required generating capacity of electricity.
- Nadi *et al.* [61] developed an adaptive architecture for modeling manufacturing processes involving several controlling variables.
- Park and Kim [62] examine different approaches to the problem of Computer Numerical Control (CNC) machining: knowledge-based expert systems, neural networks and IDs.
- Rahman *et al.* [63] discuss a methodology based on IDs to decide the dissemination of solar appliances.
- Ramamurthi *et al.* [64] have developed an ID-based expert system that performs probabilistic inference and decision making applied to the control of a drilling operation in an automated manufacturing environment.
- Schaal and Lenz [65] consider an active information system for travel industry, aimed to notify likely delays. They formulate and solve this problem modeling it as an ID.
- Sachon and Pat-Cornell [66] use an ID to quantify the effect of an airline's maintenance policy on delays, cancellations and in-flight safety.
- Vatn [67] uses IDs to provide an efficient framework for building a maintenance optimization model.

4.5 Medicine

- Bellazzi and Quaglini [68] developed a prototype for acute myeloid leukemia therapy planning. This work is used to explore the use of IDs as knowledge sources in expert systems built under blackboard architecture bases.
- Downs *et al.* [69] present two approaches to prioritizing preventive services based on expected value decision making. One of these methods uses IDs to prioritize services dynamically using individual patient data.
- Downs *et al.* [70] have developed a decision analytic approach to scoring performance on computerized patient case simulation, using expected value of information. They use IDs to represent the diagnostic problem.
- Dybowski *et al.* [71] describe a DSS for septicemia management.
- The work of Fayerweather *et al.* [72] is related with risk assessment approaches for characterizing and quantifying uncertainty. They employ an ID to estimate the likelihood of different carcinogenic risks of chemical products under specific scenarios.
- Harmanec *et al.* [73] have developed a decision analytic approach to severe head injury management. This is a medical problem extremely challenging due to the complex domain, the uncertain intervention efficacies and the time-critical setting.
- Holtzman presents Rachel [5], an advisory system for couples with fertility problems seeking medical help. Its global framework is used to automatically generate case specific IDs, which are then evaluated.

- Lehmann *et al.* [74] have worked to create a recommendation for pediatrics and other primary care providers about their role as screeners for detecting developmental dysplasia of the hip in children.
- Lehmann and Shachter [75] present a knowledge-based computer framework for interpreting randomized clinical trials. In this system statistical models are represented by IDs, whose elements are mapped into user's domain language.
- Lucas *et al.* [76] have developed a DSS capable of assisting ICU doctors in dealing with patients who are mechanically ventilated and show symptoms and signs related to the development of pneumonia.
- Lucas *et al.* [77] present a DSS for the management of primary gastric non-Hodgkin lymphoma.
- Parmigiani [78] tackles the problem of screening breast cancer. A DSS is built to address controversial questions in this problem: initial age of application, frequency of screening, impact of improved mammographic technologies, etc.
- Provan and Clarke [79] focus their work on computing diagnoses in domains with continuously changing data. This paper describes a dynamic ID construction and updating system (DYNASTY), as well as its application to constructing a decision-theoretic model to diagnose acute abdominal pain [26] and graft-versus-host disease (GVHD) [80].
- Quaglini *et al.* [81] have built a DSS for Graft-versus-host disease (GVHD), one of the major complications of allogeneic bone marrow transplantation.
- Ríos *et al.* [82] uses IDs to develop a DSS for jaundice management in newborn babies. The authors of the work have developed a multi-attribute utility function to capture the preferences of doctors and patient's parents [83].
- Sonnenberg and Gavin [84] try to develop general rules to pursue a therapeutic goal of interventional gastrointestinal endoscopy. The influences of various medical interventions on the survival of a patient are modelled by an ID.
- Sonnenberg and Vakil [85] analyze the impact of diagnostic tests on patient's health-related quality of life. An ID is used to model how non-ulcer dyspepsia and its medical care affect health-related quality of life.

4.6 Risk Analysis (not in Medicine)

- Hokstad *et al.* [86] developed an overall risk model for offshore helicopter transport, gathering factors related to maintenance, competence of crew, training, etc.
- Armacost and Pet-Edwards [87] used a set of nested IDs to summarize and characterize the important relationships and uncertainties in operational systems. This integrative approach is illustrated through a real case study performed for the U.S. Coast Guard's International Ice Patrol.

- Chee *et al.* [88] argue that IDs are a natural tool for decision making in software risk analysis. The authors use several software metrics collected during development phase and analyze the data by an ID to identify sources of errors and predict future risk items, like reliability, complexity, schedule or non reusability.
- Huseby and Skogen [89] review the trends in risk analysis used in Norway. They present a tool called DynRisk, based on IDs and using Monte Carlo Simulation as evaluation method.
- Tamimi *et al.* [90] present a risk analysis approach that makes use of IDS to build risk models.

4.7   Software Development: Information Systems Design, User Interfaces, Multi-agent Systems, Web Applications, etc

- Burgess *et al.* [91] describe the use of IDs and a risk based model to formalize the combining of factors related to software changes and enhancements management.
- Horvitz *at al.* [92] describe Lumiere project, a system designed to provide assistance to computer software users.
- Jameson *et al.* [93] deal with the construction of intelligent user interfaces, with the ability to adapt itself to user needs taking into account several factors.
- Jameson *et al.* [94] apply IDs to the problem of deciding the next action of adaptive intelligent user interfaces, as a function of perceived properties of the user and the situations.
- Koller and Milch [95] argue that traditional representations of games do not offer clearly its structure. They propose Multi-Agent IDs as a new representation language for general multi-player games.
- ADVOCATE (Madsen *et al.* [96]) is a project to design and develop new software facilities which adds intelligence to management software for unmanned underwater vehicles.
- Mudgal and Vassileva [97] apply IDs to achieve negotiation in multi-agent systems with incomplete information about the opponent.
- Pek and Poh [98] describe a research project that uses IDs for providing teachers with information on students misconceptions and students with online tutoring.
- Poole [99] embedded IDs, structured Markov decision processes and game trees to provide a representation for multi-agent decision making under uncertainty.
- Serva *et al.* [100] propose a system to tackle with the problems related to the development of information systems. They state that many of the problems in systems development have been caused bay the lack of consideration for organizational and political factors. They propose a methodology for representing this factors through the use of IDs.

- Shih and Hung [101] develop a mechanism for the construction of courseware computer systems based on IDs. This mechanism allows the instructor to analyze the relations among course units and test units.
- Zhang *et al.* [102] proposes the use of IDs for modeling multistage games.

4.8 Strategy Formulation

- Agosta [103] describes a system for optimizing a real world planning problem: oil spill emergency response.
- Al-Awadi and Harrald [104] use quantitative models and computer-based decision analysis to complex political decision processes.
- Catton and Lim [105] examine the quality of a management technique (lower cavity flooding) for nuclear plant accidents.
- Fowler [106] presents a system that integrates fully operations management within the strategy development process, which leads to the representation of an illustrative generic model of a marketing, production and selling.
- Mendoncca *et al.* [107] propose a method for using communication and computing technologies for eliciting and aggregating the knowledge of multiple, geographically separate experts on emergency management.
- Parker [108] describes an approach for analyzing force structures of the armed forces of the United States.
- Yu [109] presents two approaches for evaluating the inherent uncertainties present in accident management strategies. He considers accident management as a decision problem and uses an ID to model it.

4.9 Others

- Agosta [110] presents IES, a system for machine vision, radar signal detection and military reconnaissance.
- Binford *at al.* are the developers of SUCCESOR [111] [112], a general 3D model based vision system.
- Ezawa and Scherer [113] describe a methodology for the analysis and selection of research projects using IDs.
- Gökccay and Bilgicc [114] present a decision theoretic method that yields approximate, low cost troubleshooting plans.
- Grass and Zilberstein [115] present a system designed to control the information presented to the users, optimizing its value and giving a clear picture of the system under control.
- Horvitz and Barry [116] describe methods for managing the complexity of information displayed to people responsible for time critical decisions. The authors use this approach to the control of displays for an application at the NASA Center.
- Jenzarli [117] describes a method for explicit modeling of decisions about activity times in projects. The author extends PERT networks to include decision variables: the resulting network is called PERTID.

- Sawaragi *et al.* [118] investigates a rigorous technique that addresses the problem of bounding the reasoning cost under resource constraints and uncertainty. The authors apply this technique to an autonomous mobile robot with limited capabilities.

5 Conclusions

The use of IDs as an effective tool to represent and solve decision problems has been stated with this survey of DSSs where IDs play some role. Important works about evaluation methods, treatment of asymmetries, extensions for real time decision problems and large networks construction, among others, have situated IDs far from its initial use only as a compact representation tool for decision problems. The early papers about IDs present them as a graphic representation for decision models, complementary to decision trees. In these works decision trees were recommended for evaluating the decision models [119] [9]. However, nowadays there are much more research intereset on IDs than on decision trees, as a proof of the role of IDs as a key tool to deal with decision problems on an holistic framework.

References

1. E.J. Horvitz, J.S. Breese, and M. Henrion. Decision theory in expert systems and artificial intelligence. *International Journal of Approximate Reasoning*, 2:247–302, 1988.
2. P. Szolovits. Uncertainty and decisions in medical informatics. *Methods of Information in Medicine*, 34:111–121, 1995.
3. A.M. Thornett. Computer decision support systems in general practice. *International Journal of Information Management*, 21:39–47, 2001.
4. A. Rector. AIM: A personal view of where I have been and where we might be going. *Artificial Intelligence in Medicine*, 23:111–127, 2001.
5. S. Holtzman. *Intelligent Decision Systems*. Addison-Wesley, 1989.
6. M. Henrion, J.S. Breese, and E.J. Horvitz. Decision analysis and expert systems. *Artificial Intelligence Magazine*, 12(4):64–91, 1991.
7. E. Horvitz. Automated reasoning for biology and medicine. In R. Fortune, editor, *Advances in Computer Methods for Systematic Biology: Artificial Intelligence, Databases, and Computer Vision*. Johns Hopkins University Pres, Baltimore, 1993.
8. H.W. Gottinger and P. Weimann. Intelligent decision support systems. *Decision Support Systems*, 8:317–332, 1992.
9. R.A. Howard and J.E. Matheson. *The principles and applications of decision analysis*, volume II, chapter Influence diagrams, pages 720–762. Strategic Decisions Group, Menlo Parck, CA, 1984.
10. P. Haddawy. An overview of some recent developments in Bayesian problem solving techniques. *Artificial Intelligence Magazine*, 20(2):11–19, 1999.
11. M.P. Wellman, J.S. Breese, and R.P. Goldman. From knowledge bases to decision models. *The Knowledge Engineering Review*, 7(1):35–53, 1992.

12. J.S. Breese, R.P. Goldman, and M.P. Wellman. Knowledge-based construction of probabilistic and decision models (introduction). *IEEE Transactions on Systems, Man and Cybernetics*, 24(11):1577–1579, 1994.

13. D.E. Heckerman, E.J. Horvitz, and B.N. Nathwani. Toward normative expert systems: Part I, the Pathfinder project. *Methods of Information in Medicine*, 31:90–105, 1992.

14. F.A. Sonnenberg, C.G. Hagerty, and C.A. Kulikowsky. An architecture for knowledge-based construction of decision models. *Medical Decision Making*, 14:27–39, 1994.

15. M. Ramoni and A. Riva. Advances in health sciences education. *Basic science in medical reasoning: an artificial intelligence approach*, 2:131–140, 1997.

16. E. Horvitz and A. Seiver. Time-critical action: representations and application. In *Proceedings of the 13th Conference on Uncertainty in Artificial Intelligence*, pages 250–257, 1997.

17. H.A. Heathfield and J. Wyatt. Philosophies for the design and development of clinical decision support systems. *Methods of Information in Medicine*, 32:1–8, 1993.

18. J.W. Egar and M.A. Musen. Graph-grammar assistance for automated generation of influence diagrams. In D. Heckerman and A. Mamdani, editors, *Proceedings of the 9th Conference on Uncertainty in Artificial Intelligence*, pages 235–242. Morgan & Kaufmann, 1993.

19. S.M. Mahoney and K.B. Laskey. Network engineering for complex belief networks. In E. Horvitz and F.V. Jensen, editors, *Proceedings of the 12th Conference on Uncertainty in Artificial Intelligence*, pages 389–395. Morgan & Kaufmann, 1996.

20. K.L. Poh and E.J. Horvitz. Reasoning about the value of decision-model refinement: Methods and application. In D. Heckerman and A. Mamdani, editors, *Proceedings of the 9th Conference on Uncertainty in Artificial Intelligence*, pages 174–182. Morgan & Kaufmann, 1993.

21. R.P. Goldman and J.S. Breese. Integrating model construction and evaluation. In D. Dubois, M.P. Wellman, B. D'Ambrosio, and P. Smets, editors, *Proceedings of the 8th Conference on Uncertainty in Artificial Intelligence*, pages 104–111. Morgan & Kaufmann, 1992.

22. W. Gilks, A. Thomas, and D.J. Spiegelhalter. A language and program for complex bayesian modelling. *The Statiscian*, 43:169–178, 1993.

23. D. Nilsson and S.L. Lauritzen. Evaluating Influence Diagrams using LIMIDS. In C. Boutilier and M. Goldszmidt, editors, *Proceedings of the 16th Conference on Uncertainty in Artificial Intelligence*, pages 436–445. Morgan Kaufmann Publishers, 2000.

24. K.L. Poh and X. Wu. Computer aided decision modeling using multilevel influence diagrams. In *Proceedings of the Third International Conference on Systems Science and Systems Engineering*, pages 49–52. Scientific and Technical Documents Publishing House, 1998.

25. G.M. Provan. Tradeoffs in constructing and evaluating temporal influence diagrams. In *Proceedings of the 9th Conference on Uncertainty in Artificial Intelligence*, pages 40–47, 1993.

26. G.M. Provan and J.R. Clarke. Dynamic network construction and updating techniques for the diagnosis of acute abdominal pain. *IEEE Transactions on Pattern Analysis and Machine Intelligence*, 15(3):299–307, 1993.

27. M. Ramoni. Ignorant influence diagrams. In *Proceedings of the International Joint Conference on Artificial Intelligence*, pages 1808–1814, San Mateo, CA, 1995. Morgan & Kaufmann.

28. M. Ramoni, A. Riva, M. Stefanelli, and V. Patel. Medical decision making using ignorant influence diagrams. In *Tenth Conference on Artificial Intelligence in Medicine*, pages 231–243, New York, 1995. Springer Verlag.

29. S. Renooij and L.C. van der Gaag. Decision making in qualitative influence diagrams. In K. van Marcke and W. Daelemans, editors, *Proceeding of the Ninth Dutch Conference on AI*, pages 93–102, Belgium, 1997. University of Antwerp.

30. Y. Xiang and K.L. Poh. Time-critical dynamic decision making. In H. Prade K.B Laskey, editor, *Proceedings of the 15th Conference on Uncertainty in Artificial Intelligence*, pages 688–695, San Francisco, 1999. Morgan Kaufmann Publishers.

31. Y. Xiang and K.L. Poh. Time-critical dynamic decision modeling in medicine. *Computers in Biology and Medicine*, 32(2):85–97, March 2002.

32. M. Höhle, E. Jørgensen, and D. Nilsson. Modeling with LIMIDS - Exemplified by disease treatment in slaughter pigs. In *Proceedings from International Symposium on Pig Herd Management Modelization*, pages 17–26, Lleida, Spain, 2000.

33. A.L. Jensen and F.V. Jensen. MIDAS - An influence diagram for management of Mildew in winther wheat. In E. Horvitz and F.V. Jensen, editors, *Proceedings of the 12th Conference on Uncertainty in Artificial Intelligence*, pages 349–356, San Francisco, 1996. Morgan & Kaufmann.

34. E. Jørgensen. Influence diagrams: potential use and current limitations. Technical Report 52, Dina Foulum, Dpt. of Biometry and Informatics, Research Center Foulum, August 1996.

35. F. Sahin and J.S. Bay. A biological decision theoretic intelligent agent solution to a herding problem in the context of distributed multi-agent systems. In *Proceedings of the IEEE International Conference on Systems, Man and Cybernetics*, volume 1, pages 137–142, 2000.

36. Z. Xuan, R.J. Aspinall, and R.G. Healey. ILUDSS: a knowledge-based spatial decision support system for strategic land-use planning. *Computers and Electronics in Agriculture*, 15(4):279–301, October 1996.

37. J.H. Ahn and K.J. Ezawa. Decision support for real-time telemarketing operations through Bayesian network learning. *Decision Suport Systems*, 21(1):17–27, September 1998.

38. D.A. Freund and R.S. Dittus. Principles of pharmacoeconomic analysis of drug therapy. *PharmacoEconomics*, 1(1):20–31, January 1992.

39. P.J. Gmytrasiewicz and C.C.C. Tseng. Refining influence diagram for stock portfolio selection. Technical Report 241, Society for Computational Economics, 2001.

40. K.L. Poh. An intelligent decision support system for investment analysis. *Knowledge and Information Systems*, 2:340–358, 2000.

41. P.J. Regan and S. Holtzman. R & D Decision Advisor: an interactive approach to normative decision system model construction. *European Journal of Operational Research*, 84(1):116–133, July 1995.

42. G. Apostolakis and D.C. Bell. Modelling the public as a stakeholder in environmental risk decision making. In *IEEE International Conference on Systems, Man and Cybernetics*, volume 2, pages 1021–1026, 1995.

43. E. Casman, B. Fischhoff, M. Small, H. Dowlatabadi, J. Rose, and M.G. Morgan. Climate change and cryptosporidiosis: a qualitative analysis. *Climatic Change*, 50:219–249, 2001.
44. A.S. Heger and J.E. White. Using influence diagrams for data worth analysis. *Reliability Engineering and System Safety*, 55:195–202, 1997.
45. S. Kuikka, H. Gislason, S. Hansson, M. Hilden, H. Sparhilt, and O. Varis. Environmentally driven uncertainties in Baltic cod management-modelling by Bayesian influence diagrams. *Canadian Journal of Fisheries and Aquatic Sciences*, 56:629–641, 1999.
46. S. Kuikka. *Uncertainty analysis in fisheries management science-Baltic Sea applications*. PhD thesis, Faculty of Agriculture and Forestry, Helsinki, 1998.
47. S. Rahman and A. de Castro. A framework for incorporating environmental factors into the utility planning process. *IEEE Transactions on Power System*, 9(1):352–358, February 1994.
48. S: Siddhaye and P. Sheng. Environmental impact and design parameters in electronics manufacturing: a sensitivity analysis approach. In *Proceedings of the 2000 IEEE International Symposium on Electronics and the Environment*, pages 39–45, 2000.
49. K. Sistonen. *Decision Analysis of rehabilitation of fisheries in River Kyrnjoki*. M.sc., Laboratory of Water Resources, University of Technology, Espoo, Finland, 1995.
50. O. Varis. A belief network approach to optimization and parameter estimation: application to resource and environmental management. *Artificial Intelligence*, 101:135–163, 1998.
51. O. Varis, B. Klove, and J. Kettunen. Evaluation of a real-time forecasting system for river water quality - a trade-off between risk attitudes, costs and uncertainty. *Environmental Monitoring and Assessment*, 28:201–213, 1993.
52. O. Varis, J. Kettunen, and H. Sirvio. Bayesian influence diagram approach to complex environmental management including observational design. *Computational Statistics and Data Analysis*, 9:77–91, 1990.
53. O. Varis, J. Kettunen, and H. Sirviö. Bayesian influence diagram approach to complex environmental management including observational design. *Computational Statistics & Data Analysis*, 9(1):77–91, January 1990.
54. O. Varis and S. Kuikka. Analysis of sardine fisheries management on Lake Kariba, Zimbabwe and Zambia - structuring a Bayesian influence diagram model. Technical Report WP-90-48, International Institute for Applied Systems Analysis, Laxenburg, Austria, 1990.
55. B.W. Ang, J.P. Huang, and K.L. Poh. Break-even price of distributed generation under uncertainty. *Energy*, 24:579–589, 1999.
56. B. Attoh-Okine. Potential application of bayesian influence diagram in a pavement management system. In *Proceedings of the Second International Symposium on Uncertainty Modelling and Analysis*, pages 379–386, 1993.
57. D.M. Buede. Engineering design using decision analysis. In *IEEE International Conference on Humans, Information and Technology*, volume 2, pages 1868–1873. IEEE Systems, Man and Cybernetics, 1994.
58. P. Cox, J. Niewöhner, N. Pidgeon, S. Gerrard, B. Fischhoff, and D. Riley. The use of mental models in chemical risk protection: developing a generic workplace methodology. *Risk Analysis*, 23(2):311–324, April 2003.
59. Y. Kim. A framework for an on-line diagnostic expert system for intelligent manufacturing. *Expert Systems with Applications*, 9(1):55–61, 1995.

60. A. Lehtilä, P. Silvennoinen, and J. Vira. A belief network model for forecasting within the electricity sector. *Technological Forecasting and Social Change*, 38(2):135–150, September 1990.

61. F. Nadi, A.M. Agogino, and D.A. Hodges. Use of influence diagrams and neural networks in modeling semiconductor manufacturing processes. *IEEE Transactions on Semiconductor Manufacturing*, 4(1):52–58, February 1991.

62. K.S. Park and S.H. Kim. Artificial intelligence approaches to determination of CNC machining parameters in manufacturing: a review. *Articial Intelligence in Engineering*, 12:127–134, 1998.

63. S. Rahman, A. de Castro, and Y. Teklu. Evaluating the impact of solar-assisted appliances in demand-side management. In *Proceedings of the Second International Conference on Advances in Power System Control, Operation and Management*, volume 2, pages 956–961, December 1993.

64. K. Ramamurthi, D.P. Shaver, and A.M. Agogino. Real time expert system for predictive diagnostics and control of drilling operation. In *Proceedings of the Sixth Conference on Artificial Intelligence for Applications*, volume 1, pages 62–69, 1990.

65. M. Schaal and H.J. Lenz. Best time and content for delay notification. In *Proceedings of the Eight International Symposium on Temporal Representation and Reasoning*, pages 75–80, 2001.

66. M. Sachon and E. Pat-Cornell. Delays and safety in airline maintenance. *Reliability Engineering & System Safety*, 67:301–309, 2000.

67. Jørn Vatn. Maintenance optimisation from a decision theoretical point of view. *ress*, 58:119–126, 1997.

68. R. Bellazzi and S. Quaglini. Reusable influence diagrams. *Artificial Intelligence in Medicine*, 6(6):483–500, December 1994.

69. S.M. Downs, L.R. Barrios-Cohen, and H. Uner. Expected value priorizations of prompts and reminders. In *Proceedings of the American Medical Informatics Association 2002*, pages 215–219, 2002.

70. S.M. Downs, F. Marasaigan, V. Abraham, B. Wildemurth, and C.P. Friedman. Scoring performance on computer-based patient simulation: beyond value of information. In *Proceedings of the American Medical Informatics Association 1999*, pages 520–524, 1999.

71. R. Dybowski, W.R. Gransden, and I. Phillips. Towards a statistically oriented decision support system for the management of septicaemia. *Artificial Intelligence in Medicine*, 5(6):489–502, December 1993.

72. W.E. Fayerweather, J.J. Collins, A.R. Schnatter, F.T. Hearne, R.A. Menning, and D.P. Reyner. Quantifying uncertainty un a risk assessment using human data. *Risk Analysis*, 19(6):1077–1090, December 1999.

73. D. Harmanec, T.Y. Leong, S. Sundaresh, K.L. Poh, T.T. Yeo, I. Ng, and T.W. Lew. Decision analytic approach to severe head injury management. In *Proceedings of the American Medical Informatics Association 1999*, pages 271–275, 1999.

74. H.P. Lehmann, R. Hinton, P. Morello, and J. Santoli. Developmental dysplasia of the hip pratice guideline. *Pediatrics*, 105(4):E57, April 2000.

75. H.P. Lehmann and R.D. Shachter. A physician-based architecture for the construction and use of statistical models. *Methods of Information in Medicine*, 33(4):423–432, October 1994.

76. P. Lucas, N.C. de Brujin, K. Schurink, and A. Hoepelman. A probabilistic and decision-theoretic approach to the management of infectious disease at the ICU. *Artificial Intelligence in Medicine*, 19:251–279, 2000.

77. P. Lucas, H. Boot, and B. Taal. Computer-based decision support in the management of primary gastric non-hodgkin lymphoma. *Methods of Information in Medicine*, 37:206–219, 1998.

78. G. Parmigiani. Decision models in screening for breast cancer. *Bayesian Statistics*, 6:525–546, 1999.

79. G.M. Provan. Model selection for diagnosis and treatment using temporal influence diagrams. In *Proceedings of the International Workshop on Artificial Intelligence and Statistics*, pages 469–480, January 1993.

80. J.R. Clarke and G.M. Provan. A dynamic decision-theoretic approach to the management of GVHD. In *Proceedings of the American Association for Artificial Intelligence Spring Symposium 1992*, 1992.

81. S. Quaglini, R. Bellazi, F. Locatelli, M. Stefanelli, and C. Salvaneschi. An influence diagram for assessing GVHD prophylaxis after bone marrow transplantation in children. *Medical Decision Making*, 14:223–235, 1994.

82. S. Ríos-Insua, C. Bielza, M. Gómez, J.A. Fernández del Pozo, M. Sánchez Luna, and S. Caballero. *Applied Decision Analysis*, chapter An intelligent decision system for jaundice management in newborn babies, pages 133–144. Kluwer, 1998.

83. M. Gómez, S. Ríos-Insua, C. Bielza, and J.A. Fernández del Pozo. Multiattribute utility analysis in the IctNeo system. *Research and Practice in Multiple Criteria Decision Making. Lecture Notes in Economics and Mathematical Systems*, 487:81–92, 1998.

84. A. Sonnenberg and M.W. Gavin. General principles to enhance practive patterns in gastrointestinal endoscopy. *Alimentary Pharmacology & Therapeutics*, 16(5):1003–1009, May 2002.

85. A. Sonnenberg and N. Vakil. The benefit of negative tests in non-ulcer dyspepsia. *Medical Decision Making*, 22(3):199–207, June 2002. abstract.

86. P. Hokstad, E. Jersin, and T. Sten. A risk influence model applied to North Sea transport. *Reliability Engineering & System Safety*, 74(3):311–322, December 2001.

87. R.L. Armacost and J. Pet-Edwards. Integrative risk and uncertainty analysis for complex public sector operational systems. *Socio-Economic Planning Sciences*, 33:105–130, 1999.

88. C.L. Chee, V. Vij, and C.V. Ramamoorthy. Using influence diagrams for software risk analysis. In *Proceedings of the Seventh International Conference on Tools with Artificial Intelligence*, pages 128–131, November 1995.

89. A.B. Huseby and S. Skogen. Dynamic risk analysis: the dynrisk concept. *International Journal of Project Management*, 10(3):160–164, August 1992.

90. S. Tamimi, J. Dickman, and R. Bauman. Risk analysis using influence diagrams. In *Proceedings of the First International Symposium on Uncertainty Modeling and Analysis*, pages 348–353, 1990.

91. C.J. Burgess, I. Dattani, G. Hughes, and J.H.R. May. Using influence diagrams in software change management. *Requirements Engineering*, 6(4):175–182, 2001.

92. E. Horvitz, J. Breese, D. Heckerman, D. Hovel, and K. Rommelse. The Lumiere project: Bayesian user modeling for inferring the goals and needs of

software users. In G.F. Cooper and S. Moral, editors, *Proceedings of the 14th Conference on Uncertainty in Artificial Intelligence*, pages 256–265. Morgan & Kaufmann, 1998.

93. A. Jameson, B. Großmann-Hutter, L. March, R. Rummer, T. Bohnenberger, and F. Wittig. When actions have consequences: empirically based decision making for intelligent user interfaces. *Knowledge-Based Systems*, 14:75–92, 2001.

94. A. Jameson, B. Großmann-Hutter, L. March, and R. Rummer. Creating an empirical basis for adaptation decisions. In H. Lieberman, editor, *Proceedings of IUI 2000: International Conference on Intelligent User Interfaces*, pages 149–156, New York, 2000.

95. D. Koller and B. Milch. Multi-agent influence diagrams for representing and solving games. In *Proceedings of the Seventeenth International Conference on Artificial Intelligence*, pages 1027–1034, Seattle, Washington, 2001.

96. K.G. Olesen A.L. Madsen and S.L. Dittmer. Practical modeling in bayesian decision problems - exploiting deterministic relations. *IEEE Transactions on Systems, Man and Cybernetics*, 32(1):32–38, 2002.

97. C. Mudgal and J. Vassileva. *Cooperative Information Agents*, chapter Bilateral Negotiation with Incomplete and Uncertain Information: A Decision-Theoretic Approach Using a Model of the Opponent, pages 107–118. 2000.

98. P. Pek and K. Poh. Using decision networks for adaptative tutoring. In S.S. Young, J. Greer, H. Maurer, and Y.S. Chee, editors, *Proceedings of the International Conference on Computers in Education / International Conference on Computer-Assisted Instruction*, pages 1076–1084, 2000.

99. D. Poole. The independent choice logic for modelling multiple agents under uncertainty. *Artificial Intelligence*, 94:7–56, 1997.

100. M.A. Serva, J. Cooprider, and R. Lee. Using influence diagrams for representing organizational and political factors in information system design. In *Proceedings of the Twenty-Fifth Hawaii International Conference on Systems Science*, volume 3, pages 424–433, January 1992.

101. T.K. Shih and L. Hung. Multimedia courseware development using influence diagram. In *Proceedings of the 2002 International Conference on Multimedia and Expo*, volume 2, pages 377–380, 2002.

102. Z. Zhang anf W. Liu and W. Li. Dynamic multi-agent influence diagrams for modeling multistage games. In *Proceedings of the International Conference on Machine Learning anf Cybernetics, 2002*, volume 3, pages 1184–1188, 2002.

103. J.M. Agosta. Constraining influence diagram structure by generative planning: an application to the optimization of oil spill response. In E. Horvitz and F.V. Jensen, editors, *Uncertainty in Artificial Intelligence, 6*, pages 11–19. Morgan Kaufmann Publishers, 1996.

104. M.A. Al-Awadi and J.R. Harrald. A political model using influence diagrams and the analytic hierarchy process: the Camp David negotiations. *Socio-Economic Planning Sciences*, 23(3):145–159, 1989.

105. I. Catton and H. Lim. The impact of phenomenological uncertainties on an accident management strategy. *Reliability Engineering & System Safety*, 45(1-2):175–194, 1994.

106. A. Fowler. Systems modelling, simulation and the dynamics of strategy. *Journal of Business Research*, 56(2):135–144, February 2003.

107. D. Mendoncca, R. Rush, and W.A. Wallace. Timely knowledge elicitation from geographically separate, mobile experts during emergency response. *Safety Science*, 35:193–208, 2000.

108. S.R. Parker. Military force structure and realignment "sharpening the edge" through dynamic simulation. In *Proceedings of the Winter Simulation Conference*, pages 1288–1295, 1995.

109. D. Yu. Modeling and measuring the effects of imprecision in accident management. *Annals of Nuclear Energy*, 29(7):821–833, May 2002.

110. J.M. Agosta. The structure of Bayes networks for visual recognition. In L.N. Kanal, R.D. Shachter, T.S. Levitt, and J.F. Lemmer, editors, *Uncertainty in Artificial Intelligence, 4*, volume 9, pages 397–405, Amsterdan, The Netherlands, 1990. Elsevier Science Publishers B. V.

111. T.O. Binford, T.S. Levitt, and W.B. Mann. Bayesian inference in model based machine vision. In L.N. Kanal, T.S. Levitt, and J.F. Lemmer, editors, *Uncertainty in Artificial Intelligence, 3*, pages 73–95, Amsterdam, The Netherlands, 1989. Elsevier Science Publishers B. V.

112. T.S. Levitt, J.M. Agosta, and T.O. Binford. Model based influence diagrams for machine vision. In L.N. Kanal, M. Henrion, R.D. Shachter, and J. Lemmer, editors, *Uncertainty in Artificial Intelligence, 5*, pages 371–388, Amsterdam, The Netherlands, 1990. Elsevier Science Publishers B. V.

113. K.J. Ezawa and J.B. Scherer. Technology management using computer-aided decision engineering tool. In *Proceedings of Technology Managament: the new International Language*, pages 319–322, 1991.

114. K. Gökccay and T. Bilgicc. Troubleshooting using probabilistic networks and value of information. *International Journal of Approximate Reasoning*, 29:107–133, 2002.

115. J. Grass and S. Zilberstein. Planning information gathering under uncertainty. Technical Report UM-CS-1997-032, Computer Science Department, University of Massachusetts, 1997.

116. E. Horvitz and M. Barry. Display of information for time-critical decision making. In P. Besnard and S. Hanks, editors, *Proceedings of the 11th Conference on Uncertainty in Artificial Intelligence*, pages 296–305, San Francisco, 1995. Morgan Kaufmann Publishers.

117. A. Jenzarli. Using influence diagrams for PERT. In *Portland International Conference on Management and Technology. Innovation in Technology Management: the key to Global Leadership*, pages 395–400, 1997.

118. T. Sawaragi, O. Katai, and D. Iwai. Decision-theoretic selection of reasoning scheme for an autonomous robot under resource constraints. In *Proceedings of the 1993 IEEE/RSJ International Conference on Intelligent Robots and Systems*, volume 2, pages 1449–1456, 1993.

119. A.C. Miller, M.W. Merkhofer, R.A. Howard, J.E. Matheson, and T.R. Rice. Development of decision aids for decision analysis. Technical Report DO-27742, Stanford Research Institute, Menlo Park, CA., 1976.

Learning Bayesian Networks by Floating Search Methods

Rosa Blanco[1], Iñaki Inza[1], and Pedro Larrañaga[1]

Intelligent System Group
Deparment of Computer Science and Artificial Intelligence
University of Basque Country
P.O. Box 649, 20080 San Sebastián – Donostia, Spain
email:{rosa, inza, ccplamup}@si.ehu.es

Abstract. In this work, two novel sequential algorithms for learning Bayesian networks are proposed. The presented sequential search methods are an adaptation of a pair of algorithms proposed to feature subset selection: *Sequential Forward Floating Selection* and *Sequential Backward Floating Selection*. As far as we know, these algorithms have never been used for learning Bayesian networks. An empirical comparison among the results of the proposed algorithms and the results of two sequential algorithm (the classical B-algorithm and its extension, the B3 algorithm) is carried out over four databases from literature. The results show promising results for the floating approach to the learning Bayesian network problem.

1 Introduction

During the last few years, probabilistic graphical models are increasingly becoming a suitable tool for structuring a set of domain variables. Bayesian networks are a popular kind of probabilistic graphical model. Bayesian networks [9,30,37] are used in a large amount of different domains including diagnosis, expert systems and debugging. When an expert, which traces the probabilistic relationship of a domain, does not exist, the automatic learning of a Bayesian network from a set of samples is a useful and widely accepted alternative.

A Bayesian network has two components: a network structure and a set of conditional probabilities. The network structure is a directed acyclic graph which depicts the relationships between the variables or nodes using arrows. The conditional probability measures represent the uncertainty of the domain. As each random discrete variable X_i $(i = 1, \ldots, n)$ follows a multinomial probability distribution, in a Bayesian network the joint probability distribution $p(x_1, \ldots, x_n)$ can be factorized by:

$$p(x_1, \ldots, x_n) = \prod_{i=1}^{n} p(x_i \mid \boldsymbol{pa}(x_i))$$

where x_i represents the value of the random variable X_i, and $\boldsymbol{pa}(x_i)$ represents a combination of the values of the random variable parents of X_i in the directed acyclic graph.

The automatic learning of a Bayesian network reflects the conditional (in)dependences [17] that implicitly appear in the domain database. Although the first automatic approaches have tried to produce a list of conditional (in)dependences by the use of statistical test [43], another automatic approach has strongly emerged in the last few years: the score+search approach [8,25]. The main idea is to perform an intelligent search in a specific search space, evaluating each proposed Bayesian network, and trying to find the network that best fits the databse.

Learning a Bayesian network from data is proved to be a NP-hard task when the Bayesian Dirichlet equivalent (BDe) metric is the objective function [12]. However, it is assumed that when other scores are used, the complexity of the problem is not reduced. In this way, the use of heuristic search procedures in order to solve the learning Bayesian network problem is justified.

In the field of pattern recognition, a successful pair of sequential algorithms were proposed several years ago and they have been largely used in many pattern recognition applications. These algorithms [39], known as *Sequential Forward Floating Selection* (SFFS) and *Sequential Backward Floating Selection* (SBFS), were originally presented to select the 'best' subset of m variables from a set of M domain variables ($m < M$). In this work, an adaptation of SFFS and SBFS search procedures is presented with the aim of learning Bayesian networks. Instead of searching for the 'best' subset of variables, we search for the 'best' subset of arcs to form a Bayesian network structure.

The paper is structured as follows: Section 2 introduces the framework to perform the search of Bayesian network structures, including the scoring function, search methods and search spaces. In Section 3, the original floating methods are presented and their adapted versions for learning Bayesian network structures are proposed. In Section 4, an empirical comparison between the adapted floating methods and the B and B3 algorithms is carried out over four databases from the literature. Finally, a brief set of conclusions are presented.

2   Learning Bayesian Networks by a Score+search Approach

When a problem is faced up as a score+search approach, three target must be stated: the scoring function, the search engine and the search space.

2.1   Families of Scores

In order to introduce the notation used in this section, Figure 1 shows a structure, the local probabilities and factorization for a Bayesian network.

$X = (X_1, \ldots, X_n)$ denotes an n-dimensional random variable and $x = (x_1, \ldots, x_n)$ represents one of its possible instances. If the variable X_i has

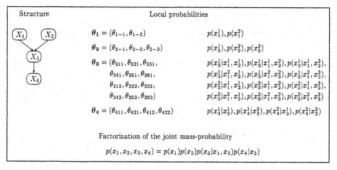

Structure	Local probabilities	

$\theta_1 = (\theta_{1-1}, \theta_{1-2})$ $p(x_1^1), p(x_1^2)$

$\theta_2 = (\theta_{2-1}, \theta_{2-2}, \theta_{2-3})$ $p(x_2^1), p(x_2^2), p(x_2^3)$

$\theta_3 = (\theta_{311}, \theta_{321}, \theta_{331},$ $p(x_3^1|x_1^1, x_2^1), p(x_3^1|x_1^1, x_2^2), p(x_3^1|x_1^1, x_2^3),$

$\theta_{341}, \theta_{351}, \theta_{361},$ $p(x_3^1|x_1^2, x_2^1), p(x_3^1|x_1^2, x_2^2), p(x_3^1|x_1^2, x_2^3),$

$\theta_{312}, \theta_{322}, \theta_{332},$ $p(x_3^2|x_1^1, x_2^1), p(x_3^2|x_1^1, x_2^2), p(x_3^2|x_1^1, x_2^3),$

$\theta_{342}, \theta_{352}, \theta_{362})$ $p(x_3^2|x_1^2, x_2^1), p(x_3^2|x_1^2, x_2^2), p(x_3^2|x_1^2, x_2^3)$

$\theta_4 = (\theta_{411}, \theta_{421}, \theta_{412}, \theta_{422})$ $p(x_4^1|x_3^1), p(x_4^1|x_3^2), p(x_4^2|x_3^1), p(x_4^2|x_3^2)$

Factorization of the joint mass-probability

$$p(x_1, x_2, x_3, x_4) = p(x_1)p(x_2)p(x_3|x_1, x_2)p(x_4|x_3)$$

Fig. 1. Structure, local probabilities and resulting factorization for a Bayesian network with four variables (X_1, X_3 and X_4 with two possible values, and X_2 with three possible values).

r_i possible values, $x_i^1, \ldots, x_i^{r_i}$, the local distribution, $p(x_i \mid pa_i^{j,S}, \theta_i)$ is an unrestricted discrete distribution, $p(x_i^{\,k} \mid pa_i^{j,S}, \theta_i) = \theta_{x_i^k | pa_i^j} \equiv \theta_{ijk}$, where $pa_i^{1,S}, \ldots, pa_i^{q_i,S}$ denotes the values of $\boldsymbol{Pa}_i^S$, the set of parents of the variable X_i in the structure S. The term q_i denotes the number of possible different instances of the parent variables of X_i. Thus, $q_i = \prod_{X_g \in \boldsymbol{Pa}_i} r_g$. The local parameters are given by $\theta_i = ((\theta_{ijk})_{k=1}^{r_i})_{j=1}^{q_i})$. The parameter θ_{ijk} represents the conditional probability of variable X_i being in its kth value, knowing that the set of its parent variables is in its jth value. We represent a database with N cases by $D = \{\boldsymbol{x}_1, \ldots, \boldsymbol{x}_N\}$. The information contained in D is used to learn the Bayesian network structure, S.

Using the maximum likelihood estimate for θ_{ijk}, $\widehat{\theta}_{ijk} = \frac{N_{ijk}}{N_{ij}}$, where N_{ijk} denotes the number of cases in D in which the variable X_i has the value x_i^k and $\boldsymbol{Pa}_i$ has its jth value, and $N_{ij} = \sum_{k=1}^{r_i} N_{ijk}$. When a certain form of penalty model complexity into the maximized likelihood is included, we obtain a general formula for the *penalized maximum likelihood* score as:

$$\sum_{i=1}^{n}\sum_{j=1}^{q_i}\sum_{k=1}^{r_i} N_{ijk} \log \frac{N_{ijk}}{N_{ij}} - f(N) \sum_{i=1}^{n} q_i(r_i - 1)$$

where $f(N)$ is a non-negative penalization function. A usual choice for the penalty function is the Jeffreys-Schwarz criterion, sometimes called the Bayesian Information Criterion (BIC) [41], where $f(N) = \frac{1}{2} \log N$.

In the Bayesian approach to the Bayesian network model induction from data, we express our uncertainty on the model (structure and parameters) by defining a variable whose states correspond to the possible network structure hypothesis S^h and assessing the probability $p(S^h)$. In the approach known as the Bayesian model selection, we select the model whose logarithm of the relative posterior probability: $\log p(S, D)$ is maximum. Taking this $\log p(S \mid D) \propto \log p(S, D) = \log p(S) + \log p(D \mid S)$ into account, and under the assumption that the prior distribution over the structure is uniform, an equivalent criterion is the log *marginal likelihood* ($\log p(D \mid S)$) of the data given the structure. It is possible to compute the marginal likelihood efficiently and in a closed form under a number of general assumptions [14,27]. For instance, Cooper and Herskovits (1992) proved that if the cases occur independently, there are no missing values, and if the density of the parameters given the structure is uniform, then:

$$p(D \mid S) = \prod_{i=1}^{n} \prod_{j=1}^{q_i} \frac{(r_i - 1)!}{(N_{ij} + r_i - 1)!} \prod_{k=1}^{r_i} N_{ijk}!.$$

This score is known as the K2 metric. A number of modifications of the K2 score metric were introduced in [6] in order to control the tendency towards simpler network structures.

The literature includes several scores that, inspired on the information theory [42], are able to calculate the *entropy* of a probability distribution represented by a Bayesian network. It is shown that the entropy of the distribution represented by a Bayesian network structure S is [28]:

$$H_S = \sum_{i=1}^{n} \sum_{j=1}^{q_i} p(\mathbf{Pa}_i = j) H_{X_i \mid \mathbf{Pa}_i = j}$$

where $H_{X_i \mid \mathbf{Pa}_i = j} = \sum_{k=1}^{r_i} p(X_i = x_i^k \mid \mathbf{Pa}_i = j) \ln p(X_i = x_i^k \mid \mathbf{Pa}_i = j)$.

2.2   Score+search Approaches to Learn Bayesian Networks

The algorithms that perform the learning of Bayesian networks from data can be divided into two main groups: algorithms based on conditional independence tests and algorithms based on a scoring function (or the score+search approach).

The methods based on conditional independence tests aim to accomplish a study of the (in)dependence relationships among the variables and try to model these relationships into a structure. The central point of this approach is the number and complexity of the tests carried out producing an effect on the efficiency of the algorithms. Recent works try to reduce the complexity, for instance Cheng et al. (2002) only requires a polynomial number of conditional independence tests in typical cases. Campos and Huete (2000) makes an intensive use of low order conditional independence tests. The research in this area proves that under conditions of complete data and given a node ordering, conditional independence tests for learning Bayesian network from data are equivalent to local log-scoring metrics [16].

A score+search approach performs a search for the structure model that 'best' represents the data according to the scoring measure. A search engine explores the search space evaluating the quality of the solutions by means of the scoring function. The type of search engine and the complexity of the score function are the essential points of this approach.

As the exhaustive enumeration of the search space to perform the search of the Bayesian network that fits the data is unfeasible, several score+search algorithms have been proposed throughout the last few years in order to examine the space of solutions. For instance, if the space of structures is the search space, the number of possible directed acyclic graphs for a domain with n variables is given by [40]:

$$f(n) = \sum_{i=1}^{n}(-1)^{i+1} \binom{n}{i} 2^{i(n-i)} f(n-i);$$

$$f(0) = 1; \quad f(1) = 1.$$

Although the search engines proposed in the literature can be divided into different axes, we classify them in two basic families: *sequential* and *population-based* algorithms.

When a single solution is recursively built, or a unique solution is kept during the optimization process, the search method can be considered as sequential. Among sequential algorithms, hill-climbing (also known as B algorithm) [7] is the most popular sequential procedure to perform the search within the space of Bayesian network structures.

It is a greedy algorithm which adds the arc that maximizes the scoring at each iteration starting with an empty solution. The well-known K2 algorithm [14] is a variation of this greedy search. It requires an ordering among the domain variables and looks for the best set of parents of each variable within the previous subset of variables. Another variation of the hill-climbing method is the iterated hill-climbing approach [13]. This procedure is motivated by the strong hill-climbing dependence of the starting point. It proposes to start the search at several points of the search space and it returns the best solution found so far.

Variable neighbourhood search, also proposed as a modification of the hill-climbing procedure [20], uses a systematic change of the neighbourhood space within a randomized local search algorithm. Other 'semi' hill-climbing approaches to learn the structure of a Bayesian network from data have been proposed in the literature: simulated annealing [2,13], branch-and-bound [46] or beam search [21].

All the previous algorithms relax the constraints of the classic greedy search. They can be seen as extensions or approximations of the hill-climbing approach. A different point of view to perform the search of a Bayesian network structure in a sequential way is given by the Markov Chain Monte Carlo [22] algorithm. It constructs a Markov Chain over orders. To tackle the problem of learning Bayesian networks, it samples entire networks from the posterior probability given the order.

In opposition to sequential methods, population-based approaches maintain a set of solutions and combine them to obtain a new set of solutions. All population-based methods are *stochastic* algorithms, that is, the same initial conditions do not produce the same final solution. Within this category, the following methods appear in the literature: genetic algorithms [33,34,36], ant colonies optimization [18] and estimation of distributions algorithms [5,38]. The main problem of population-based algorithms is the possible non-validity of new individuals requiring a repair method. This is motivated by the non-closure with respect to the search space of the majority of the used basic operators. Novel operators to preserve the closure in a specific search space are proposed in the literature [15,24].

2.3   The Search Space

The most usual approach to perform the search of the Bayesian network model is to implement it in the *space of directed acyclic graphs*. The search process in this space comes across difficulties when the search strategy has an evolutionary nature due to the non-acyclicity of the operators of these strategies. On the other hand, a sequential approach only needs to construct the solution verifying the non-violation of the constraint of acyclicity.

The BDe metric [26] assesses two Bayesian networks with the same value reflecting the same set of conditional independences. In this way, the search can also be performed in the *space of equivalence classes* (classes that reflect the same set of conditional independences) [11]. Recent works [23] note the relationship between the cardinality of Bayesian network structures space and the equivalence of classes space: this can be interpreted as a deceleration in the popularization of this promising approach. There are two basic reasons for this deceleration: the cardinality of this space is not largely reduced and the search process in this space has a high computational cost. However, with a convenient graphical representation for the equivalence class structure and a set of suitable operators to move within the space, a score+search framework in this space slightly outperforms the results of a traditional approach [11].

Although the search of Bayesian network structures is usually performed in a search space, literature reports attempts to combine the characteristics of both search space: directed acyclic graphs and equivalence classes [32].

The literature also proposes to perform the search within the *space of skeletons* [45]. The advantage of this space for heuristics inspired on evolutionary computation is that the operations performed with the old population to create a new ones are closed: valid individuals are always generated in the offspring without the need of repair-operators. However, it implies the necessity of finding the orientation of the edges.

Other authors [22,33] propose to perform the search within the *space of orders* of the n variables of the problem. The motivation for the birth of this approach is that several structure learning algorithms need the n variables to be ordered as well as a smoother posterior landscape.

3 Floating Methods to Learn Bayesian Networks Structures

3.1 The Original SFFS and SBFS Search Methods

Floating search methods appeared some years ago to solve the feature selection problem. In Feature Subset Selection (FSS) [31], the main goal is to obtain a subset of domain variables. This subset is formed by the 'best' domain variables with respect to a score metric. In pattern recognition, the number of variables to be selected is usually previously fixed as a parameter m, and the method stops when m variables from M ($m < M$) are selected.

The first proposed algorithms for FSS were the *sequential forward selection* (SFS) and the *sequential bacward elimination* (SBE). These methods are classical greedy algorithms. Both algorithms suffer from the 'nesting' problem: when a decision is taken, it cannot be reconsidered. In this case, when a feature is added (or removed), it can not be discarded (or considered) later. To avoid this effect, other sequential but more sophisticated algorithms were proposed such as the *step-wise* [47] and the *l-r* [44]. The *step-wise*, in each iteration, decides whether to add or to remove a single variable, depending on the best score metric achieved. The *l-r* is similar to *step-wise*, but it considers whether to add l variables or to remove r variables. Although these algorithms try to avoid, in part, the nesting problem, there is not way to determine the best values of l and r in the case of the *l-r* algorithm. The procedure proposed by *sequential forward floating selection* (SFFS) and *sequential backward floating selection* (SBFS) [39] allows l and r to intelligently 'float', and not have to be previously fixed.

Before explaining the original algorithms and their corresponding adaptation to the learning of Bayesian networks, a number of preliminaries are required. Bearing in mind that we are talking about the FSS problem, let $\mathcal{X}_k = \{X_i \in \mathcal{Y}; 1 \le i \le k\}$, where $\mathcal{Y}$ is the total set of domain variables, and let $J(X_i)$ the value of the objective function, only for the X_i variable, that is, the set $\mathcal{X}_k = \{X_i\}$, which can be considered as the *individual significance* of the variable X_i. Obviously, $J(\mathcal{X}_k)$ is the value of the the $\mathcal{X}_k$ subset of variables. Then, the *significance $S_{k-1}(X_j)$ of the variable $X_j \in \mathcal{X}_k$*, that is, when X_j is a variable of the set $\mathcal{X}_k$ and X_j is the candidate to be excluded from the set, is defined by:

$$S_{k-1}(X_j) = J(\mathcal{X}_k) - J(\mathcal{X}_k \setminus X_j).$$

The *significance $S_{k+1}(X_j)$ of the variable $X_j \in \mathcal{Y} \setminus \mathcal{X}_k$*, that is, when X_j is a variable out of the set $\mathcal{X}_k$ and X_j is candidate to be included in the set, with respect to the set $\mathcal{X}_k$ is defined by:

$$S_{k+1}(X_j) = J(\mathcal{X}_k \cup X_j) - J(\mathcal{X}_k).$$

Figure 2 shows the original SFFS algorithm as proposed in Pudil et al. (1994). It is assumed that a previous subset of k variables is selected before

Step 1: Inclusion
> Using the basic greedy method, select feature X_{k+1} from the set of available measurements, $\mathcal{Y} \setminus \mathcal{X}_k$ to form feature set $\mathcal{X}_{k+1}$, i.e., the most significant feature X_{k+1} with respect to the set $\mathcal{X}_k$ is added to $\mathcal{X}_k$. Therefore $\mathcal{X}_{k+1} = \mathcal{X}_k \cup X_{k+1}$.

Step 2: Conditional exclusion
> Find the least significant feature in the set $\mathcal{X}_{k+1}$. If X_{k+1} is the least significant feature in the set $\mathcal{X}_{k+1}$, i.e. $J(\mathcal{X}_{k+1} \setminus X_{k+1}) \geq J(\mathcal{X}_{k+1} \setminus X_j)$ $\forall j = 1, 2, \ldots, k$, then set $k = k + 1$ and return to Step 1, but if X_r $1 \leq r \leq k$, is the least significant feature in the set $\mathcal{X}_{k+1}$, i.e. $J(\mathcal{X}_{k+1} \setminus X_r) > J(\mathcal{X}_k)$ then exclude X_r from $\mathcal{X}_{k+1}$ to form a new feature set $\mathcal{X}'_k$, i.e $\mathcal{X}'_k = \mathcal{X}_{k+1} \setminus X_r$. Note that now $J(\mathcal{X}'_k) > J(\mathcal{X}_k)$. If $k = 2$, the set $\mathcal{X}_k = \mathcal{X}'_k$ and $J(\mathcal{X}_k) = J(\mathcal{X}'_k)$ and return to Step 1, else go to Step 3.

Step 3: Continuation of conditional exclusion
> Find the least significant feature X_s in the set $\mathcal{X}'_k$. If $J(\mathcal{X}'_k \setminus X_s) \leq J(\mathcal{X}_{k-1})$ then set $\mathcal{X}_k = \mathcal{X}'_k$, $J(\mathcal{X}_k) = J(\mathcal{X}'_k)$ and return to Step 1. If $J(\mathcal{X}'_k \setminus X_s) > J(\mathcal{X}_{k-1})$ then exclude X_s from $\mathcal{X}'_k$ to from a newly reduced set $\mathcal{X}'_{k-1}$, i.e. $\mathcal{X}'_{k-1} = \mathcal{X}'_k \setminus X_s$. Set $k = k - 1$. Now if $k = 2$, the set $\mathcal{X}_k = \mathcal{X}'_k$ and $J(\mathcal{X}_k) = J(\mathcal{X}'_k)$ and return to Step 1, else repeat Step 3.

Fig. 2. Original Sequential Forward Floating Selection Algorithm

the process begins, usually $k = 2$ variables are selected in a greedy way by means of SFS. The SFFS starts with an *inclusion* step. In this stage, a candidate feature, to be part of the solution, is selected by SFS and added to $\mathcal{X}_k$ obtaining $\mathcal{X}_{k+1}$, and the SFFS goes to the exclusion step. At the *exclusion* step, the least significant feature from $\mathcal{X}_{k+1}$ is proposed to be removed from the solution. If the least significant feature is the last inserted, the deletion is rejected and the algorithm returns to the inclusion step. If not, the least significant variable from $\mathcal{X}_{k+1}$ is deleted to form $\mathcal{X}'_k$ and the algorithm goes to the *continuation of the conditional exclusion*. At this continuation of the conditional exclusion step, the algorithm asks for the least significant variable of $\mathcal{X}'_k$. If the value of the scoring function of $\mathcal{X}'_k$ without the least significant feature is less or equals the value of the scoring function of $\mathcal{X}_{k-1}$ (the last subset of variables with the same cardinality), it rejects the deletion and returns to the inclusion step. If not, the variable is excluded from $\mathcal{X}'_k$ to form $\mathcal{X}'_{k-1}$ and the methods repeats the exclusion step.

Note that the inclusion of variables is carried out while the least significant variable in the set is the one last added. The exclusion of variables is performed while the objective function is higher than the objective function of the last subset with the same cardinality. Due to this fact, the number of included (excluded) variables 'float' during the search process.

In Figure 3, the corresponding 'up-bottom' approach to the SFFS is depicted. Note that all the steps are similar and opposed to each other. When the SFFS includes (excludes) a feature in the candidate subset, the SBFS procedure proposes the exclusion (inclusion) of a feature.

Step 1: Exclusion
    Using the basic greedy method, select feature X_{k+1} from the set of
    available measurements, $\mathcal{X}_k$ to remove from the current set, i.e., the least
    significant feature X_{k+1} is deleted from the set $\mathcal{X}_k$ to form $\mathcal{X}_{k+1}$.
Step 2: Conditional inclusion
    Find the most significant feature in the set of excluded features $\mathcal{Y} \setminus \mathcal{X}_{k+1}$.
    If X_{k+1} is the most significant feature in the set $\mathcal{X}_{k+1}$, i.e.
    $J(\mathcal{X}_{k+1} \cup X_{k+1}) \geq J(\mathcal{X}_{k+1} \cup X_j) \forall j = 1, 2, \ldots, k$, then set $k = k + 1$ and
    return to *Step 1*, but if X_r $1 \leq r \leq k$, is the most significant feature in the
    set $\mathcal{X}_{k+1}$, i.e. $J(\mathcal{X}_{k+1} \setminus X_r) > J(\mathcal{X}_k)$ then include X_r to $\mathcal{X}_{k+1}$ to form a
    new feature set $\mathcal{X}'_k$, i.e $\mathcal{X}'_k = \mathcal{X}_{k+1} \cup X_r$. Note that now $J(\mathcal{X}'_k) > J(\mathcal{X}_k)$. If
    $k = 2$, the set $\mathcal{X}_k = \mathcal{X}'_k$ and $J(\mathcal{X}_k) = J(\mathcal{X}'_k)$ and return to *Step 1*, else go
    to *Step 3*.
Step 3: Continuation of conditional inclusion
    Find the most significant feature X_s in the set of excluded features $\mathcal{Y} \setminus \mathcal{X}'_k$.
    If $J(\mathcal{X}'_k \cup X_s) \leq J(\mathcal{X}_{k-1})$ then set $\mathcal{X}_k = \mathcal{X}'_k$, $J(\mathcal{X}_k) = J(\mathcal{X}'_k)$ and return
    to *Step 1*. If $J(\mathcal{X}'_k \cup X_s) > J(\mathcal{X}_{k-1})$ then include X_s to the set $\mathcal{X}'_k$ to from
    a newly extended set $\mathcal{X}'_{k-1}$, i.e. $\mathcal{X}'_{k-1} = \mathcal{X}'_k \cup X_s$. Set $k = k - 1$. Now if
    $k = 2$, the set $\mathcal{X}_k = \mathcal{X}'_k$ and $J(\mathcal{X}_k) = J(\mathcal{X}'_k)$ and return to *Step 1*, else
    repeat *Step 3*.

Fig. 3. Original Sequential Backward Floating Selection Algorithm

3.2 Adapted SFFS and SBFS to Learn Bayesian Networks Structures

In the Bayesian network learning problem, a structure is a graph $\mathcal{G}_k = \{\mathcal{V}, \mathcal{E}_k\}$ where $\mathcal{V} = \{X_1, X_2, \ldots, X_n\}$ is the set of nodes and $\mathcal{E}_k \subset \mathcal{A} = \{a_{11}, a_{12}, \ldots, a_{ij}, \ldots, a_{nn}\}$. $\mathcal{A}$ denotes the set of all possible arcs between the nodes of $\mathcal{V}$.

Let $g(\mathcal{E}_k)$ be the score metric for the complete structure $\mathcal{E}_k$. The *significance* $S_{k-1}(a_{ij}^k)$ of the arc $a_{ij}^k \in \mathcal{E}_k$ between the nodes X_i and X_j, that is, the arc candidate to be excluded from the structure, is determined as:

$$S_{k-1}(a_{ij}^k) = g(\mathcal{E}_k) - g(\mathcal{E}_k \setminus a_{ij}^k).$$

The *significance* $S_{k+1}(a_{ij}^k)$ of the arc $a_{ij}^k \in \mathcal{A} \setminus \mathcal{E}$ between the nodes X_i and X_j, that is, the arc candidate to be included in the structure, is determined as:

$$S_{k+1}(a_{ij}^k) = g(\mathcal{E}_k \cup a_{ij}^k) - g(\mathcal{E}_k).$$

Figure 4 and Figure 5 show the proposed sequential algorithms in order to learn Bayesian network structures. The adapted SFFS and SBFS try to avoid the main problem of all the greedy methods, including the B-algorithm: the nesting problem.

The adapted floating search algorithms follow the general scheme of the original methods. For instance, the forward algorithm starts with the *inclusion* step, where the most significant arc is proposed to be part of the network structure $\mathcal{E}_k$ to form $\mathcal{E}_{k+1}$. At the *conditional exclusion*, the least significant arc of $\mathcal{E}_{k+1}$ is proposed to exclusion. If it is the last introduced arc, the exclusion is refused. If not, the arc is deleted to form $\mathcal{E}'_k$ and the *continuation*

Fig. 4. Adapted Sequential Forward Floating Selection Algorithm

of the conditional exclusion is carried out. At this stage, the algorithm asks for the least significant arc of the network structure $\mathcal{E}_k'$. If the value of the scoring function of the structure without the least significance arc is lower than the last time the structure had the same number of arcs, the exclusion is rejected and the method returns to the inclusion step. If not, the arc is deleted to form $\mathcal{E}_{k-1}'$ and another arc to be deleted is proposed.

In contrast to other sequential algorithms, with these methods, the number of added arcs do not need to be previously fixed, 'float' during the search.

It must be noted that a Bayesian network structure is represented by means of a directed acyclic graph. When adding an arc, if the resulting graph is cyclic, the arc is rejected and another arc is selected. For simplicity, this refusal of an arc when cyclicity appears is not written in Figure 4 and Figure 5.

In the original SFFS and SBFS, the improvement of the total score, neither in addition nor in deletion, is considered. It is possible that, by adding or deleting a feature, the score metric gets worse. In adapted algorithms, a threshold of this worsening is fixed. This threshold is used as follows: when an arc is added (or removed), if the current score is worse than the fixed threshold, the arc is rejected and another arc is proposed to carry out the movement. However, it is possible that a small deterioration in the score leads to a high improvement in several movements. The threshold is decreased with the number of visited solution candidates. For simplicity, the use of the score is not presented in Figure 4 and Figure 5. It must be noted that when the threshold is fixed to 0, the adapted algorithms turn into classic greedy methods.

Step 1: Exclusion
   Using the basic greedy method, select an arc a_{ij}^{k+1} from the set $\mathcal{E}_k$ to remove from the current set, i.e., the least significant arc a_{ij}^{k+1} is deleted from the set $\mathcal{E}_k$ to form a graph $\mathcal{E}_{k+1} = \mathcal{E}_k \setminus a_{ij}^{k+1}$. If its not possible to exclude an arc, stop.

Step 2: Conditional inclusion
   Find the most significant arc in the set of excluded arcs $A \setminus \mathcal{E}_{k+1}$. If a_{ij}^{k+1} is the most significant arc in the set $\mathcal{E}_{k+1}$, i.e.
   $g(\mathcal{E}_{k+1} \cup a_{ij}^{k+1}) \geq g(\mathcal{E}_{k+1} \cup a_{ij}^l) \forall l = 1, 2, \ldots, k$, the set $k = k + 1$ and return to *Step 1*, but if a_{ij}^r $1 \leq r \leq k$, is the most significant arc in the set $\mathcal{E}_{k+1}$, i.e. $g(\mathcal{E}_{k+1} \cup -a_{ij}^r) > g(\mathcal{E}_k)$ then include a_{ij}^r to $\mathcal{E}_{k+1}$ to form a new graph $\mathcal{E}_k'$, i.e $\mathcal{E}_k' = \mathcal{E}_{k+1} \cup a_{ij}^r$. Note that now $g(\mathcal{E}_k') > g(\mathcal{E}_k)$. If $k = 2$, the set $\mathcal{E}_k = \mathcal{E}_k'$ and $g(\mathcal{E}_k) = g(\mathcal{E}_k')$ and return to *Step 1*, else go to *Step 3*.

Step 3: Continuation of conditional inclusion
   Find the most significant arc a_{ij}^s in the set of excluded arcs $A \setminus \mathcal{E}_k'$. If $g(\mathcal{E}_k' \cup a_{ij}^s) \leq g(\mathcal{E}_{k-1})$ then set $\mathcal{E}_k = \mathcal{E}_k'$, $g(\mathcal{E}_k) = g(\mathcal{E}_k')$ and return to *Step 1*. If $g(\mathcal{E}_k' \cup a_{ij}^s) > g(\mathcal{E}_{k-1})$ then include a_{ij}^s to the set $\mathcal{E}_k'$ to from a new graph $\mathcal{E}_{k-1}'$, i.e. $\mathcal{E}_{k-1}' = \mathcal{E}_k' \cup a_{ij}^s$. Set $k = k - 1$. Now if $k = 2$, the set $\mathcal{E}_k = \mathcal{E}_k'$ and $g(\mathcal{E}_k) = g(\mathcal{E}_k')$ and return to *Step 1*, else repeat *Step 3*.

Fig. 5. Adapted Sequential Backward Floating Selection Algorithm

4   Experimental Results

In order to compare the behaviour of both floating methods, they are tested using the K2 [14] scoring function over four Bayesian network databases from the literature: *Alarm* [3], *Insurance* [4], *Hailfinder* [1] and *Mildew* [29]. The *Alarm* network is considered a benchmark to evaluate Bayesian network learning algorithms. It has 37 nodes, 46 arcs and it is related with the diagnosis in medical domain. Whereas the experimentation with *Alarm* is performed over the first 3,000 cases of the well-known database proposed by Cooper and Herskovits (1992), the experiments with *Insurance, Hailfinder* and *Mildew* networks are carried out over 10,000 cases simulated from the real Bayesian networks. The *Insurance* network has 27 nodes and 52 arcs and it evaluates the risk in car insurance. *Hailfinder* has 56 variables, 66 arcs and it was designed to forecast severe weather in North-eastern Colorado, USA. Finally, the *Mildew* Bayesian network, composed of 35 nodes and 46 arcs, decides on the amount of fungicides to be used against an attack of mildew on wheat.

   With the purpose of comparing the results of the adapted floating methods with the results of a 'standard' and 'benchmark' sequential algorithm, a comparison with the B algorithm [7] is performed. Moreover, the B3 algorithm is proposed in order to make an extensive comparison. This algorithm is a sequential greedy search engine, as with the B algorithm, that performs a deterministic local search taking three operators at each step into account: arc addition, arc deletion and arc reversal. The SFFS is compared with the *forward* B and B3 algorithms which, starting with an empty solution, iteratively construct by means of the operators a Bayesian networks structure. The SBFS is compared with the *backward* B and B3 algorithms. In this case, the search engine starts with a network solution, and while B3 works as in the forward case, the B algorithm removes the arc from the structure the arc with the major deterioration of the score at each step.

To perform the comparison of the backward methods, 10 complete structures are generated. To create each of these structures, a randomly generated ancestral ordering is needed.

Moreover, due to the fact that a partial solution may be required to start the original floating algorithms, the experimentation is enlarged for the adapted floating methods, starting the search process in the middle of the search space between the empty and the full structures. These structures are generated by means of two different processes. First, the networks are randomly generated as follows: starting with an empty solution structure, an arc is added to the network structure with a fixed probability. This initialization is called *Random* initialization, and the probability of the appearance of an arc is fixed to 0.05. The random generation of structures may produce non-valid Bayesian networks. To overcome this difficulty, a simple 'repairing' operator is used: when a cycle is detected in the structure, one arc from the cycle is deleted randomly. 10 independent structures are generated as the initialization network of floating methods. The improvement threshold is fixed to $0.10n$ where n is the number of variables of the domain.

The second approach to generate the initial structures is based on the network induced by the forward B algorithm called *B-noise* initialization. When a forward floating algorithm is used, if an arc is selected by the B algorithm, it can be deleted with a certain probability. When a backward floating algorithm is used, if an arc is not selected by the B algorithm, it can be added with a certain probability. The idea is to produce networks with fewer and more arcs respectively. In the case of the forward method, the probability of deletion is fixed to 0.10. In the case of backward method, the probability of addition is fixed taking the expected number of deleted arcs for the forward algorithm to be the same as the expected number of added arcs for backward engine into account.

Tables 1, 2, 3 and 4 show the results in terms of score and number of evaluated solutions during the search process. When more than one run is performed, the average score (and their standard deviation) and the average number of evaluated solutions (and their standard deviation) are displayed. The empty blocks on the tables are unfeasible combinations of initial structures and search engines. It must be noted that for the *Hailfinder* domain, only when the search engines start with an empty solution, the problem is computationally treatable.

In order to carry out a more thorough analysis of the results, the statistical significance of the obtained differences is studied when several runs are performed. The Mann-whitney [35] test is carried out to determine the

Table 1. Score results and number of visited solutions of the networks induced by the search algorithms over 10 runs (when feasible) in a simulation of the *Alarm* network. The real value of the K2 scoring function is -14412.69.

		Forward			Backward		
		B	B3	SFFS	B	B3	SBFS
Empty	score	-14520.2	-14440.2	-14472.9			
	n.eval	61832	69228	255999			
Full	score				$-20259.2 \pm 496.1^\dagger$	-16288.51 ± 299.1	-18149.7 ± 488.7
	n.eval				66359.3 ± 2341.5	$247788.5 \pm 8761.7^\dagger$	$157540.0 \pm 3108.9^\dagger$
Random	score			-14790.2 ± 225.8			-14885.7 ± 396
	n.eval			203378.3 ± 9203.1			205335.7 ± 9391.1
B-noise	score			-14473.0 ± 0.2			$-15256.2 \pm 389.5^\dagger$
	n.eval			$244762.8 \pm 912.5^\dagger$			187655.0 ± 9929.7

Table 2. Score results and number of visited solutions of the networks induced by the search algorithms over 10 runs (when feasible) in a simulation of the *Insurance* network. The real value of the K2 scoring function is -59257.60.

		Forward			Backward		
		B	B3	SFFS	B	B3	SBFS
Empty	score	-58964.3	-58641.6	-58964.3			
	n.eval	28435	36000	75294			
Full	score				$-67162.1 \pm 2111.63^\dagger$	-59783.3 ± 159.9	$-64881.0 \pm 762.1^\dagger$
	n.eval				22032.5 ± 182.4	$67009.8 \pm 4052.7^\dagger$	$52255.8 \pm 1341.2^\dagger$
Random	score			-60068.6 ± 669.0			$-60585.5 \pm 383.4^\dagger$
	n.eval			69245.0 ± 2404.2			69488.1 ± 4112.9
B-noise	score			-59932.4 ± 241.8			-60502.1 ± 231.6
	n.eval			$61924.2 \pm 2815.8^\dagger$			54594.9 ± 4917.4

significance of the differences in Tables 1, 2, 3 and 4. For each initialization, statistically significant differences with respect to the best attained results are marked. The symbol '$\dagger$' denotes statistically significant differences at the 0.05 confidence level with respect to the best results reached by an algorithm in a certain initialization scheme.

The comparison can be made in three steps: (i) forward methods with an 'empty' initialization, (ii) backwards methods with a 'full' initialization and (iii) floating methods with 'middle' initialization.

For the forward methods starting with an empty initialization, the *Alarm*, *Insurance*, *Hailfinder* and *Mildew* domains show a similar behaviour. The floating method, requiring a higher number of evaluated solutions, improves the scoring function values of the B algorithm. However, the proposed B3 algorithm attains slightly better fitness results than the floating algorithm.

The obtained results by backward methods, starting with a full initalization, should be studied for each problem domain separately.

For the *Alarm* domain, although the B algorithm requires the lowest number of evaluated structures with statistically significant differences, it attains the lowest score value. The B3 algorithm achieves the best scoring results without significant difference with respect to the floating method.

In the *Insurance* and *Mildew* domains, the B algorithm needs the lowest number of evaluated solutions (with statistically significant differences) to

Table 3. Score results and number of visited solutions of the networks induced by the search algorithms over 10 runs (when feasible) in a simulation of the *Hailfinder* network. The real value of the K2 scoring function is -217663.39.

| | | Forward | |
	B	B3	SFFS
Empty score	-217105.1	-217072.6	-217085.64
n.eval	208227	214974	304616

Table 4. Score results and number of visited solutions of the networks induced by the search algorithms over 10 runs (when feasible) in a simulation of the *Mildew* network. The real value of the K2 scoring function is -205668.23.

| | | Forward | | | Backward | | |
		B	B3	SFFS	B	B3	SBFS
Empty	score	-202685	-202199	-202655			
	n.eval	49650	61208	210159			
Full	score				$-239320 \pm 5242^{\dagger}$	-205443 ± 2268	$-212818 \pm 2280^{\dagger}$
	n.eval				12266 ± 234	$90096 \pm 2009^{\dagger}$	$155897 \pm 4289^{\dagger}$
Random	score			-203862 ± 1363			-203584 ± 1082
	n.eval			202755 ± 9537			201431 ± 13504
B-noise	score			$-203639 \pm 2726^{\dagger}$			-202660 ± 102
	n.eval			180420 ± 8625			$213539 \pm 9731^{\dagger}$

reach the lowest values of the score function. The best value of the scoring measure is obtained by the B3 algorithm with statistically significant differences with respect to floating methods and the B algorithm.

When the floating methods are compared, the forward floating seems to attain better results with a lower number of evaluated solutions. In this way, for the *Alarm* domain, when the B-noise initialization is used, statistically significant differences are not found for scoring values and required evaluated solutions. However, when the Random initialization is proposed, the forward method requires a statistically significant higher number of evaluated structures to reach a statistically significant best scoring function. A similar behaviour can be observed in the *Insurance* domain. Although the forward method needs a statistically significant higher number of evaluated networks for the B-noise initialization, it attains better but not statistically significant scoring results with respect to the backward algorithm. Finally, in the *Mildew* problem domain, for the B-noise initialization, a statistically significant worse value of the scoring measure is attained by the forward method with a statistically significant lower number of evaluated solutions. For the Random initialization, the score results and the number of evaluated structures are so close that any statistically significant difference can be found.

In order to perform a qualitative study of the learnt solution by the *forward* floating method, Figures 6, 7, 8 and 9 show the original network structures and the network structure induced by the forward floating algorithm.

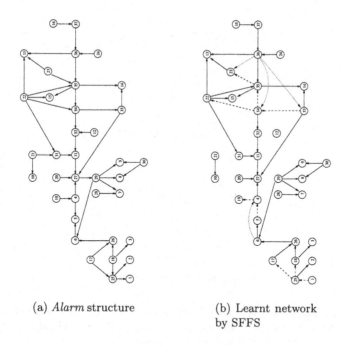

(a) *Alarm* structure

(b) Learnt network
by SFFS

Fig. 6. The real *Alarm* network has 37 nodes and 46 arcs. The network structure learnt by SFFS, starting with an empty solution, has 47 arcs: 4 extra arcs (dotted line), 8 reverse arcs (dashed line) and 3 missing arcs.

5   Conclusions

A novel search engine to learn Bayesian networks from data in a score+search framework has been proposed. Two algorithm proposed in the field of the feature subset selection for patter recognition are adapted to perform the search of the Bayesian network structure that best fits the data.

In an empirical comparison with respect to the results of the classic well-known sequential B algorithm promising results are achieved with the *Alarm*, *Insurance*, *Hailfinder* and *Mildew* networks.

However, a comparison with the proposed B3 algorithm (an adaptation of the classic hill-climbing) are not as good as expected. This fact may be due to the number of operators used by different algorithms. Where floating methods only perform both the arc addition and arc deletion operations, the B3 algorithm performs the arc addition, arc deletion and arc reversal. This reversing operator seems to provide an extra power in order to explore the search space. To confirm these suspicions, a comparison of floating methods with respect to a hill-climbing with arc addition and arc deletion should be carried out.

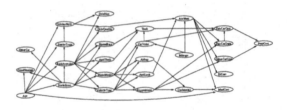

(a) *Insurance* structure

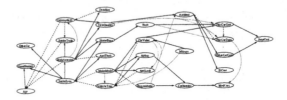

(b) Learnt network by SFFS

Fig. 7. The real *Insurance* network has 27 nodes and 52 arcs. The network structure learnt by SFFS, starting with an empty solution, has 49 arcs: 13 extra arcs (dotted line), 11 reverse arcs (dashed line) and 14 missing arcs.

Acknowledgments

This work is partially supported by the Department of Education, University and Research of the Basque Government and by the Ministry of Science and Technology under grants PI 1999-40 and TIC2001-2973-C05-03 respectively.

References

1. B. Abramson, J. M. Brown, W. Edwards, A. Murphy, and R. L. Winkler, 'Hailfinder: A Bayesian system for forecasting severe weather', *International Journal of Forecasting*, **12**, 57–71, (1996).
2. S. Acid, L. M. de Campos, and J. F. Huete, 'The search of causal orderings: a short cut for learning belief networks', in *Proceedings of the Sixth European Conference on Symbolic and Quantitative Approache to Reasoning with Uncertainty*, pp. 228–239, (2001).
3. I.A. Beinlinch, H.J. Suermondt, R.M. Chavez, and G.F. Coooper, 'The ALARM monitoring system: a case study with two probabilistic inference techniques for belief networks', in *Proceedings of the Second European Conference on Artificial Intelligence in Medicine*, pp. 247–257, (1989).
4. J. Binder, D. Koller, S. Russell, and K. Kanazawa, 'Adaptive probabilistic networks with hidden variables', *Machine Learning*, **29**(2–3), 213–244, (1997).

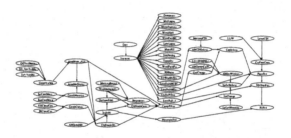

(a) *Hailfinder* structure

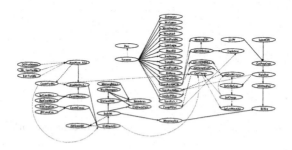

(b) Learnt network by SFFS

Fig. 8. The real *Hailfinder* network has 56 nodes and 66 arcs. The network structure learnt by SFFS, starting with an empty solution, has 69 arcs: 17 extra arcs (dotted line), 13 reverse arcs (dashed line) and 14 missing arcs.

5. R. Blanco, I. Inza, and P. Larrañaga, 'Learning Bayesian networks in the space of structures by estimation of distribution algorithms', *International Journal of Intelligent Systems*, **18**, 205–220, (2003).

6. C. Borgelt and R. Kruse, 'An empirical investigation of the K2 metric', in *Proceedings of the Sixth European Conference on Symbolic and Quantitative Approache to Reasoning with Uncertainty*, pp. 240–251, (2001).

7. W. Buntine, 'Theory refinement in Bayesian networks', in *Proceedings of the Seventh Conference on Uncertainty in Artificial Intelligence*, pp. 52–60, (1991).

8. W. Buntine, 'A guide to the literature on learning probabilistic networks from data', *IEEE Transactions on Knowledge and Data Engineering*, **8**(2), 195–210, (1996).

9. E. Castillo, J.M. Gutiérrez, and A.S. Hadi, *Expert Systems and Probabilistic Network Models*, Springer-Verlag, New York, 1997.

10. J. Cheng, R. Greiner, J. Kelly, D. Bell, and W. Liu, 'Learning Bayesian networks from data: an information-theory based approach', *Artificial Intelligence*, **137**(1–2), 43–90, (2002).

11. D.M. Chickering, 'Learning equivalence classes of Bayesian network structures', *Journal of Machine Learning Research*, **2**, 445–498, (2002).

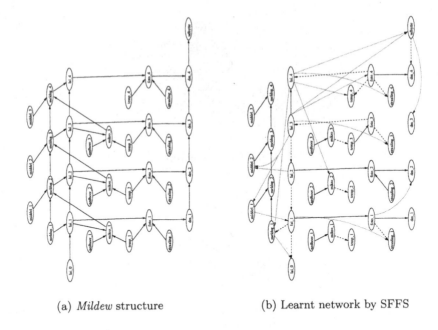

(a) *Mildew* structure   (b) Learnt network by SFFS

Fig. 9. The real *Mildew* network has 35 nodes and 46 arcs. The network structure learnt by SFFS, starting with an empty solution, has 46 arcs: 16 extra arcs (dotted line), 14 reverse arcs (dashed line) and 16 missing arcs.

12. D.M. Chickering, D. Geiger, and D. Heckerman, 'Learning Bayesian networks is NP-hard', Technical Report MSR-TR-94-17, Microsoft Research, Advanced Technology Division, Microsoft Corporation, Redmond, WA, (1994).

13. D.M. Chickering, D. Geiger, and D. Heckerman, 'Learning Bayesian networks: Search methods and experimental results', in *Preliminary Papers of the Fifth International Workshop on Artificial Intelligence and Statistics*, pp. 112–128, (1995).

14. G.F. Cooper and E.A. Herskovits, 'A Bayesian method for the induction of probabilistic networks from data', *Machine Learning*, **9**, 309–347, (1992).

15. C. Cotta and J. Muruzábal, 'Towards more efficient evolutionary induction of Bayesian networks', in *Parallel Problem Solving From Nature VIII. Lecture Notes in Computer Science 2439*, pp. 730–739, (2002).

16. R.G. Cowell, 'Conditions under which conditional independence and scoring methods lead to identical selection of Bayesian network models', in *Proceedings of the Seventeenth Conference on Uncertainty in Artificial Intelligence*, pp. 91–97, (2001).

17. A.P. Dawid, 'Conditional independence in statistical theory', *Journal of the Royal Statistics Society, Series B*, **41**, 1–31, (1979).

18. L.M. de Campos, J.M. Fernández-Luna, J.J. Gámez, and J.M. Puerta, 'Ant colony optimization for learning Bayesian networks', *International Journal on Artificial Reasoning*, **31**(3), 109–136, (2002).

19. L.M. de Campos and J.F. Huete, 'A new approach for learning belief networks using independence criteria', *International Journal of Approximate Reasoning*, **24**(1), 11–37, (2000).

20. L.M. de Campos and J.M. Puerta, 'Stochastic local algorithms for learning belief networks: searching in the space of the orderings, symbolic and quantitative approaches to reasoning with uncertainty', *Lecture Notes in Artificial Intelligence 2143*, 228–239, (2001).

21. N. Friedman, M. Goldszmidt, and A. Wyner, 'On the application of the bootstrap for computing confidence measures on features of induced Bayesian networks', in *Proceedings of the Seventh International Workshop on Artificial Intelligence and Statistics*, (1999).

22. N. Friedman and D. Koller, 'Being Bayesian about network structure: A Bayesian approach to structure discovery in Bayesian networks', *Machine Learning*, **50**(1-2), 95–125, (2003).

23. S. Gillispie and M. Perlman, 'Enumerating Markov equivalence classes of acyclic digraph models', in *Proceedings of the Seventeenth Conference on Uncertainty in Artificial Intelligence*, pp. 171–177, (2001).

24. J. Habrant, 'Structure learning of Bayesian networks from databases by genetic algorithms', in *Proceedings of the International Conference on Enterprise Information Systems*, (1999).

25. D. Heckerman, 'A tutorial on learning with Bayesian networks', Technical Report MSR-TR-95-06, Microsoft Advanced Technology Division, Microsoft Corporation, Seattle, Washington, (1995).

26. D. Heckerman and D. Geiger, 'Likelihoods and parameters priors for Bayesian networks', Technical Report MST-TR-95-54, Microsoft Advanced Technology Division, Microsoft Corporation, Seattle, Washington, (1995).

27. D. Heckerman, D. Geiger, and D.M. Chickering, 'Learning Bayesian networks: The combination of knowledge and statistical data', *Machine Learning*, **20**, 197–243, (1995).

28. E.A. Herskovits and G.F. Cooper, 'Kutató: An entropy-driven system for construction of probabilistic expert systems from database', in *Proceedings of the Sixth Conference on Uncertainty in Artificial Intelligence*, pp. 54–62, (1990).

29. A.L. Jensen and F.V. Jensen, 'MIDAS – An influence diagram for management of mildew in winther wheat', in *Proceedings of the Twelfeth Conference on Uncertainty in Artificial Intelligence*, pp. 349–356, San Francisco, (1996). Morgan Kaufmann.

30. F.V. Jensen, *Bayesian Networks and Decision Graphs*, Springer Verlag, 2001.

31. J. Kittler, 'Feature set search algorithms', in *Pattern Recognition and Signal Processing*, ed., C.H. Chen, pp. 41–60. Sithoff and Noordhoff, (1978).

32. T. Kočka and R. Castelo, 'Improved learning of Bayesian networks', in *Proceedings of the Seventeenth Conference on Uncertainty in Artificial Intelligence*, pp. 269–276, (2001).

33. P. Larrañaga, C.M.H. Kuijpers, R.H. Murga, and Y. Yurramendi, 'Learning Bayesian network structures by searching for the best ordering with genetic algorithms', *IEEE Transactions on Systems, Man, and Cybernetics - Part A: Systems and Humans*, **26**(4), 487–493, (1996).

34. P. Larrañaga, M. Poza, Y. Yurramendi, R.H. Murga, and C.M.H. Kuijpers, 'Structure learning of Bayesian networks by genetic algorithms: A performance analysis of control parameters', *IEEE Transactions on Pattern Analysis and Machine Intelligence*, **18**(9), 912–926, (1996).

35. H.B. Mann and D.R. Whitney, 'On a test of whether one of two random variables is stochastically larger than the other', *Annals of Mathematical Statistics*, **18**, 50–60, (1947).

36. J.W. Myers, K.B. Laskey, and T. Levitt, 'Learning Bayesian networks from incomplete data with stochastic search algorithms', in *Proceedings of the Fifteenth Conference on Uncertainty in Artificial Intelligence*, pp. 476–485, (1999).

37. J. Pearl, *Probabilistic Reasoning in Intelligent Systems*, Morgan Kaufmann, 1988.

38. J.M. Peña, J.A. Lozano, and P. Larrañaga, 'Unsupervised learning of Bayesian networks via estimation of distribution algorithms', in *Proceedings of the First European Workshop on Probabilistics Graphical Models*, pp. 144–151, (2002).

39. P. Pudil, J. Novovicova, and J. Kittler, 'Floating search methods in feature selection', *Pattern Recognition Letters*, **15**(1), 1119–1125, (1994).

40. R.W. Robinson, 'Counting unlabelled acyclic digraphs', in *Lecture Notes in Mathematics: Combinatorial Mathematics V*, pp. 28–43. Springer-Verlag, (1977).

41. G. Schwarz, 'Estimating the dimension of a model', *Annals of Statistics*, **7**(2), 461–464, (1978).

42. C.E. Shannon, 'A mathematical theory of communication', *The Bell System Technical Journal*, **27**, 379–423, (1948).

43. P. Spirtes, C. Glymour, and R. Scheines, *Causation, Prediction, and Search*, Lecture Notes in Statistics 81, Springer-Verlag, 1993.

44. S.D. Stearns, 'On selecting features for pattern classifiers', in *Third International Conference on Pattern Recognition*, pp. 71–75, (1976).

45. H. Steck, 'On the use of skeletons when learning in Bayesian networks', in *Proceedings of the Sixteenth Conference on Uncertainty in Artificial Intelligence*, pp. 558–565, (2000).

46. J. Tian, 'A branch and bound algorithm for MDL learning Bayesian networks', in *Proceedings of the Sixteenth Conference on Uncertainty in Artificial Intelligence*, pp. 580–588, (2000).

47. G. Wahba, 'Comments on 'monotone regression splines in action'', *Statistical Science*, **3**, 456–458, (1988).

A Graphical Meta-Model for Reasoning about Bayesian Network Structure

Luis M. de Campos[1], José A. Gámez[2], and J. Miguel Puerta[2]

[1] Dpto. de Ciencias de la Computación e I.A., ETSII - Universidad de Granada, 18071 - Granada.

[2] Dpto. de Informática, EPSA - Universidad de Castilla-La Mancha, 02071 - Albacete.

Abstract. When the amount of available data is small with respect to the problem size, many Bayesian networks can account for the data in a similar way. In these cases model averaging offers a framework which allows us to make better predictions than model selection. Selective model averaging directly uses a subset of these networks with high probability mass to reason over the probability of the structural features (arcs) which can appear in the problem domain network model. In this paper, instead of using this subset of networks to reason about the domain network structure, we propose to use the probability distribution over the structural features induced by these networks, in order to learn a new Bayesian network whose variables are the structural features which can appear in the problem domain network. This network can be considered as a higher level model or meta model, because it can be seen as an approximation of the whole space of Bayesian networks defined over the problem domain variables. This meta-model (Bayesian network) can be used to reason about the probability of structural features, as selective model averaging, but also as a decision support tool to be used by an expert in the problem domain.

1 Introduction

Bayesian Networks (BNs), also known as Probabilistic Belief Networks or Causal Networks, are knowledge representation tools capable of efficiently manage the dependence/independence relationships among the random variables that compose the problem domain we wish to model. This representation has two components: a) a graphical structure, or more precisely a directed acyclic graph (DAG), and b) a set of parameters, which together specify a joint probability distribution over the random variables [15]. In Bayesian networks, the graphical structure represents dependence and independence relationships. The parameters are a collection of conditional probability measures, which shape the relationships.

In the last decade the problem of learning or estimating a Bayesian network from data has received a great of deal of research into the uncertainty in artificial intelligence community. The most common approach to discover the structure of a BN is the combination of a machine learning algorithm (to induce networks from a data file) with the use of model selection (to select the network with the highest score). From the obtained model, an expert in the domain might study some relations between the problem variables, either by confirming some of the present arcs as causal relations or by disagreeing with them. This approach is reasonable when the amount of available data with respect to the problem size is substantial, because in these cases the highest score model is orders of magnitude more likely than any other [10]. On the other hand, when we have a small amount of data, many very different models can account for the data reasonably well. Then, instead of selecting a single model, it is more cautious to deal with a subset of high score models and to study the structural relations between variables from this subset, using model averaging.

In this paper we propose a new approach for evaluating the posterior probability of certain structural Bayesian network properties. We also use a subset of high score BN models but go a step further: we build a Bayesian network meta-model which tries to represent the interactions governing the structural relations between the problem variables. This meta-model can be seen as an approximated representation of the posterior probability on the whole space of BN models, given the available data.

The rest of the paper is structured as follows: we begin in Section 2 with the preliminaries, where we briefly describe some of the concepts and methods about learning Bayesian networks and Selective Model Averaging. In Section 3 we give a description of the proposed methodology for computing several structural network features. In Section 4 we present two examples to illustrate the methodology developed in the paper. Finally, Section 5 contains the concluding remarks.

2   Learning Bayesian Networks

A BN is a directed acyclic graph $G = (\mathbf{V}, \mathbf{E})$, where the set of nodes $\mathbf{V} = \{x_1, x_2, \ldots, x_n\}$ represents the system variables and $\mathbf{E}$, a set of arcs, represents the direct dependence relationships among the variables. A set of parameters is also stored for each variable in $\mathbf{V}$, which are usually conditional probability distributions. For each variable $x_i \in \mathbf{V}$ we have a family of conditional distributions $P(x_i|Pa(x_i))$, where $Pa(x_i)$ represents the parent set of the variable x_i. From these conditional distributions we can recover the joint distribution over $\mathbf{V}$:

$$P(x_1, x_2, \ldots, x_n) = \prod_{i=1}^{n} P(x_i|Pa(x_i)) \tag{1}$$

This expression represents a decomposition of the joint distribution. The dependence/independence relationships which make this decomposition possible are graphically encoded (through the d-separation criterion [15]) by means of the presence or absence of direct connections between pairs of variables.

Given a joint probability distribution P over $\mathbf{V}$, X and Y are probabilistically independent given Z, denoted as $I_P(X, Y|Z)$, if and only if the following statement holds: $P(\mathbf{x}|\mathbf{yz}) = P(\mathbf{x}|\mathbf{z})$, $\forall \mathbf{x}, \mathbf{y}, \mathbf{z}$ such that $P(\mathbf{yz}) > 0$, where $\mathbf{x}$, $\mathbf{y}$ and $\mathbf{z}$ denote a configuration for the subsets of variables X, Y and Z, respectively. If we denote by $\langle X, Y|Z \rangle_G$ the fact that Z d-separates X and Y in a DAG G, then G is said to be an *I-map* of a probability distribution P if $\langle X, Y|Z \rangle_G \implies I_P(X, Y|Z)$, and is *minimal* if no arc can be eliminated from G without violating the I-map condition. G is a *D-map* of P if $\langle X, Y|Z \rangle_G \impliedby I_P(X, Y|Z)$. When a DAG G is both an I-map and a D-map of P, it is said that G and P are *isomorphic* models. It is always possible to build a minimal I-map of any given probability distribution P, but some distributions do not admit an isomorphic model [15].

The problem of learning a BN can be stated as follows: given a *training set* $D = \{\mathbf{v}^1, \ldots, \mathbf{v}^m\}$ of instances of $\mathbf{V}$, find the BN that best matches D. The common approach to this problem is to introduce a scoring function, f, that evaluates each network with respect to the training data, and then to search for the best network according to this score. Different Bayesian and non-Bayesian scoring metrics can be used [1,9,12].

A desirable and important property of a metric is its decomposability in the presence of full data, i.e, the scoring function can be decomposed in the following way:

$$f(G : D) = \sum_{i=1}^{n} f(x_i, Pa(x_i)) \qquad (2)$$

where $f(x_i, Pa(x_i))$ depends on the count of the variables x_i and $Pa(x_i)$ in D, i.e, the number of instances in D that match each possible instantiation of x_i and $Pa(x_i)$. The decomposition of the metric is very important for the learning task: a local search procedure that changes one arc at each move can efficiently evaluate the improvement obtained by this change, because it can reuse most of the computations made in previous stages (only the statistics corresponding to the variables whose parent sets have been modified need to be recomputed). One example is a greedy hill-climbing method that at each step performs the local change yielding the maximal gain, until it reaches a local maximum of the scoring function. The usual choices for local changes in the space of DAGs are arc addition, arc deletion and arc reversal [9]. As this procedure is trapped in the first local maximum it reaches, several methods for avoiding this situation have been used, such as stochastic hill-climbing (with random restart [9]), Variable Neighborhood Search [6,7], Genetic Algorithms [13], Simulated Annealing [9], Tabu Search [2], Ant Colonies [5], etc.

2.1   Model Selection

When a substantial amount of data is available for the learning process, then we should take the BN with the highest score as our model for the structural representation of the problem domain. In this case, and due to the presence of a great amount of data, we/the expert should agree with the relations shown in the selected model (BN). In this framework, known as *Model Selection*, we use the selected model as the *true* model for the data D.

2.2   Selective Model Averaging

When the previous hypothesis does not hold, i.e., when we have a small amount of data with respect to the size of the model, then there are probably many models which may account for the data in a reasonable way. In these cases model selection does not seem to be the right choice. Model averaging provides a framework for reasoning about certain hypotheses over data. If F is the hypothesis of interest, then its posterior distribution given the data D is:

$$P(F|D) = \sum_{m=1}^{M} P(F|G_m, D)P(G_m|D) \qquad (3)$$

This is the average of the posterior distributions of the hypothesis under each of the possible BNs, weighted by the posterior model probabilities. In equation 3, $\mathcal{G} = \{G_1, G_2, \ldots, G_M\}$ is the set of all the possible BNs with n variables. Several authors have used this framework in order to improve the predictive reasoning task with respect to the model selection framework [14]. On the other hand, this framework has also been used for reasoning over the structure of the model [10,8]. In this paper we are only interested in the second process, i.e., model averaging-based structural reasoning.

For example, if we have a prior hypothesis F about the structure of our network, as could be the absence or presence of a direct relation between variables x_i and x_j, then we could be interested in obtaining the probability $P(F|D)$, being $F = \{x_i - x_j\}$ the possibility of having a direct relation between the two variables x_i and x_j. This relation can be:

$$\{x_i - x_j\} = \{x_i \leftrightarrow x_j, x_i \leftarrow x_j, x_i \rightarrow x_j\},$$

that is to say, F is a discrete variable with three exhaustive states: ($\leftrightarrow$) no direct connection exists between the two variables; ($x_i \rightarrow x_j$) there is a direct connection from x_i to x_j; and ($x_i \leftarrow x_j$) indicating the presence of a direct connection from x_j to x_i. Of course, the probability of having a direct connection between the two variables can be calculated as the sum of the probabilities for the last two states of F ($\leftarrow, \rightarrow$).

Following the notation introduced by [8] we will refer to this kind of hypotheses (arcs) as *structural features*. Therefore, Equation 3 can be rewritten as:

$$P(F|D) = \sum_{m=1}^{M} F(G_m)P(G_m|D) \qquad (4)$$

where $F(G_m)$ is 1 if the feature F appears in G_m and 0 otherwise.

Exhaustive enumeration over the set of possible BNs structures is feasible only for tiny domains. In the rest of cases, several authors have proposed to approximate this exhaustive enumeration by finding a subset $\mathcal{G}_h \subset \mathcal{G}$ of structures with a high score, and then the posterior distribution of F is estimated as follows:

$$\hat{P}(F|D) = \sum_{G_m \in \mathcal{G}_h} F(G_m)\hat{P}(G_m|D) \tag{5}$$

where

$$\hat{P}(G_i|D) = \frac{P(G_i|D)}{\sum_{G_m \in \mathcal{G}_h} P(G_m|D)} \tag{6}$$

This approach is known as *Selective Model Averaging*. Its implementation presents several difficulties, being the hard one that of searching for the subset $\mathcal{G}_h$ of BNs with high score.

By using a Bayesian scoring metric (which measures network's posterior probabilities), then Equation 6 can be rewritten as follows:

$$\hat{P}(G_i|D) = \frac{f(G_i : D)}{\sum_{G_m \in \mathcal{G}_h} f(G_m : D)} \tag{7}$$

Example 1: Let us consider a problem with three variables: x_1, x_2 and x_3. Thus, the set of possibles structural features (arcs) is: $F_1 = \{x_1 - x_2\}$, $F_2 = \{x_1 - x_3\}$ and $F_3 = \{x_2 - x_3\}$. Suppose also that we have a database D over these variables and a metric $f(G : D)$. Then, we can obtain an empirical distribution over $\mathcal{G}$, $\hat{P}(G|D)$ from Equation 7.

$\hat{P}(G|D)$	F_1	F_2	F_3
0.20	$x_1 \leftarrow x_2$	$x_1 \not\leftrightarrow x_3$	$x_2 \leftarrow x_3$
0.20	$x_1 \rightarrow x_2$	$x_1 \not\leftrightarrow x_3$	$x_2 \rightarrow x_3$
0.05	$x_1 \rightarrow x_2$	$x_1 \not\leftrightarrow x_3$	$x_2 \leftarrow x_3$
0.20	$x_1 \leftarrow x_2$	$x_1 \not\leftrightarrow x_3$	$x_2 \rightarrow x_3$
0.15	$x_1 \rightarrow x_2$	$x_1 \rightarrow x_3$	$x_2 \leftarrow x_3$
0.15	$x_1 \rightarrow x_2$	$x_1 \leftarrow x_3$	$x_2 \leftarrow x_3$
0.01	$x_1 \not\leftrightarrow x_2$	$x_1 \leftarrow x_3$	$x_2 \leftarrow x_3$
0.02	$x_1 \rightarrow x_2$	$x_1 \leftarrow x_3$	$x_2 \not\leftrightarrow x_3$
0.02	$x_1 \rightarrow x_2$	$x_1 \not\leftrightarrow x_3$	$x_2 \not\leftrightarrow x_3$
0.0	*	*	*

Table 1. A possible approximation for the probability distribution $\hat{P}(G|D)$. The last tuple indicates that the probability of the remaining graph configurations is 0.0

From this distribution and using equation 5 we can compute the probability of any structural feature, or even the combination of some of them. For example, we can compute $\hat{P}(F_1 = \{x_1 \leftarrow x_2\}|D) = 0.40$.

This is the most frequent use of selective model averaging; however, we think that this model/distribution can be exploited in order to obtain more information about the possible arcs in the BN structure. As an example, let us suppose that the following independence assertion is supported by D: x_1 *and* x_3 *are conditionally independent given* x_2. After the selective model averaging process, we can compute $P(x_1 - x_2|D)$, but what if we know that the correct relation is $x_1 \to x_2$? In this case, either the arc $x_2 \to x_3$ should increase (and the arc $x_3 \to x_2$ should decrease) its probability, or the existence of arcs linking x_1 and x_3 as well x_2 and x_3 should be more probable, in order to assure that we have a *minimal I-map* of D.

In the next section we show how this kind of information can be obtained and exploited by using a Graphical Meta-Model.

3   Learning a Graphical Meta-Model

Our idea is to see the distribution $\hat{P}(G|D)$ as a joint distribution over a new set of variables, which are just the structural features: Note that the set $\mathcal{G}^*$ of directed graphs with n variables can be expressed as the Cartesian product of $n(n-1)/2$ structural features, $\mathcal{G}^* = \mathcal{F} = F_1 \times F_2, \ldots, \times F_{n(n-1)/2}$, each variable F_k referring to a different pair of variables. Thus, the set of DAGs with n variables, $\mathcal{G}$, is a subset of $\mathcal{F}$. Therefore, instead of using Equation 4, we can express $\hat{P}(F|D)$ in the following way:

$$\hat{P}(F|D) = \sum_{F \in \mathbf{F}} \hat{P}(\mathbf{F}|D) \tag{8}$$

i.e., we sum over all the configurations $\mathbf{F}$ of $\mathcal{F}$ that contain the feature F. Now, we are going to search for some structure (a Bayesian network structure) within the joint distribution $\hat{P}(\mathbf{F}|D)$. In this way we could compute efficiently the probabilities of different structural features.

As we have previously said, the enumeration of all possible BNs with n variables is not feasible, so we will consider for the averaging process just a subset of graphs with a high score $f(G : D)$. In our opinion, this set of graphs could be enough to extract the main structural features among the variables, and then to use this information to approximate the joint probability distribution $P(\mathbf{F}|D)$.

If we use a *heuristic* search engine during our BN learning process then, while searching across the search space, different BN structures will be visited. From this set of visited networks, we are interested in those with a high score and having significant structural differences among them. We can save these DAGs and their associated probabilities in a table like a database of cases. In fact, if we observe the example in Table 1, we realize that this table looks like a pseudo meta-database which has structural features as possible columns or variables. This higher level database can be used to learn a BN having structural features as variables, which will work as a meta-model encoding relations among the possible arcs linking the domain variables.

Thus, the process can be divided in three stages:

1. From our problem domain database D we obtain a set of networks $\mathcal{G}_h$ with high score by using a heuristic learning method.
2. From $\mathcal{G}_h$ we compute $\hat{P}(\mathbf{F}|D)$, and using this probability distribution we obtain a higher level database. The number of cases for each instance G_i in this new database will be directly proportional to its probability $\hat{P}(G_i|D)$. As an example, if we consider the probability distribution expressed in Table 1, the DAG $x_1 \leftarrow x_2 \leftarrow x_3$ should have 20 cases of a total of 100.
3. Taking the high-level generated database as the input, we learn (by using any BN learning algorithm) the network M_{BN}, which represents our meta-model. Notice that when learning M_{BN} we are looking for the BN which best matches the probability distribution $\hat{P}(\mathbf{F}|D)$. Figure 1 shows a possible meta-model for the problem described in example 1.

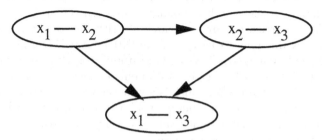

Fig. 1. Meta-Bayesian Network

Once we have built the meta-model M_{BN}, it can be used for reasoning about the structure of the Bayesian network defined over the problem domain variables. Thus, by using standard (exact or approximate) propagation algorithms, we can efficiently compute the probability of any structural feature $P(F_i|D)$ and, also, answer other kinds of queries, as $P(F_i|D, \mathbf{F_j})$, where $\mathbf{F_j} \subset \mathcal{F}$.

In this way, an expert in the problem domain could see this meta Bayesian network as a tool for reasoning about structural features, and to use it to analyse the probabilities of some arcs or to study the impact in the structure of the BN defined over the domain variables, when entering some arcs as evidence.

For example, if the domain expert agrees with some structural features $\mathbf{F_o}$, s/he could instantiate them as evidence, and analyse the marginal probability of the remaining structural features given by the meta-model. In our example, as we have

$$\hat{P}(x_1 - x_2|D) = \begin{cases} \hat{P}(x_1 \leftarrow x_2|D) = 0.4 \\ \hat{P}(x_1 \rightarrow x_2|D) = 0.59 \\ \hat{P}(x_1 \nleftrightarrow x_2|D) = 0.01 \end{cases}$$

the expert could agree with the relation $x_1 \leftarrow x_2$ and enter it as evidence, then the meta-model M_{BN} can compute the probability $\hat{P}(x_2 - x_3 | x_1 \leftarrow x_2, D)$ or any other structural feature.

3.1   Computational issues

A tool has been implemented in order to carry out the kind of analysis described in the previous section. The tool has been implemented by using the Elvira software (http://leo.ugr.es/~elvira), although it works as an independent application. In this subsection we will describe the more relevant aspects about the implemented tool.

The search engine used in the learning stages is an algorithm based on the *Variable Neighbourhood Search* (VNS) meta-heuristic, and it is focused on learning in the space of orderings [7]. As VNS is a multiple restarts local search, we will obtain several local maxima with structural differences. On the other hand, as the search is mainly focused on the space of orderings, the movements carried out by the search algorithm in this space will yield, in general, very different structures in the DAGs visited. During this search stage all the individuals/DAGs visited are stored (together with their metric value) in a hash table. The metric used in our score+search learning process has been the BDeu metric [9], which is a well known *decomposable* and *score equivalent* metric. However, in order to avoid precision problems when dealing with BDeu, we divide the obtained value by using the constant $\log N$, being N the number of cases in the database.

The next task to be accomplished is the one of selecting the subset $\mathcal{G}_h$ of DAGs. As an initial approach we have considered a simple strategy: if v is the score of the best DAG included in the hash table, then we include in $\mathcal{G}_h$ every DAG G_i such that the absolute value of its metric value, $|v_i|$, is in the interval $[|v| \cdot (1 - \epsilon), |v|]$. In our experiments we have chosen $\epsilon = 0.005$. In practice we have found that this simple selection yields a reasonable number of DAGs, with significant structural differences among them.

The next stage in the discovering process consists in learning the meta-model M_{BN} from the set of selected DAGs $\mathcal{G}_h$. First, we should note that if our problem domain has n variables, then the meta-model will have $n(n-1)/2$, i.e., $O(n^2)$. Thus, the new learning problem will probably be computationally quite expensive. In our experiments we realized that, in general, in this higher level learning process, as the amount of data is small, a great number of relations (arcs) are not supported by the data, i.e., for this kind of relations we get an extreme probability ($\hat{P}(x_i \leftrightarrow x_j | D) = 1$). Because of this, for this learning process we only take into account those variables whose entropy is greater than 0 ($H(P(x_i - x_j | D)) > 0$) (otherwise this variable is conditionally independent from the remaining variables). By carrying out this preselection of variables, the size of the problem of learning the meta-model can be reduced drastically. For example, in the ALARM domain [4], which has 37 variables, we would have 666 variables involved in the higher level learning task; however, by using the previous preselection only around 80 variables are considered.

Once we have selected the set of variables (structural features) to be considered in the meta-model learning process, we have to choose the learning algorithm to be used. As we have already commented, any BN learning algorithm could be used in this stage. In our case, we have chosen as our main algorithm a greedy hill-climbing method with the classical operators of deletion, addition and reversal of arcs. However, if the number of variables is too high we approximate the structure of the meta-model by a tree-shaped Bayesian network, which is learned by using the Chow-Liu algorithm [3].

Finally, propagation algorithms running over the learned meta-model are used to answer queries about the network structure. In this case and due to the variety of available algorithms in the Elvira software, exact and approximate propagation can be carried out depending on the meta-model complexity. Two kinds of inference tasks are used in our tool: probability propagation and abduction (maximum a posteriori explanation, MPE). Probability propagation is used to study the marginal probability of each structural feature after entering some evidence. Abduction/MPE is used to obtain the network structure which better explains the evidence.

4   Examples

In this section we will use two toy examples to illustrate the proposed methodology. In these two examples we focus on certain relations between the variables and then we monitorize the evolution of the probabilities of these relations when some evidence is entered.

Example 4.1: We have edited a BN with four variables A, B, C, D, and with the chain structure shown in figure 2. The conditional probability distributions have been fixed such that there is a strong dependence between the variables linked by an arc. *Probabilistic Logic Sampling* [11] has been used to generate a database with 100 cases.

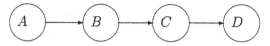

Fig. 2. Original BN for the Example 4.1.

After the learning process and the selection of $\mathcal{G}_{\mathbf{h}}$ the meta-model M_{BN} containing 6 variables (which is displayed in Figure 3) is learned. Without any extra information (evidence) the *a priori* probabilities of each structural feature are shown in the first column of Table 2.

By looking in M_{BN} for the configuration of highest probability (MPE) we obtain the network shown in Figure 4. As we can see, in this case we obtain

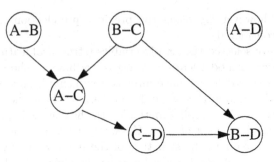

Fig. 3. Meta-model for the Example 4.1.

$\hat{P}(F_i\|D)$		$\hat{P}(F_i\|\mathbf{F_j}, D)$	
$A - B = \begin{cases} \leftrightarrow = 0.00 \\ \rightarrow = 0.48 \\ \leftarrow = 0.52 \end{cases}$		$A - B = \begin{cases} \leftrightarrow = 0.00 \\ \rightarrow = 1.00 \\ \leftarrow = 0.00 \end{cases}$	
$A - C = \begin{cases} \leftrightarrow = 0.76 \\ \rightarrow = 0.08 \\ \leftarrow = 0.16 \end{cases}$		$A - C = \begin{cases} \leftrightarrow = 0.49 \\ \rightarrow = 0.17 \\ \leftarrow = 0.34 \end{cases}$	
$A - D = \begin{cases} \leftrightarrow = 1.00 \\ \rightarrow = 0.00 \\ \leftarrow = 0.00 \end{cases}$		$A - D = \begin{cases} \leftrightarrow = 1.00 \\ \rightarrow = 0.00 \\ \leftarrow = 0.00 \end{cases}$	
$B - C = \begin{cases} \leftrightarrow = 0.00 \\ \rightarrow = 0.49 \\ \leftarrow = 0.51 \end{cases}$		$B - C = \begin{cases} \leftrightarrow = 0.00 \\ \rightarrow = 0.49 \\ \leftarrow = 0.51 \end{cases}$	
$B - D = \begin{cases} \leftrightarrow = 0.75 \\ \rightarrow = 0.17 \\ \leftarrow = 0.08 \end{cases}$		$B - D = \begin{cases} \leftrightarrow = 0.75 \\ \rightarrow = 0.17 \\ \leftarrow = 0.08 \end{cases}$	
$C - D = \begin{cases} \leftrightarrow = 0.00 \\ \rightarrow = 0.53 \\ \leftarrow = 0.47 \end{cases}$		$C - D = \begin{cases} \leftrightarrow = 0.00 \\ \rightarrow = 0.58 \\ \leftarrow = 0.42 \end{cases}$	

Table 2. Example 4.1. Marginal prior probabilities and marginal posterior probabilities. $\mathbf{F_j} = \{A \rightarrow B\}$

a network which is different from the original one, but has the same set of conditional independence relations.

Now, let us suppose that an expert in the domain introduces the arc $A \rightarrow B$ as evidence, because s/he is sure of this causal relation. The right column in Table 2 shows the posterior probabilities for the structural features, after entering the evidence and propagating it over M_{BN}. The main change in these probabilities occurs in the arc $A - C$, decreasing considerably the probability of not including an arc between these two variables ($\leftrightarrow$). In this case, if we get the problem domain network by picking the state of maximum probability independently for each structural feature, then we get the network $A \rightarrow B \leftarrow C \rightarrow D$, which clearly does not satisfy the I-map property because of the marginal independence relation between A and C. However, if we act in a global way by looking for the MPE in M_{BN}, then we get the structure of the original network (Figure 2).

Finally, if we also observe $A \leftrightarrow C$, then the posterior probability of arc $B \rightarrow C$ is 1.0. In this case the MPE is still the original BN.

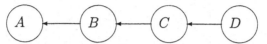

Fig. 4. Example 4.1. Prior maximum configuration of $\hat{P}(\mathbf{F}|D)$ (Maximum configuration without evidence).

Example 4.2: In this case we will consider the network in Figure 5, containing five variables A, B, C, D, E.

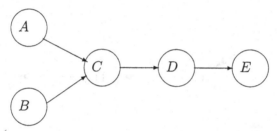

Fig. 5. Original BN for the Example 4.2.

The learned meta-model is displayed in Figure 6. It is interesting to note that the marginal independences represented in the meta-model BN could give us some useful information. In this case the only marginal independence is between the edges $A - B$ and $D - E$, which could be related with the "distance" between these elements in the true network. Probably in larger BNs we would find much more marginal independence relationships in the meta-model network. We conjecture that a large and reasonably sparse BN would give rise to a meta-model metwork composed of overlapping clusters of edges, which do not influence very much among each other.

The first column of Table 3 shows the a priori marginal probabilities for the structural features. In this case the network obtained by looking for the MPE in the learned meta-model, which does not coincide with the original network, is shown in Figure 7.

If we proceed as in the previous example and $B \rightarrow C$ is entered as evidence, then the impact of this observation can be seen in the second column of Table 3. Again, using abductive reasoning in order to look for the structure which best supports the entered evidence, we obtain the original BN structure.

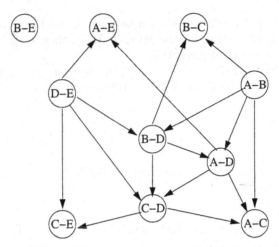

Fig. 6. Meta-model for the Example 4.2.

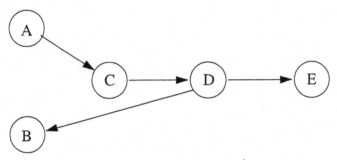

Fig. 7. Example 4.2. Prior maximum configuration of $\hat{P}(\mathbf{F}|D)$ (Maximum configuration without evidence).

Going further in this example, we force a complex structure which is not present in the original network. Thus, we fixed $\{A \rightarrow C, B \rightarrow C, D \rightarrow C\}$ as evidence (notice that we are reversing the arc $C \rightarrow D$ with respect to the original network). Table 4 shows the probabilities of the considered structural features before and after propagating the evidence. As we can see in Figure 8, which displays the MPE network, the system adds the arcs $\{A \rightarrow D, B \rightarrow D\}$. In this way, it seems us that the system tries to modify the structure in order to adapt it in the best way to the given evidence. In this case it tries to preserve the minimal I-map condition and the effect of this change is the same as an arc reversal operation.

5   Concluding Remarks

In this paper we have considered the problem of Bayesian network structural discovering from data. We start from the basis of having a non substantial

$\hat{P}(F_i|D)$			$\hat{P}(F_i|\mathbf{F_j}, D)$		
$A - C =$	$\leftrightarrow = 0.36$		$A - C =$	$\leftrightarrow = 0.00$	
	$\rightarrow = 0.30$			$\rightarrow = 0.50$	
	$\leftarrow = 0.34$			$\leftarrow = 0.50$	
$B - C =$	$\leftrightarrow = 0.84$		$B - C =$	$\leftrightarrow = 0.00$	
	$\rightarrow = 0.10$			$\rightarrow = 1.00$	
	$\leftarrow = 0.06$			$\leftarrow = 0.00$	
$C - D =$	$\leftrightarrow = 0.13$		$C - D =$	$\leftrightarrow = 0.00$	
	$\rightarrow = 0.38$			$\rightarrow = 0.79$	
	$\leftarrow = 0.49$			$\leftarrow = 0.21$	
$A - B =$	$\leftrightarrow = 0.49$		$A - B =$	$\leftrightarrow = 1.00$	
	$\rightarrow = 0.23$			$\rightarrow = 0.00$	
	$\leftarrow = 0.28$			$\leftarrow = 0.00$	
$A - D =$	$\leftrightarrow = 0.41$		$A - D =$	$\leftrightarrow = 0.91$	
	$\rightarrow = 0.32$			$\rightarrow = 0.06$	
	$\leftarrow = 0.27$			$\leftarrow = 0.03$	
$B - D =$	$\leftrightarrow = 0.52$		$B - D =$	$\leftrightarrow = 0.74$	
	$\rightarrow = 0.15$			$\rightarrow = 0.26$	
	$\leftarrow = 0.33$			$\leftarrow = 0.00$	
$D - E =$	$\leftrightarrow = 0.00$		$D - E =$	$\leftrightarrow = 0.00$	
	$\rightarrow = 0.72$			$\rightarrow = 0.95$	
	$\leftarrow = 0.28$			$\leftarrow = 0.05$	
$A - E =$	$\leftrightarrow = 0.91$		$A - E =$	$\leftrightarrow = 1.00$	
	$\rightarrow = 0.07$			$\rightarrow = 0.00$	
	$\leftarrow = 0.02$			$\leftarrow = 0.00$	
$B - E =$	$\leftrightarrow = 1.00$		$B - E =$	$\leftrightarrow = 1.00$	
	$\rightarrow = 0.00$			$\rightarrow = 0.00$	
	$\leftarrow = 0.00$			$\leftarrow = 0.00$	
$C - E =$	$\leftrightarrow = 0.92$		$C - E =$	$\leftrightarrow = 0.96$	
	$\rightarrow = 0.06$			$\rightarrow = 0.03$	
	$\leftarrow = 0.02$			$\leftarrow = 0.01$	

Table 3. Example 4.2. Marginal prior probabilities and marginal posterior probabilities. $\mathbf{F_j} = \{B \rightarrow C\}$

amount of data with respect to the problem size. Selective model averaging is adopted in order to have an approximation of the probability distribution of structural features given the data. The main novelty of the paper resides on the fact of learning or computing a meta-model (or meta-BN) which constitutes an approximation of that probability distribution. From this meta-model we can perform structural discovery by using standard BN propagation techniques, as probability propagation and abductive inference. The meta-model can be used not only for analysing the prior o posterior probabilities of the structural features, but also to study how the network structure changes in order to absorb the entered evidence (arcs). Searching for the MPE is used to accomplish this task.

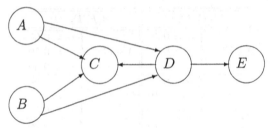

Fig. 8. Example 4.2. Maximum a posteriori configuration of $\hat{P}(\mathbf{F} \setminus \mathbf{F_j} | \mathbf{F_j}, D)$, where $\mathbf{F_j} = \{A \to C, B \to C, D \to C\}$

For the future we will carry out a serious experimental evaluation of the models obtained from the process presented in the paper rather than toy examples. To do this, we have to consider other ways of selecting the models in $\mathcal{G}_h$ and other searching techniques. In this latter aspect, we plan to use Markov Chain Monte Carlo methods in the search process, because of the theoretical advantages exhibited by these methods [8]. Besides, we think that it could be of interest to be able of computing the probability of more complex structural features, as the Markov Blanket of a given variable. Finally, we also plan to work in the scalability of learning algorithm when a great number of variables has to be considered, as it could occur when learning the meta-model.

Acknowledgments

This work has been supported by the Dirección General de Investigación e Innovación de la Junta de Comunidades de Castilla-La Mancha under Project num. PBC-02-002.

References

1. S. Acid and L.M. de Campos. A hybrid methodology for learning belief nertworks: Benedict. *International Journal of Approximate Reasoning*, 27(3):235–262, 2001.
2. R. R. Bouckaert. *Bayesian Belief Networks: From Construction to Inference*. PhD thesis, University of Utrecht, 1995.
3. C. Chow and C. Liu. Approximating discrete probability distributions with dependence trees. *IEEE transactions on Information Theory*, 14:462–467, 1968.
4. G.F. Cooper and E. Herskovits. A Bayesian method for the induction of probabilistic networksmfrom data. *Machine Learning*, 9:309–347, 1992.
5. L.M. de Campos, J.A. Gámez and J.M. Puerta. Ant colony optimization for learning Bayesian networks. *International Journal of Approximate Reasoning*, 31(3):291–311, 2002.
6. L.M. de Campos and J.M. Puerta. Stochastic local and distributed search algorithms for learning belief networks. In *Proceedings of the Third International Symposium on Adaptive Systems: Evolutionary Computation and Probabilistic Graphical Models*, pages 109–115, 2001.

$\hat{P}(F_i\|D)$		$\hat{P}(F_i\|\mathbf{F_j}, D)$	
$A - C =$	$\leftrightarrow = 0.36$ $\rightarrow = 0.30$ $\leftarrow = 0.34$	$A - C =$	$\leftrightarrow = 0.00$ $\rightarrow = 1.00$ $\leftarrow = 0.00$
$B - C =$	$\leftrightarrow = 0.84$ $\rightarrow = 0.10$ $\leftarrow = 0.06$	$B - C =$	$\leftrightarrow = 0.00$ $\rightarrow = 1.00$ $\leftarrow = 0.00$
$C - D =$	$\leftrightarrow = 0.13$ $\rightarrow = 0.38$ $\leftarrow = 0.49$	$C - D =$	$\leftrightarrow = 0.00$ $\rightarrow = 0.00$ $\leftarrow = 1.00$
$A - B =$	$\leftrightarrow = 0.49$ $\rightarrow = 0.23$ $\leftarrow = 0.28$	$A - B =$	$\leftrightarrow = 1.00$ $\rightarrow = 0.00$ $\leftarrow = 0.00$
$A - D =$	$\leftrightarrow = 0.41$ $\rightarrow = 0.32$ $\leftarrow = 0.27$	$A - D =$	$\leftrightarrow = 0.00$ $\rightarrow = 0.67$ $\leftarrow = 0.33$
$B - D =$	$\leftrightarrow = 0.52$ $\rightarrow = 0.15$ $\leftarrow = 0.33$	$B - D =$	$\leftrightarrow = 0.00$ $\rightarrow = 1.00$ $\leftarrow = 0.00$
$D - E =$	$\leftrightarrow = 0.00$ $\rightarrow = 0.72$ $\leftarrow = 0.28$	$D - E =$	$\leftrightarrow = 0.00$ $\rightarrow = 1.00$ $\leftarrow = 0.00$
$A - E =$	$\leftrightarrow = 0.91$ $\rightarrow = 0.07$ $\leftarrow = 0.02$	$A - E =$	$\leftrightarrow = 1.00$ $\rightarrow = 0.00$ $\leftarrow = 0.00$
$B - E =$	$\leftrightarrow = 1.00$ $\rightarrow = 0.00$ $\leftarrow = 0.00$	$B - E =$	$\leftrightarrow = 1.00$ $\rightarrow = 0.00$ $\leftarrow = 0.00$
$C - E =$	$\leftrightarrow = 0.92$ $\rightarrow = 0.06$ $\leftarrow = 0.02$	$C - E =$	$\leftrightarrow = 1.00$ $\rightarrow = 0.00$ $\leftarrow = 0.00$

Table 4. Example 4.2. Marginal prior probabilities and marginal posterior probabilities. $\mathbf{F_j} = \{B \rightarrow C, D \rightarrow C, A \rightarrow C\}$

7. L.M. de Campos and J.M. Puerta. Stochastic local search algorithms for learning belief networks: Searching in the space of orderings. In *ECSQARU – Lecture Notes in Artificial Intelligence*, volume 2143, pages 228–239. Springer-Verlag, 2001.
8. N. Friedman and D. Koller. Being Bayesian about network structure. In *Proceedings of the 16th Conference on Uncertainty in Artificial Intelligence*, pages 201–210, San Mateo, 2000. Morgan Kaufmann.
9. D. Heckerman, D. Geiger, and D.M. Chickering. Learning bayesian networks: The combination of knowledge and statistical data. *Machine Learning*, 20:197–243, 1995.
10. D. Heckerman, C. Meek, and G. Cooper. A Bayesian approach to causal discovery. Technical Report MSR-TR-97-05, Microsoft Research, Advanced Tech-

nology Division, 1997.

11. M. Henrion. Propagating uncertainty by logic sampling in bayes networks. In J. Lemmer and L.N. Kanal, editors, *Uncertainty in Artificial Intelligence, 2*, pages 149–164. North Holland, Amsterdam, 1988.

12. W. Lam and F. Bacchus. Learning bayesian belief networks. an approach based on the mdl principle. *Computational Intelligence*, 10(4):269–293, 1994.

13. P. Larrañaga, M. Poza, Y. Yurramendi, R. Murga, and C. Kuijpers. Structure learning of Bayesian networks by genetic algorithms. A perfomance analysis of control parameters. *IEEE Transactions on Pattern Analysis and Machine Intelligence*, 18(9):912–926, 1996.

14. D. Madigan and E. Raftery. Model selection and accounting for model uncertainty in graphical models using Occam's window. *Journal Americal Statistical Association*, 89:1535–1546, 1994.

15. J. Pearl. *Probabilistic reasoning in intelligent systems: networks of plausible inference*. Morgan Kaufmann, San Mateo, 1988.

Restricted Bayesian Network Structure Learning

Peter J.F. Lucas

Institute for Computing and Information Sciences, University of Nijmegen,
Toernooiveld 1, 6525 ED Nijmegen, The Netherlands
E-mail: peterl@cs.kun.nl

Abstract. Learning the structure of a Bayesian network from data is a difficult problem, as its associated search space is superexponentially large. As a consequence, researchers have studied learning Bayesian networks with a fixed structure, notably naive Bayesian networks and tree-augmented Bayesian networks, which involves no search at all. There is substantial evidence in the literature that the performance of such restricted networks can be surprisingly good. In this paper, we propose a restricted, polynomial time structure learning algorithm that is not as restrictive as both other approaches, and allows researchers to determine the right balance between classification performance and quality of the underlying probability distribution. The results obtained with this algorithm allow drawing some conclusions with regard to Bayesian-network structure learning in general.

1 Introduction

Health care is currently in the process of being transformed, albeit slowly, by the introduction of information technology into patient care. It is expected that one of the most significant consequences of this will be the future availability of huge quantities of clinical data, at the moment still hidden in paper records, for the purpose of data mining and knowledge discovery. This will allow exploiting these data in the construction of decision-support systems; these systems may play a role in improving the quality of patient care. Many see Bayesian networks as the appropriate tools in this context, as they are intuitive, even to the novice, and allow for incorporating qualitative knowledge of (patho)physiological mechanisms as well as of statistical information that is amply available in medicine.

In the past decade, much emphasis has been put on building Bayesian networks based on (medical) expert knowledge [1,15,16,21,22]. Unfortunately, building a Bayesian network for a realistic medical problem in this way may take years. With the future availability of huge quantities of clinical data in mind, learning Bayesian-network structures from data becomes appealing, raising the important question as to when structure learning will pay off. One would expect that the more faithful a Bayesian-network's structure is in reflecting the statistical independences hidden in the underlying data, the better its performance. However, previous research by Domingos and Pazzani has shown that, however crude as Bayesian networks, naive Bayesian classifiers often outperform more sophisticated network structures as well as other types of classifiers [10]. In addition, Friedman et al. [11], and Cheng and Greiner [4] have shown that so-called *tree-augmented Bayesian networks* (TANs), which in comparison to naive Bayesian classifiers incorporate extra dependences among features in the form of a tree structure often outperform naive Bayesian classifiers. These results explain why naive Bayesian classifiers and TANs are looked upon by researchers as being state-of-the-art classifier models.

There is a problem, though, with this research; most of the conclusions are based on experimental results obtained with datasets from the UCI Machine Learning Repository. It is not easy to judge the quality of these datasets, but since the majority of these datasets are medical in nature, and as the author is a medical doctor it is at least possible to say to what extent these medical datasets can be considered to be characteristic for the field of medicine. One observation is that even for such complicated disorders as diabetes and breast cancer, the available datasets contain only a few attributes (8 and 10, respectively). It appears that many of these datasets are biased as medically significant variables have often not been included. For the purpose of studying Bayesian-network learning these datasets are therefore less suitable, because many of the relationships between variables that participate in the disorder's causative mechanisms cannot be revealed as the relevant variables are missing. Unfortunately, when it comes to comparing results with other work, using datasets from the UCI repository is almost inevitable.

In this paper, we investigate the hypothesis that both naive Bayesian networks and TANs can be seen as end-points in a more general construction process, and that somewhere between these two extremes better models are to be found. In addition, the effects of entering partial evidence on the performance of networks are studied. This mirrors the practical situation that when a model is actually used in medical practice, not all patient data will be readily available.

For learning and evaluation purposes, we used a clinical research dataset of diseases of the liver and biliary tract [27]. This dataset is uncommon in that it contains data of a significant number of patients, with each patient described by a significant number of attributes. This dataset has been put together with great care. In order to be able to compare our results with results obtained by others, two datasets (the lymphoma and hepatitis datasets) were selected from the UCI Machine Learning Repository. These two were considered to be the best medical datasets available in this repository, even though still troublesome from a medical viewpoint.

Bayesian networks and their variants are introduced in the next section in the context of biomedical research, as are the clinical datasets that were used in this study. Next, the algorithm which is studied in this paper is described in Section 3. The results achieved with this algorithm are summarised in Section

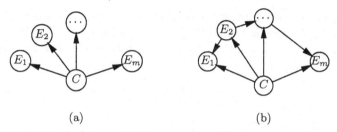

(a)   (b)

Fig. 1. Naive Bayesian network (a); and tree-augmented Bayesian network (b).

4. These are subsequently discussed against the backdrop of Bayesian-network structure learning in Section 5.

2 Background

In this section, the methods and datasets used in this study are reviewed. In addition, the medical context of this research is sketched.

2.1 Bayesian Network Classifiers

A *Bayesian network* $\mathcal{B}$ is defined as a pair $\mathcal{B} = (G, \Pr)$, where G is a directed, acyclic graph $G = (V(G), A(G))$, with a set of vertices $V(G) = \{V_1, \ldots, V_q\}$, representing a set of discrete stochastic variables $\mathcal{V}$, and a set of arcs $A(G) \subseteq V(G) \times V(G)$, representing conditional and unconditional stochastic independences among the variables, modelled by the absence of arcs among vertices. The basic property of a Bayesian network is that any variable corresponding to a vertex in the graph G is conditionally independent of its non-descendants given its parents; this is called the *local Markov property* [6]. On the variables $\mathcal{V}$ is defined a joint probability distribution $\Pr(V_1, \ldots, V_q)$, taking into account the conditional independence relationships modelled by the network, i.e. the following equality holds:

$$\Pr(V_1, \ldots, V_q) = \prod_{k=1}^{q} \Pr(V_k \mid \pi(V_k))$$

here, $\pi(V_k)$ stands for the set of parents of vertex V_k. In the following, V_k or V refers to a (free) variable; a specific value of a variable is denoted by v_k or v. Furthermore, expressions of the form $\bigcirc_X f(X)$ are abbreviated notations of $f(x_1) \circ \cdots \circ f(x_p)$, where $x_1, \ldots, x_p$ are elements in the domains of the variables $X_1, \ldots, X_p$, and $\circ$ is a binary operator.

Bayesian-network models conforming to the topology shown in Figure 1(a) correspond to the situation where a distinction is made between *evidence (feature) variables* E_i and a *class variable* C, with the evidence variables assumed to be conditionally independent given the class variable. In the following such networks will be called *naive Bayesian networks* by way of analogy with the special form of Bayes' rule, nicknamed "naive Bayes' rule", for which the same assumptions hold. A naive Bayesian network is normally used to determine the class value with maximum a posteriori probability, i.e.

$$c_{max} = \text{argmax}_C \{\Pr(C \mid \mathcal{E})\}$$

with given evidence $\mathcal{E} \subseteq \{E_1, \ldots, E_m\}$.

A naive Bayesian network lacks important probabilistic dependence information, but has the advantage that the assessment of the required probabilities $\Pr(E_j \mid C)$ and $\Pr(C)$ is straightforward. Determination of the a posteriori probability distribution $\Pr(C \mid \mathcal{E})$ is computationally speaking trivial. One would expect that adopting such strong simplifying assumptions may be at the expense of reduced performance. However, Domingos and Pazzani have convincingly shown that naive Bayesian networks yield surprisingly powerful classifiers, much more robust than previously thought [10]. This can be explained by noting that when classifying cases based on, for example, a single class and a number of feature variables, a structured Bayesian network may be optimal in the sense that it may fit the underlying probability distribution of the data best, but this does not necessarily imply that its performance in terms of percentage of correctly classified cases is optimal as well [29]. This explains why the naive Bayes' rule is becoming increasingly popular, after having fallen into disgrace two decades ago.

Building upon work from the late 1960s by Chow and Liu [5], Friedman et al. [11], subsequently showed that when the evidence variables are linked together as a directed tree, and these variables are then connected to the class variable as in a naive Bayesian network, the resulting network, called a *tree-augmented naive (TAN) Bayesian network*, or TAN for short (see Figure 1(b)), often outperforms a naive Bayesian network. The method uses a minimum-cost spanning tree algorithm in selecting branches for the TAN, where the used cost measure is the negative value of the mutual information between variables E_i, E_j, $i \neq j$: $I_{\Pr}(E_i, E_j)$, also referred to as the Kullback-Leibler divergence [19], where Pr is a probability distribution estimate based on the data. Friedman et al. suggest using mutual information between variables E_i, E_j *conditioned* on the class variable C [11]:

$$I_{\Pr}(E_i, E_j \mid C) = \sum_{E_i, E_j, C} \Pr(E_i, E_j, C) \cdot \log \frac{\Pr(E_i, E_j \mid C)}{\Pr(E_i \mid C) \Pr(E_j \mid C)} \quad (1)$$

which offers advantages when focusing on building a classifier, as the conditional mutual information takes class influences into account.

2.2   Bayesian Models in Medicine

As early as in the 1960s, computer programs were already developed to investigate the applicability of a computerised version of the naive Bayes' rule to the problem of medical management of disease, in particular diagnosis [35]. Diagnosis of liver disease and congenital heart disease were among the first subjects for which computer-based Bayesian models were constructed [26,37]. Since then, the construction and validation of such computer-based systems have been undertaken by several research groups in various medical domains [3,8,12,14,17,24,36]. A frequently cited medical application from the early 1970s to the end of the 1980s is the 'acute abdominal pain program' developed by De Dombal et al., a program capable of diagnosing causes of abdominal pain, such as perforated peptic ulcer [7–9]. As in most of these early systems, it was assumed that the elements in the diagnostic class were mutually exclusive and exhaustive, and that the observable findings that constitute the evidence are conditionally independent given the class variable [23]. In some cases, the underlying probability distributions were based on a clinical dataset, whereas in others subjective estimates by experienced clinicians were taken [34].

As soon as Bayesian-network technology became available at the end of the 1980s, biomedical researchers started developing Bayesian networks, usually using expert knowledge as a foundation. Examples of early Bayesian-network systems include Pathfinder [14–16], a system aimed at supporting pathologists in the diagnosis of white-blood-cell tumours, and MUNIN, a system meant to assist neurologists in the interpretation of electromyograms [1]. There is also some work from the early 1990s were researchers compared various other representation formalisms, such as classification rules and prototype representation, to Bayesian networks in an attempt to gain insight into the pros and cons of exploiting probability theory for medical decision support [18,28,33]. The last word about this topic has not yet been said [30].

Modern Bayesian networks in medicine not only concern diagnostic applications, but are also able to assist in the prediction of prognosis and in the selection of optimal treatment if a Bayesian network is augmented with decision theory. Examples are a system that assists in the prediction of the outcome of treatment for non-Hodgkin lymphoma of the stomach, and in the selection of optimal treatment for this disorder [21], and a system that assists in the diagnosis of mechanically ventilated pneumonia of patients in the ICU and in the selection of optimal antibiotic treatment for this disorder [22]. As mentioned above, machine learning and statistics increasingly play a part in the construction process of network models in medicine.

2.3   The Datasets

We review the three datasets that have been used in the evaluation of the algorithm.

The Copenhagen Computer Icterus (COMIK) group has been working for more than a decade on the development of a system for diagnosing disease of the liver and biliary disease, known as the Copenhagen Pocket

Chart [24,25]. It has been based on the analysis of data of 1002 jaundiced patients. The Copenhagen Pocket Chart classifies a given jaundiced patient into one of four different diagnostic categories: acute non-obstructive, chronic non-obstructive, benign obstructive, and malignant obstructive jaundice, based on the values of 21 variables to be filled in by the clinician. Table 1 shows the Pocket Chart, where 'no' and 'ob' stand for non-obstructive and obstructive, respectively, 'ac' and 'ch' stand for acute and chronic, respectively, and 'be' and 'ma' stand for 'benign' and 'malignant'. For the selection of relevant variables, the order of the likelihood ratios $\lambda_j = \Pr(e_j \mid c)/\Pr(e_j \mid \neg c)$ was used to select 24 relevant variables from 107 initially given variables E_j; this subset was reduced to 21 relevant variables by using Bayes' rule, by which the impact on the classification performance of omitting additional variables was studied.

The chart offers a compact representation of three logistic regression equations [2,13]: $S_c = \sum_k^{n_c} \omega_k^c e_k^c$, $c \in \{$non-obstructive, acute, benign$\}$, with resulting a posteriori probability distributions:

$$\Pr(c \mid \mathcal{E}) = [1 + \exp - S_c]^{-1}$$

Assuming stochastic independence, the probability of acute obstructive jaundice, for example, is computed as follows: $\Pr($acute obstructive jaundice $\mid \mathcal{E}) = \Pr(\text{ac} \mid \mathcal{E}) \cdot \Pr(\text{ob} \mid \mathcal{E})$.

The performance and usefulness of this classification scheme has been extensively investigated by research groups in several countries, such as in Sweden [20] and the Netherlands [32], using retrospective data from patients. These studies showed that, when taking the diagnostic conclusions of expert clinicians as a point of reference, the system is able to produce a correct conclusion (one of the four possible diagnostic categories) in about 75–77% of jaundiced patients.

The other two datasets used in this study concerning lymphoma and hepatitis, respectively, are popular medical datasets in machine-learning research, and were originally donated to the research community by a research group located in Ljubljana. The lymphoma dataset contains 148 records concerning 19 variables; the hepatitis dataset contains 155 records concerning 20 variables. The lymphoma dataset has, in contrast to the other datasets, no missing values. We have also experimented with the hepatitis dataset after removing records with missing data, leaving 80 records.

In this paper, we study the three datasets. As our research is not concerned with variable selection, we took either all variables (lymphoma and hepatitis data), or those selected by the COMIK group.

3   FANs: Forest-Augmented Bayesian Networks

The following algorithm, which is a variant of the modification by Friedman et al. of the Chow and Liu algorithm [11] is studied in this paper. It allows

	No vs. Ob	Ac vs. Ch	Be vs. Ma		No vs. Ob	Ac vs. Ch	Be vs. Ma
Age: 31 – 64 years	+7	+5		Physical			
≥ 65 years	+12	+5		examination:			
Previous history:				Spiders	-6	+11	
Jaundice due to	-7	+8		Ascites	-3	+6	
cirrhosis				Liver surface nodular		+5	
Cancer in GI-tract,				Gall bladder:			
pancreas, bile	+10		+7	Courvoisier	+16		+11
system, or breast				firm or tender	+5		
				Clinical chemistry:			
Leukaemia or	-13			bilirubin ≥ 200μmol/l	+5	-5	+5
malignant							
lymphoma							
Previous biliary				Alkaline phosphatase:			
colics or proven				400 – 1000 U/l	+6		
gallstones	+3	+7	-7	> 1000 U/l	+11		+6
In treatment for							
congestive heart							
failure		-5					
Present history:				ASAT:			
				40 – 319 U/l		+5	
≥ 2 weeks			+7	≥ 320 U/l	-10	+1	+6
Upper abdominal pain:				Clotting factors:			
sever	+9		-6	≤ 0.55		+8	+5
slight or moderate	+4			0.56 – 0.70		+5	+5
Fever:				LDH ≥ 1300 U/l		-5	+7
without chills		-3	-5				
with chills		-6	-10				
Intermittent jaundice	+5		-5				
Weight loss (≥ 2 kg)			+4				
Alcohol:							
1 – 4 drinks per day	-4						
≥ 5 drinks per day	-4	+4		SUM left			
SUM left				CONSTANTS	-19	-21	-8
				TOTAL SCORE			

Table 1. Pocket Diagnostic Chart [27].

for exploring the search space of Bayesian-network models bounded by naive Bayesian networks and TANs.

FAN Algorithm: Let $k \geq 0$; assume that evidence variables E_i, $i = 1, \ldots, m$, a class variable C, and a dataset D, with $|D| = n$, are given.

(1) The conditional mutual information $I_{\mathrm{Pr}}(E_i, E_j \mid C)$ for all pairs of evidence variables E_i, E_j, $i \neq j$, are computed using formula (1).

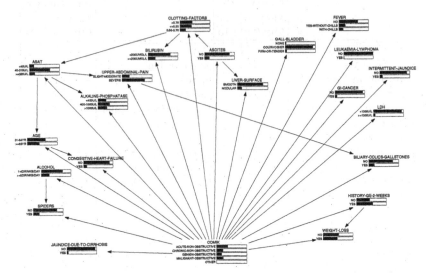

Fig. 2. COMIK FAN with 11 arcs added to its naive Bayesian network backbone.

(2) An undirected complete graph with vertices E_i, $i = 1, \ldots, m$, is built, with costs attached to the edges defined by $-I_{\Pr}(E_i, E_j \mid C)$.

(3) A minimum-cost spanning forest for the undirected cost graph is constructed, containing exactly k edges.

(4) The undirected forest is transformed into a directed forest by choosing a root vertex for every tree in the forest, and by adding an outward direction to the branches encountered on the paths from the root to every other vertex in the tree.

(5) The directed forest is transformed into a connected directed graph by adding an arc (directed edge) from the class vertex C to every evidence vertex E_i in the forest. The resulting directed graph is called a *forest-augmented network* model, or *FAN* model for short.

(6) The conditional probability distributions of the FAN model are learnt from the data in the dataset.

All operations are polynomial time, with step (1) being the most expensive, $\mathcal{O}(nm^2)$, one. Figure 3 gives a summary of the most important steps in the algorithm.

The joint probability distribution of the FAN models were learnt using Bayesian updating with Dirichlet priors based on the datasets of approximately 900 (COMIK), 135 (lymphoma) and 140 (hepatitis) cases, each time using the remainder of each dataset for testing (see below) [31]. Thus, the conditional probability distribution for each variable V_i was computed as the weighted average of a probability estimate and the Dirichlet prior, as follows:

$$Pr(V_i \mid \pi(V_i), D) = \frac{n}{n + n_0} \widehat{\Pr}_D(V_i \mid \pi(V_i)) + \frac{n_0}{n + n_0} \Theta_i$$

where $\widehat{\Pr}_D$ is the probability distribution estimate based on a given dataset D, and Θ_i is the Dirichlet prior. We choose Θ_i to be a uniform probability distribution. The parameter n_0 is equal to the number of past cases on which the contribution of Θ_i is based; here we took after experimentation $n_0 = 5$.

As an example, consider the FAN model shown in Figure 2, which includes every variable mentioned in the Pocket Diagnostic Chart, with the evidence vertices forming a forest of 10 trees, of which 3 contained more than one vertex. A similar FAN model for lymphoma is shown in Figure 4 and for hepatitis in Figure 5.

4   Evaluation

4.1   Methods

Using the FAN algorithm described above, 21 Bayesian networks for the COMIK dataset, 18 networks for lymphoma and 19 networks for hepatitis were constructed. These included naive Bayesian networks, with an empty set of added branches, and TAN models, which contained a forest consisting of a single tree with 20 branches for the COMIK dataset, 17 branches for the lymphoma dataset and 18 branches for the hepatitis dataset. The performance of each network was evaluated using tenfold cross-validation, i.e. the dataset was split up into 10 (almost) equal parts, and the performance of each network was determined by evaluating the results for each of the parts, after its underlying joint probability distribution was learnt from the other 9 parts.

Three additive components make up the error in classifying: (1) the intrinsic error due to noise in the data, (2) the statistical bias in the model, and (3) the variance (model's sensitivity to the characteristics of the dataset) [2,10,13]. One would expect a large statistical bias for the naive Bayesian classifier, as its independence assumptions are almost always unjustified, and a somewhat lower one for FAN and TAN models. On the other hand, experimental evidence shows that the naive Bayesian classifier has a low variance [10]. Models with a greater representational power have a greater ability to respond to the dataset, i.e. they have a large *information-storage capacity*, and tend to have a lower bias and higher variance [2]. Tenfold cross-validation

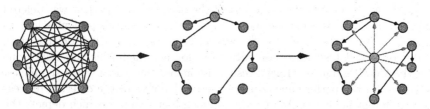

Fig. 3. The three phases (from left to right) of the FAN algorithm.

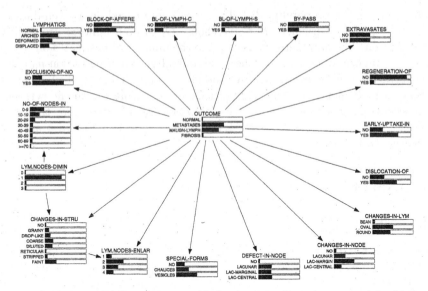

Fig. 4. Lymphoma FAN with 3 arcs added to its naive Bayesian network backbone.

offers a good balance between the bias and variance of learning results, and this was also confirmed experimentally [13].

The performance of the networks was measures by comparing the clinical diagnosis with the class value with maximum a posteriori probability. The resulting measure is called the *success rate*. The success rate conveys information about the quality of classification, but it offers only rough information about how close the a posteriori probability distribution is to reality. More subtle effects can be uncovered by determining for each patient case $r_k \in D$, with actual class value c_k, the entropy

$$E_k = - \ln \Pr(c_k \mid \mathcal{E})$$

which has the informal meaning of a penalty: when the probability $\Pr(c_k|\mathcal{E}) = 1$, then $E_k = 0$ (actually observing c_k generates no information); otherwise, $E_k > 0$. The total score for dataset D is now defined as the sum of the individual scores: $E = \sum_{k=1}^{n} E_k$.

In order to obtain insight into the effects of partial data on the conclusions drawn by each network, a part of the data for each patient was deleted at random, with the percentage of data deleted for each patient equal to 12.5%, 25%, 37.5%, 50%, 67.5%, and 75%, respectively. The deleted data only concerned test data, not data used for learning. This was only done for the COMIK dataset. Initial results indicated that there was a significant variation in the performance due to the randomness of the deletion process over different networks. As an example, consider Figure 6 which depicts the variation in the performance results of a FAN model with a forest containing

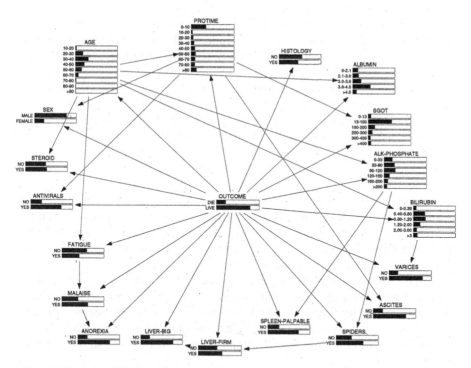

Fig. 5. Hepatitis FAN with 17 arcs added to its naive Bayesian network backbone.

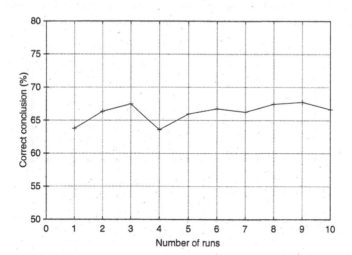

Fig. 6. Variation in the performance of the same FAN model when 50% of the evidence in each patient record is deleted at random for each of the 10 runs of the cross-validation procedure.

14 arcs, where each time 50% of the evidence was deleted at random from the patient record. As a consequence, the random deletion process was repeated 20 times for each network and deletion percentage, in order to average out this random variation. The number 20 was determined experimentally; after 10 runs with each network and deletion percentage, the average results started to converge. The same set of random numbers was used for different networks. Making the deletion process completely random would have made the averaging out process computationally intractable.

4.2   Results

The results for the 21 FAN models of the COMIK dataset are given in Figures 7 and 8. The plot in Figure 7 clearly indicates that including more arcs into a FAN model has no obvious beneficial effects on the performance; indeed, the differences between the various network models for any given evidence deletion percentage are very small.

Figure 8 makes clear that even though adding arcs has only a slight effect on the classification performance of a network, this is less true so for the underlying probability distribution, which is affected to various degrees. Figure 8 also shows that adding arcs has almost no effect after 11 arcs have been added. In addition, almost no effect can be noticed if less than 50% of the available evidence is entered.

Figures 9 and 10 summarise the results obtained for the lymphoma dataset, whereas Figures 11 and 12 do the same for the FAN models regarding hepatitis. For these models, there were significant differences in performance for

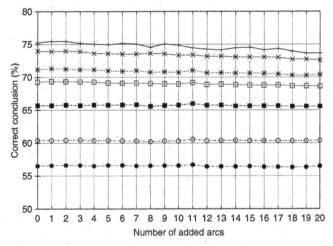

Fig. 7. Success rate for different Bayesian-network topologies after random deletion of 0% (+), 12.5% (×), 25% (⋆), 37.5% (□), 50% (■), 67.5% (○), 75% (●) of the evidence for each patient.

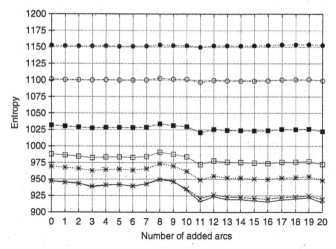

Fig. 8. Entropy results for different Bayesian-network topologies after random dele-tion of 0% (+), 12.5% (×), 25% (⋆), 37.5% (□), 50% (■), 67.5% (○), 75% (●) of the evidence for each patient.

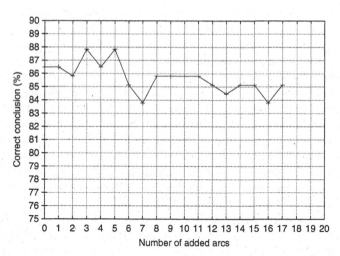

Fig. 9. Success rate for different Bayesian network topologies using lymphoma data.

different network topologies. The best performing Bayesian network of lym-phoma included three arcs, and is depicted in Figure 4; the best performing models concerning hepatitis included 17 arcs when records with missing data were deleted, and 16 arcs otherwise.

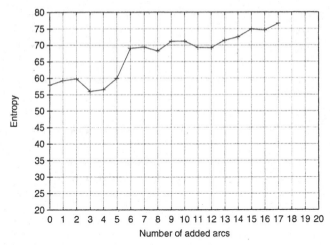

Fig. 10. Entropy results for different Bayesian-network topologies using lymphoma data.

5 Discussion

The first conclusion that can be drawn is that if one is merely interested in using a Bayesian network for classification purposes, developing a naive Bayesian network may suffice, even though it may not always yield the best model possible. This confirms previous research by others. For only one of the datasets we were able to confirm the finding that TANs outperform naive Bayesian classifiers. For the COMIK dataset, changes in topology gave rise to only very small changes in performance. As the COMIK dataset can be viewed upon as a large, good-quality clinical dataset, these conclusions, which are statistically significant due to the dataset's size, are worth noting. With the two medical datasets of low quality from the UCI Repository, we were indeed able to discover differences in performance, which indicates that there are certain datasets for which FAN learning will be worthwhile, even if we are only interested in classification performance.

It was also observed that adding dependence information to a Bayesian network may improve the quality of the underlying probability distribution. In the case of the lymphoma and hepatitis datasets this quality improvement was to some extent reflected in the performance figures as well. This means that local improvements in the quality of a Bayesian network do sometimes translate into global quality improvement, but not as a straightforward function of the number of arcs added. The central conclusion of this work is therefore that if one wishes to learn a Bayesian network that offers good performance for both classification and regression problems, learning a FAN model where the number of arcs is a parameter determined by the problem at hand, may be an appropriate solution.

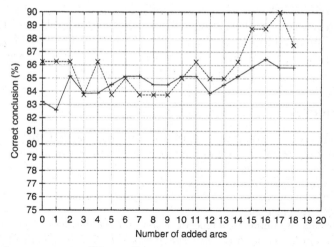

Fig. 11. Success rate for different Bayesian network topologies using hepatitis data with missing values (+), and hepatitis data without missing values (×).

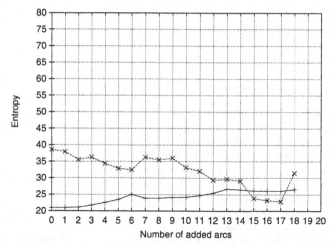

Fig. 12. Entropy results for different Bayesian-network topologies using hepatitis data with missing values (+), and hepatitis data without missing values (×).

The FAN algorithm is an example of a restricted Bayesian-network structure learning algorithm [4]. By putting restrictions on the topology of the network to be learnt using local information, it is possible to learn a structure in polynomial time. As the Bayesian-network topology space is superexponentially large, other researchers usually resort to using heuristic search methods such as hill climbing and tabu search. However, the results achieved with these algorithms are thus far rather disappointing for real-life datasets.

We now believe that this research can only be expected to yield positive results for regression problems, and not for classification problems. The FAN algorithm proposed in this paper can thus be regarded as providing a base level for such more sophisticated algorithms. The future will learn whether these algorithms are able to keep up to the expectations.

References

1. S. Andreassen, M. Woldbye, B. Falck and S.K. Andersen. MUNIN — A Causal Probabilistic Network for Interpretation of Electromyographic Findings, in: *Proceedings of the 10th International Joint Conference on Artificial Intelligence*, Milan, Italy, 1987, pp. 366–372.
2. M. Berthold and D.J. Hand (Eds.). *Intelligent Data Analysis: an Introduction.* Springer, Berlin, 1999.
3. F. Burbank. A computer diagnostic system for the diagnosis of prolonged undifferentiated liver disease. *American Journal of Medicine* 46 (1996) 401–415.
4. J. Cheng and R. Greiner. Comparing Bayesian network classifiers. In: *Proceedings of UAI'99)*. Morgan Kaufmann, San Francisco, CA, 1999, pp. 101–107.
5. C.K. Chow and C.N. Liu. Approximating discrete probability distributions with dependence trees. *IEEE Trans. on Info. Theory* 14 (1968) 462–467.
6. R.G. Cowell, A.P. Dawid, S.L. Lauritzen and D.J. Spiegelhalter. *Probabilistic Networks and Expert Systems.* Springer, New York, 1999.
7. F.T. de Dombal. Computers and decision-making: an overview for gastroenterologists. *Frontiers in Gastrointestinal Research* 7 (1984) 119–133.
8. F.T. de Dombal, D.J. Leaper, J.R. Staniland, A.P. McAnn and J.C. Horrocks. Computer-aided diagnosis of acute abdominal pain. *British Medical Journal* ii (1972) 9–13.
9. F.T. de Dombal, V. Dallos and W.A. McAdam. Can computer-aided teaching packages improve clinical care in patients with acute abdominal pain? *British Medical Journal* 302 (1991) 1495–1497.
10. P. Domingos and M. Pazzani. On the optimality of the simple Bayesian classifier under zero-one loss. *Machine Learning* 29 (1997) 103–130.
11. N.I.R. Friedman, D. Geiger and M. Goldszmidt. Bayesian network classifiers. *Machine Learning* 29 (1997) 131-163.
12. G.A. Gorry and G.O. Barnett. Experience with a model of sequential diagnosis. *Computers and Biomedical Research* 1 (1968) 490–507.
13. T. Hastie, R. Tibshirani and J. Friedman. *The Elements of Statistical Learning: Data Mining, Inference, and Prediction.* Springer, New York, 2001.
14. D.E. Heckerman. *Probabilistic Similarity Networks.* The MIT Press, Cambridge, Massachusetts, 1992.
15. D.E. Heckerman, E.J. Horvitz and B.N. Nathwani. Towards normative expert systems: part I – The Pathfinder project. *Methods of Information in Medicine* 31 (1992) 90–105.
16. D.E. Heckerman and B.N. Nathwani. Towards normative expert systems: part II – probability-based representations for efficient knowledge acquisition and inference. *Methods of Information in Medicine* 31 (1992) 106–116.
17. R.P. Knill-Jones, R.B. Stern, D.H. Girmes, J.D. Maxwell, R.P.H. Thompson and R. Williams. Use of a sequential Bayesian model in the diagnosis of jaundice. *British Medical Journal* i (1973) 530–533.

18. M. Korver and P.J.F. Lucas. Converting a rule-based expert system into a belief network. *Medical Informatics* 18(3) (1993) 219–241.
19. S. Kullback and R. Leibler. On information and sufficiency. *Annals of Mathematical Statistics* 22 (1951) 79–86.
20. G. Lindberg, C. Thomson, A. Malchow-Møller, P. Matzen and J. Hilden. Differential diagnosis of jaundice: applicability of the Copenhagen Pocket Diagnostic Chart proven in Stockholm patients. *Liver* 7 (1987) 43–9.
21. P.J.F. Lucas, H. Boot and B.G. Taal. Computer-based decision-support in the management of primary gastric non-Hodgkin lymphoma. *Methods of Information in Medicine* 37 (1998) 206–219.
22. P.J.F. Lucas, N.C. de Bruijn, K. Schurink and I.M. Hoepelman. A Probabilistic and decision-theoretic approach to the management of infectious disease at the ICU. *Artificial Intelligence in Medicine* 19(3) (2000) 251–279.
23. P.J.F. Lucas and L.C. van der Gaag. *Principles of Expert Systems*. Addison-Wesley, Wokingham, 1991.
24. A. Malchow-Møller, C. Thomson, P. Matzen et al. Computer diagnosis in jaundice: Bayes' rule founded on 1002 consecutive cases. *Journal of Hepatology* 3 (1986) 154–163.
25. A. Malchow-Møller, L. Mindeholm, H.S. Rasmussen and B. Rasmussen et al. Differential diagnosis of jaundice: junior staff experience with the Copenhagen pocket chart. *Liver* 7 (1987) 333–338.
26. W.B. Martin, P.C. Apostolakos and H. Roazen. Clinical versus actuarial prediction in the differential diagnosis of jaundice. *American Journal of Medical Science* 240 (1960) 571–578.
27. P. Matzen, A. Malchow-Møller, J. Hilden et al. Differential diagnosis of jaundice: a pocket diagnostic chart. *Liver* 4 (1984) 360–71.
28. B. Middleton, M.A. Shwe, D.E. Heckerman, M. Henrion, E.J. Horvitz, H.P. Lehmann and G.F. Cooper. Probabilistic diagnosis using a reformulation of the INTERNIST-1/QMR knowledge base, II. Evaluation of diagnostic performance, *Methods of Information in Medicine* 30 (1991) 256–267.
29. S. Monti and G.F. Cooper. A Bayesian network classifier that combined a finite mixture model and a naive Bayes model, in: K. Blackmond Laskey and H. Prade (Eds.), *Proceeding of UAI'98*. Morgan Kaufmann, San Francisco, CA, 1999, pp. 447–456.
30. A. Onísko, P.J.F. Lucas and M. Druzdzel. Comparison of rule-based and Bayesian network approaches in medical diagnostic systems, in: S. Quaglini, P. Barahona, S, Andreassen, *Proceedings of the 8th Conference on Artificial Intelligence in Medicine in Europe, AIME 2001*. Springer, Berlin, 2001, pp. 283–292.
31. M. Ramoni and P. Sebastiani. Bayesian methods, in: M. Berthold and D.J. Hand (Eds.), *Intelligent Data Analysis: an Introduction*. Springer, Berlin, 1999, pp. 130–166.
32. R.W. Segaar, J.H.P. Wilson, J.D.F. Habbema, A. Malchow-Møller, J. Hilden, and P.J. van der Maas. 'Transferring a Diagnostic Decision Aid for Jaundice'. *Netherlands Journal of Medicine* 33 (1988) 5–15.
33. M.A. Shwe, B. Middleton, D.E. Heckerman, M. Henrion, E.J. Horvitz, H.P. Lehmann and G.F. Cooper. Probabilistic diagnosis using a reformulation of the INTERNIST-1/QMR knowledge base, I. The probabilistic model and inference algorithms. *Methods of Information in Medicine* 30 (1991) 241–255.

34. D.J. Spiegelhalter, R.C.G. Franklin and K. Bull. Assessment, criticism and improvement of imprecise subjective probabilities for a medical expert system, in: M. Henrion, R.D. Shachter, L.N. Kanal and J.F. Lemmer (Eds.), *Uncertainty in Artificial Intelligence 5*. North-Holland, Amsterdam, 1990, pp. 285–294.

35. D.J. Spiegelhalter and R.P. Knill-Jones. Statistical and knowledge-based approaches to clinical decision-support systems, with an application in gastroenterology. *Journal of the Royal Statistical Society* 147 (1984) 35–77.

36. B.S. Todd and R. Stamper. *The Formal Design and Evaluation of a Variety of Medical Diagnostic Programs*. Technical Monograph PRG-109, Oxford University Computing Laboratory, Oxford University, 1993.

37. H.R. Warner, A.F. Toronto, L.G. Veasey, R. Stephenson. A mathematical approach to medical diagnosis – applications to congenital heart disease. *Journal of the American Medical Association* 177 (1961) 177–184.

Scaled Conjugate Gradients for Maximum Likelihood: An Empirical Comparison with the EM Algorithm

Kristian Kersting[1] and Niels Landwehr[1]

Institute for Computer Science, Machine Learning Lab
Albert-Ludwigs-University, Georges-Köhler-Allee, Gebäude 079,
D-79085 Freiburg i. Brg., Germany

Abstract. To learn Bayesian networks, one must estimate the parameters of the network from the data. EM (Expectation-Maximization) and gradient-based algorithms are the two best known techniques to estimate these parameters. Although the theoretical properties of these two frameworks are well-studied, it remains an open question as to when and whether EM is to be preferred over gradients. We will answer this question empirically. More specifically, we first adapt scaled conjugate gradients well-known from neural network learning. This accelerated conjugate gradient avoids the time consuming line search of more traditional methods. Secondly, we empirically compare scaled conjugate gradients with EM. The experiments show that accelerated conjugate gradients are competitive with EM. Although, in general EM is the domain independent method of choice, gradient-based methods can be superior.

1 Introduction

Bayesian networks are one of the most important, efficient and elegant frameworks for representing and reasoning with probabilistic models. They specify joint probability distributions over finite sets of random variables, and have been applied to many real-world problems in diagnosis, forecasting, automated vision, sensor fusion and manufacturing control. Over the past years, there has been much interest in the problem of learning Bayesian networks from data to avoid the problems of knowledge elicitation. For learning Bayesian networks, parameter estimation is a fundamental task not only because of the inability of humans to reliably estimate the parameters, but also because it forms the basis for the overall learning problem [10].

A classical method for parameter estimation is *maximum likelihood* parameter estimation. Here, one selects those parameters maximizing the likelihood which is the probability of the observed data as a function of the unknown parameters with respect to the current model. Unfortunately in many real-world domains, the data cases available are incomplete, i.e. some values may not be observed. For instance in medical domains, a patient rarely gets all of the possible tests. In presence of missing data, the parameter estimation becomes much more difficult [6]. In fact, the maximum likelihood estimate typically cannot be written in closed form. It is a numerical optimization problem, and all known algorithms involve nonlinear optimization and multiple calls to a Bayesian network inference as subroutines. The latter ones have been proven to be NP-hard [7]. The most prominent techniques within Bayesian networks are the Expectation-Maximization (EM) algorithm and gradient-based approaches.

The comparison between EM and (advanced) gradient techniques is not well understood. Both methods perform a greedy local search which is guaranteed to converge to stationary points. They both exploit expected sufficient statistics as their primary computation step. However, there are important differences. The EM is easier to implement, converges much faster than simple gradient, and is somewhat less sensitive to starting points. Conjugate Gradients estimate the step size (see below) with a line search involving several additional Bayesian network inferences compared to EM. But, gradients are more flexible than EM, as they easily allow e.g. to learn non-multinomial parameterizations using the chain rule for derivatives [3] or to choose other scoring functions than the likelihood [13]. The EM algorithm may slowly converge when close to a solution, but can be sped up near the maximum using gradient-based approaches [26,17,1,23]. Other acceleration approaches execute only a partial maximization step of the EM algorithm based on gradient techniques leading to generalized EM algorithms [17].

In this context, we report on an experimental comparison between EM and (accelerated) conjugate gradients to maximum likelihood parameter estimation within Bayesian networks over finite random variables. In doing so, we follow Ortiz and Kaelbling's [23] suggestion for such a comparison in the context of (discrete) Bayesian network. To overcome the expensive line search, we adapted *scaled conjugate gradients* [18] from the field of learning neural networks. They avoid the line search by using a Levenberg-Marquardt approach [16] in order to scale the step size, and employ the Hessian of the scoring function to quadratically extrapolate the maximum instead of doing a line search. This type of accelerated conjugate gradients are novel in the context of Bayesian networks. Other work investigated approximated line searches only to accelerate EM [26,23,1]. From the experimental results, we will argue that scaled conjugate gradients are competitive with EM. (Accelerated) gradient methods can be superior depending on properties of the domain such as the fraction of latent parameters and the prior information encoded in the network structure, though EM seems to be the best *domain independent* choice.

We proceed as follows. After briefly introducing Bayesian networks in Section 2, we review maximum likelihood estimation via the EM algorithm and simple gradient-ascent in Section 3. In Section 4, we briefly review a well-known criterion to classify data sets into "EM-hard" and "EM-easy" ones. Section 5 reviews conjugate gradients and introduces scaled conjugate gradients. In Section 6, we experimentally compare the EM algorithm with

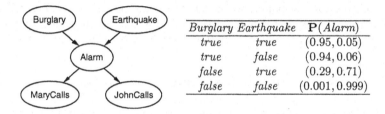

Burglary	Earthquake	$\mathbf{P}(Alarm)$
true	true	$(0.95, 0.05)$
true	false	$(0.94, 0.06)$
false	true	$(0.29, 0.71)$
false	false	$(0.001, 0.999)$

Fig. 1. An example of a Bayesian network modelling a burglary alarm system

conjugate gradients and scaled conjugate gradients. Before concluding, we discuss related work.

2   Bayesian Networks

Throughout the paper, we will use X to denote a random variable, x a state and $\mathbf{X}$ (resp. $\mathbf{x}$) a vector of variables (resp. states). We will use $\mathbf{P}$ to denote a probability distribution, e.g. $\mathbf{P}(X)$, and P to denote a probability value, e.g. $P(x)$.

A *Bayesian network* [24] represents the joint probability distribution $\mathbf{P}(\mathbf{X})$ over a set $\mathbf{X} = \{X_1, \ldots, X_n\}$ of random variables. In this paper, we restrict each X_i to have a finite set $x_i^1, \ldots, x_i^{k_i}$ of possible states. Consider Judea Pearl's famous *Alarm* network graphically illustrated in Fig. 1. The (boolean) random variables are *Burglary, Earthquake, Alarm, JohnCalls,* and *MaryCalls*. A Bayesian network is an augmented, acyclic graph, where each node corresponds to a variable X_i and each edge indicates a direct influence among the variables. For instance, *Burglary* and *Earthquake* directly affects the probability of the *Alarm* going off. We denote the parents of X_i in a graph-theoretical sense by $\mathbf{Pa}_i$, e.g. $\mathbf{Pa}_{Alarm} = \{Burglary, Earthquake\}$. The family of X_i is $\mathbf{Fa}_i := \{X_i\} \cup \mathbf{Pa}_i$, e.g. $\mathbf{Fa}_{Alarm} = \{Alarm, Burglary, Earthquake\}$. With each node X_i, a conditional probability table is associated specifying the distribution $\mathbf{P}(X_i \mid \mathbf{Pa}_i)$, cf. Fig. 1. The table entries are

$$\theta_{ijk} = P(X_i = x_i^k \mid \mathbf{Pa}_i = \mathbf{pa}_i^j), \tag{1}$$

where $\mathbf{pa}_i^j$ denotes the jth joint state of the X_i's parents, e.g. $\mathbf{pa}_{Alarm}^2 = \{Burglary = true, Earthquake = false\}$. The network stipulates the assumption that each node X_i in the graph is conditionally independent of any subset $\mathbf{A}$ of nodes that are not descendants of X_i given a joint state of its parents. Thus, the joint probability distribution over $\mathbf{X}$ factors to

$$\mathbf{P}(X_1, \ldots, X_n) = \prod_{i=1}^n \mathbf{P}(X_i \mid \mathbf{Pa}_i). \tag{2}$$

In the rest of the paper, we will represent a Bayesian network with given structure by the vector $\boldsymbol{\theta}$ consisting of all θ_{ijk}'s. For instance for the *alarm* network we have $\boldsymbol{\theta} = (\ldots, 0.95, 0.05, 0.94, 0.06, 0.29, 0.71, 0.001, 0.999, \ldots)$.

3   Basic Maximum Likelihood Estimation

Our task is to learn the numerical parameters θ_{ijk} for a Bayesian network of a given structure. More formally, we have some initial model (assignment of parameters) $\boldsymbol{\theta}$. We also have some set of data cases $\mathbf{D} = \{\mathbf{d}_1, \ldots, \mathbf{d}_N\}$. Each data case $\mathbf{d}_i$ is a (possibly) partial assignment of values to variables in the network, e.g. for the *alarm* network

ID	Burglary	Earthquake	Alarm	JohnCalls	MaryCalls
$\mathbf{d}_1$	true	false	true	true	false
$\mathbf{d}_2$	false	?	true	false	true
...	...	...	...	...	...

where the question mark denotes unobserved states. We assume that the data cases are independently sampled from identical distributions (i.i.d). We want to construct the model $\boldsymbol{\theta}^*$ which best explains the data $\mathbf{D}$.

We quantify 'which best explains the data', as usual, via the likelihood of the data $\mathbf{D}$ in $\boldsymbol{\theta}$. The likelihood $L(\mathbf{D}, \boldsymbol{\theta})$ is the probability of the data $\mathbf{D}$ as a function of the unknown parameters $\boldsymbol{\theta}$, i.e.

$$L(\mathbf{D}, \boldsymbol{\theta}) = P(\mathbf{D} \mid \boldsymbol{\theta}) . \tag{3}$$

It simplifies because of the i.i.d. assumption to

$$L(\mathbf{D}, \boldsymbol{\theta}) = \prod_{l=1}^{N} P(\mathbf{d}_l \mid \boldsymbol{\theta}) . \tag{4}$$

Thus, the search space $\mathcal{H}$ to be explored is spanned by the space over the possible values of $\boldsymbol{\theta}$. We seek to find that $\boldsymbol{\theta}^* \in \mathcal{H}$ that maximizes the likelihood, i.e., $\boldsymbol{\theta}^* = \mathrm{argmax}_{\theta \in \mathcal{H}} L(\mathbf{D}, \boldsymbol{\theta})$. Due to the monotonicity of the logarithm, this simplifies to $\boldsymbol{\theta}^* = \mathrm{argmax}_{\theta \in \mathcal{H}} \log L(\mathbf{D}, \boldsymbol{\theta}) = \mathrm{argmax}_{\theta \in \mathcal{H}} \sum_{l=1}^{N} \log P(\mathbf{d}_l \mid \boldsymbol{\theta})$, and , finally, due to the monotonicity of the logarithm, this yields

$$\boldsymbol{\theta}^* = \underset{\theta \in \mathcal{H}}{\mathrm{argmax}} \sum_{l=1}^{N} \log P(\mathbf{d}_l \mid \boldsymbol{\theta}) . \tag{5}$$

An important issue is whether the data $\mathbf{D}$ is complete, or not. In the case of complete data, i.e. the values of all random variables are observed as in data case $\mathbf{d}_1$, Lauritzen [15] showed that maximum likelihood estimation simply corresponds to frequency counting in the following way. Let $c(\mathbf{a} \mid \mathbf{D})$ denote the *counts* for a particular joint state $\mathbf{a}$ of variables $\mathbf{A}$ in the data cases, i.e.

the number of cases in which the variables in $\mathbf{A}$ are assigned the values $\mathbf{a}$, then

$$\theta^*_{ijk} = \frac{c(\mathbf{fa}_i^{jk} \mid \mathbf{D})}{c(\mathbf{pa}_i^{j} \mid \mathbf{D})} , \tag{6}$$

where $\mathbf{fa}_i^{jk}$ denotes the joint state consisting of the jth state of variable X_i and the kth joint state of $\mathbf{Pa}_i$, e.g. $\mathbf{fa}_{Alarm}^{2,3} = \{Alarm = false, Burglary = false, Earthquake = true\}$. However, in the presence of missing data as in data case $\mathbf{d}_2$, the maximum likelihood estimates typically cannot be written in closed form, and iterative optimization schemes like the EM or conjugate gradient algorithms are needed. Both approaches will be introduced in turn in the following two subsections.

Expectation-Maximization

The Expectation-Maximization algorithm [8] is a classical approach to realize maximum likelihood estimation in the presence of missing values. The basic observation underlying the Expectation-Maximization algorithm is the observation of the last section: learning would be easy if we would knew the values for all the random variables. Therefore, it iteratively performs two steps to find the maximum likelihood parameters of a model:

1. Based on the current parameters θ and the observed data $\mathbf{D}$, the algorithm computes a distribution over all possible completions of each partially observed data case.
2. Each completion is then treated as a fully-observed data case weighted by its probability. New parameters are then computed based on (6).

Lauritzen [15] showed that this idea leads to a modified (6) where *expected counts* ec instead of *counts* c are used:

$$\theta^*_{ijk} = \frac{ec(\mathbf{fa}_i^{jk} \mid \mathbf{D})}{ec(\mathbf{pa}_i^{j} \mid \mathbf{D})} . \tag{7}$$

Expected counts are defined as follows $ec(\mathbf{a}) = \sum_{l=1}^{N} P(\mathbf{a} \mid \mathbf{d}_l, \theta^n)$. To compute $P(\mathbf{a} \mid \mathbf{d}_l, \theta^n)$ any Bayesian network inference engine can be used.

Gradient-Ascent Learning

Gradient ascent, also known as *hill climbing*, is a classical method for finding a maximum of a (scoring) function. It iteratively performs two steps:

1. It computes the *gradient* vector ∇_θ of partial derivatives of the scoring function with respect to the parameters of a Bayesian network at a given point $\theta \in \mathcal{H}$.
2. Then it takes a small step in the direction of the gradient to the point $\theta + \delta \nabla_\theta$ where δ is the step-size parameter.

The algorithm will converge to a local maximum for small enough δ (cf. [16]). Thus, we have to compute the partial derivatives of $\log P(\mathbf{D} \mid \boldsymbol{\theta})$ with respect to parameters θ_{ijk}. According to Binder *et al.* [3], they are given by

$$\frac{\partial \log P(\mathbf{D} \mid \boldsymbol{\theta})}{\partial \theta_{ijk}} = \sum_{l=1}^{N} \frac{P(\mathbf{fa}_i^{jk} \mid \mathbf{d}_l, \boldsymbol{\theta})}{\theta_{ijk}} = \frac{ec(\mathbf{fa}_i^{jk} \mid \mathbf{D})}{\theta_{ijk}}. \tag{8}$$

Like for the EM algorithm, any Bayesian network inference engine can be used to compute $P(\mathbf{fa}_i^{jk} \mid \mathbf{d}_l, \boldsymbol{\theta})$. However in contrast to the EM algorithm, the described gradient-ascent method has to be modified to take into account the constraint that the parameter vector $\boldsymbol{\theta}$ consists of probability values, i.e. $\theta_{ijk} \in [0, 1]$ and $\sum_j \theta_{ijk} = 1$. As Binder *et al.* explain, there are two ways to enforce this:

1. Projecting the gradient onto the constraint surface, and
2. reparameterizing the problem so that the new parameters automatically respect the constraints on θ_{ijk} no matter what their values are.

We choose the latter approach as the reparameterized problem is fully unconstrained. More precisely, we define the parameters $\beta_{ijk} \in \mathbb{R}$ such that

$$\theta_{ijk} = \frac{\beta_{ijk}^2}{\sum_l \beta_{ilk}^2}. \tag{9}$$

where the β_{ijk} are indexed like θ_{ijk}. This enforces the constraints given above, and a local maximum with respect to β_{ijk} is also a local maximum with respect to θ_{ijk}, and vice versa. The gradient with respect to β_{ijk} can be found by computing the gradient with respect to θ_{ijk} and then deriving the gradient with respect to β_{ijk} using the chain rule of derivatives. More precisely, the chain rule of derivatives yields

$$\frac{\partial \log P(\mathbf{D} \mid \boldsymbol{\theta})}{\partial \beta_{ijk}} = \sum_{i'j'k'} \frac{\partial \log P(\mathbf{D} \mid \boldsymbol{\theta})}{\partial \theta_{i'j'k'}} \cdot \frac{\partial \theta_{i'j'k'}}{\partial \beta_{ijk}}. \tag{10}$$

Since $\partial \theta_{i'j'k'} / \partial \beta_{ijk} = 0$ unless $i = i'$ and $j = j'$, (10) simplifies to

$$\frac{\partial \log P(\mathbf{D} \mid \boldsymbol{\theta})}{\partial \beta_{ijk}} = \sum_{j'} \frac{\partial \log P(\mathbf{D} \mid \boldsymbol{\theta})}{\partial \theta_{ij'k}} \cdot \frac{\partial \theta_{ij'k}}{\partial \beta_{ijk}}$$

$$= \sum_{j'} \frac{\partial \log P(\mathbf{D} \mid \boldsymbol{\theta})}{\partial \theta_{ij'k}} \cdot \frac{\partial \beta_{ij'k}^2 / \partial \beta_{ijk} \cdot \sum_l \beta_{ilk}^2 - (\partial \sum_l \beta_{ilk}^2) / (\partial \beta_{ijk}) \cdot \beta_{ij'k}^2}{(\sum_l \beta_{ilk}^2)^2}$$

$$= \frac{2 \cdot \beta_{ijk}}{(\sum_l \beta_{ilk}^2)^2} \left(\frac{\partial \log P(\mathbf{D} \mid \boldsymbol{\theta})}{\partial \theta_{ijk}} \sum_l \beta_{ilk}^2 - \sum_l \frac{\partial \log P(\mathbf{D} \mid \boldsymbol{\theta})}{\partial \theta_{ilk}} \beta_{ilk}^2 \right) \; .$$

Now, using the identity $\partial \log P(\mathbf{D})/\partial \theta_{ijk} = ec(\mathbf{fa}_i^{jk})/(\beta_{ijk}^2) \cdot \sum_l \beta_{ilk}^2$ and (8), we get

$$\frac{\partial \log P(\mathbf{D} \mid \boldsymbol{\theta})}{\partial \beta_{ijk}} = 2 \cdot \left(\frac{ec(\mathbf{fa}_i^{jk} \mid \mathbf{D})}{\beta_{ijk}} - \theta_{ijk} \cdot \sum_l ec(\mathbf{fa}_i^{lk} \mid \mathbf{D}) \right) \; . \qquad (11)$$

Again, only local computations are involved. Moreover, many numerical optimization libraries provide advanced gradient methods which need in addition to a procedure to compute the log-likelihood only a procedure to compute the gradient of the log-likelihood function (see e.g. Numerical Recipes source code [25], Matlab [12], and Netlab [21]). Thus, implementing gradient-based maximum likelihood estimators for Bayesian network seems to be only slightly more difficult than implementing the EM algorithm. However, the used gradient-based method should be carefully selected to avoid computational complexity as we will explain in Section 5. Before doing so, let us explain when the EM algorithm is expected to converge slowly.

4 "EM-hard" and "EM-easy" Problems

As stated in the introduction, the EM algorithm converges much faster than simple gradient. In this section, we will briefly review a well-known condition on the data cases when the EM algorithm is expected to slowly converge. For more technical details we refer to [8,17].

Assume that $\langle \boldsymbol{\theta}^k \rangle$ is a sequence of parameters computed by the EM algorithm. Furthermore, assume that $\langle \boldsymbol{\theta}^k \rangle$ converges to some $\boldsymbol{\theta}^*$. Then, in the neighbourhood of $\boldsymbol{\theta}^*$, the EM algorithm is essentially a linear iteration, see e.g. [17]. However, as shown by Dempster et al. [8],

the greater the proportion of missing information (question mark entries in the data cases, see Section 3), the slower the rate of convergence (in the neighbourhood of $\boldsymbol{\theta}^*$).

Therefore, it is reasonable to classify data cases into "EM-hard" and "EM-easy" problems depending on the fraction of observed and latent parameters of the model. A variables is called *latent*, if the states of the variable are never observed. When this ratio is small, then EM exhibits fast convergence; if the ratio approaches unity, EM will exhibit slow convergence.

5 (Scaled) Conjugate Gradients

As described in Section 3, the simple gradient ascent algorithm uses a fixed step size δ when following the gradient, i.e. in every iteration the parameters are chosen according to

$$\theta^{k+1} = \theta^k + \delta \cdot \nabla_\theta \log L(\mathbf{D}, \theta^k) \,,$$

where θ^k denotes the vector of parameters in the k-th iteration. Though it will converge to a (local) maximum for small enough δ, it is not a priori clear how to choose δ. As a consequence, it often converges very slowly. Instead, it would be better to perform a series of *line searches* to choose δ in each iteration, i.e. to do a one dimensional iterative search for δ in the direction of the gradient $\nabla_{\theta^k} := \nabla_\theta \log L(\mathbf{D}, \theta^k)$ maximizing $\log L(\mathbf{D}, \theta^k + \delta \cdot \nabla_{\theta^k})$. A common line-search technique is to initially bracket a maximum and then isolate it using an exact algorithm like Brent's Method, but other methods are possible (see e.g. [25]). One of the problems with the resulting gradient ascent algorithm is that a maximization in one direction could spoil past maximizations. This problem is solved in *conjugate gradient* methods.

Conjugate Gradient

Conjugate gradients (CG) compute so-called conjugate directions $h_0, h_1, \ldots$, which are not interfering, and estimate the step size along these directions with line searches. As Nocedal [22] noted, the best variant of conjugate gradients is generally believed to be the *Polak-Ribiere* conjugate gradient. Following the schema of hill climbing, it iteratively performs two steps starting with $\theta_0 \in \mathcal{H}$ and $h_0 = \nabla_{\theta_0}$:

1. (Conjugate directions) Set the next direction h_{k+1} according to $h_{k+1} = \nabla_{\theta^{k+1}} + \gamma_k \cdot h_k$ where

$$\gamma_k = \frac{(\nabla_{\theta^{k+1}} - \nabla_{\theta^k}) \cdot \nabla_{\theta^{k+1}}}{\nabla_{\theta^k} \cdot \nabla_{\theta^k}} \,. \tag{12}$$

2. (Line search) Compute θ^{k+1} by maximizing $\log L(\mathbf{D}, \theta^k + \delta \cdot h_{k+1})$ in the direction of h_{k+1}.

For a detailed discussion, we refer to e.g. [16,22].

There are still several drawbacks of doing a line search. First, it introduces new problem-dependent parameters such as stopping criterion. Second, the line search involves several likelihood evaluations, i.e. network inferences which are known to be NP-hard [7]. Thus, the line search dominates the computational costs of conjugate gradients resulting in a serious disadvantage compared to the EM which does one inference per iteration. Therefore, it is not surprising that researchers in the Bayesian network learning community used inexact line search to reduce the complexity [26,23,1] when accelerating EM. However, the use of both exact and inexact line searches within conjugate gradients requires careful considerations [23,22]. Thus, we decided to use a variant of conjugate gradients called *scaled conjugate gradients* due to Møller [18] which led to significant speed up in the context of learning neural networks while preserving accuracy.

Scaled Conjugate Gradients

To underpin the modularity of gradient-based techniques, we will explain scaled conjugate gradients for minimizing a general (scoring) function $f : \mathbb{R}^k \mapsto \mathbb{R}$ as it is implemented e.g. in the Netlab library [21]. Using the log-likelihood as f and substituting f by $-f$ yields the corresponding maximum likelihood parameter estimation for (discrete) Bayesian network. Further details about scaled conjugate gradients can be found in [18,20].

Scaled conjugate gradients (SCGs) substitute the line search by a scaling of the step size δ depending on success in error reduction and goodness of a quadratic approximation of the likelihood. The approximation costs one additional inference per iteration compared to EM. Let $\tilde{f}_{\boldsymbol{\theta}}$ denote the quadratic approximation of f at $\boldsymbol{\theta}$, i.e.

$$\tilde{f}_{\boldsymbol{\theta}}(\mathbf{x}) = f(\boldsymbol{\theta}) + f'(\boldsymbol{\theta})^T \cdot \mathbf{x} + \frac{1}{2}\mathbf{x}^T \cdot f''(\boldsymbol{\theta}) \cdot \mathbf{x} .$$

Furthermore, let $\boldsymbol{\theta}^k$ be the last parameter estimation, and h_k the current conjugate direction. Then, the step size δ_k minimizing $\tilde{f}_{\boldsymbol{\theta}^k}(\boldsymbol{\theta}^k + \delta \cdot h_k)$ is given by

$$\delta_k = \frac{-h_k^T \cdot \tilde{f}'_{\boldsymbol{\theta}^k}(\boldsymbol{\theta}^k)}{h_k^T \cdot f''(\boldsymbol{\theta}^k) \cdot h_k} ,$$

provided that the Hessian $f''(\boldsymbol{\theta}^k)$ is positive definite. Assuming this, the next parameter estimation would be

$$\boldsymbol{\theta}^k = \tilde{f}_{\boldsymbol{\theta}^k}(\boldsymbol{\theta}^k + \delta_k \cdot h_k) .$$

However, the computation of the second order term $f''(\boldsymbol{\theta}^k) \cdot h_k$ in (5) is expensive. Therefore, it is approximated by

$$s_k = \frac{f'(\boldsymbol{\theta}^k + \sigma_k \cdot h_k) - f'(\boldsymbol{\theta}^k)}{\sigma_k} \tag{13}$$

for some small σ_k. The approximation (13) tends in the limit to the true value $f''(\boldsymbol{\theta}^k) \cdot h_k$. As Møller pointed out, the approximation (13) is poor when the current point is far from the desired minimum [9]. The Hessian becomes indefinite, and the search might fail. Møller proposed to introduce a scalar λ_k into (13) to regulate the indefiniteness of $f''(\boldsymbol{\theta}^k)$ by adding $\lambda_k \cdot h_k$ to the right-hand side of (13), i.e.,

$$s_k = \frac{f'(\boldsymbol{\theta}^k + \sigma_k \cdot h_k) - f'(\boldsymbol{\theta}^k)}{\sigma_k} + \lambda_k \cdot h_k .$$

Now, there are two cases depending on whether the Hessian is positive definite or not. To test on that, it is enough to evaluate the sign of

$$\alpha_k := h_k^T \cdot s_k .$$

In case that $\alpha_k \leq 0$, i.e. the Hessian is indefinite, the value λ_k is raised to ensure positive definiteness. The new value λ_k is set to

$$\bar{\lambda}_k = 2 \cdot \left(\lambda_k - \frac{\alpha_k}{|h_k|^2} \right) \, ,$$

where $\bar{\lambda}_k$ is the renamed, scaled λ_k. This leads to

$$\bar{\alpha}_k = -\alpha_k + \lambda_k \cdot |h_k|^2 \, .$$

Otherwise the scaling parameters λ_k and α_k remain unchanged. In both cases, the step size δ_k is given by

$$\delta_k = \frac{\mu_k}{\alpha_k} \, .$$

where $\mu_k := -h_k^T \cdot f'(\theta^k)$. The value λ_k directly scales the step size δ_k in the way, that the bigger the λ_k, the smaller the step size δ_k.

So far, we have assumed that the (artificially scaled) Hessian is a good approximation. However, even if it is positive definite, the approximation might be poor. To measure the quality of the approximation, Møller considers the relative error

$$\begin{aligned} \Delta_k &= \frac{f(\theta_k) - f(\theta_k + \delta_k \cdot h_k)}{f(\theta_k) - \tilde{f}_{\theta_k}(\delta_k \cdot h_k)} \\ &= \frac{2 \cdot \alpha_k \left(f(\theta_k) - f(\theta_k + \delta_k \cdot h_k) \right)}{-h_k^T \cdot f'(\theta^k)} \, . \end{aligned}$$

The closer Δ_k is to 1.0, the better is the approximation. Now, the scaling parameter λ_k is raised and lowered according to

$$\lambda_k \leftarrow \begin{cases} 0.25 \cdot \lambda_k & \Delta_k > 0.75 \\ \lambda_k + \frac{(1 - \Delta_k)\alpha_k}{|h_k|^2} & \Delta_k < 0.25 \end{cases} \, .$$

This leads to the overall algorithm as given in Table 1. In general, it may get stuck in local maxima, and may have problems on plateaux and ridges. Well-known techniques such as random restarts or random perturbations are possibilities to avoid them.

6   Experiments

In the experiments described below, we implemented all three algorithms, i.e., EM, conjugate-gradients, and scaled conjugate-gradients using Netica API (http://www.norsys.com) for Bayesian network inference. We adapted the conjugate gradient (CG) described in [25] to fullfil the constraints described above. Based on this code, we adapted the scaled conjugate gradient (SCG) as implemented in the Netlab library [21], see also [4]) with an upper bound on the scale parameter of $2 \cdot 10^6$. To see how sensitive the methods to (simple) priors are, we also implemented each of them using BDeu priors [11]. The prior was set to be 1 for all parameters.

Table 1. Basic scaled scaled conjugate gradient algorithm as introduced by Møller [18]. The parameters σ and λ_1 are usually set in between $0 < \sigma \leq 10^{-4}$ and $0 < \lambda_1 \leq 10^{-6}$. The *converged* condition in line 21 loop refers to some stopping criterion such as stopping when a change in log-likelihood is less than some user-specified threshold from one iteration to the next

```
0:   Initialize θ₁, σ, λ₁, and λ̄₁ = 0.0
1:   Compute gradient and the conjugate direction h₀ according to (11)
2:   success := true and k = 1
3:   If success = true then        /* calculate second oder information*/
4:      σₖ := σ/hₖ
5:      sₖ := [f'(θᵏ +σₖ · hₖ) − f'(θᵏ)] /σₖ + λₖ · hₖ
6:      αₖ := hₖᵀ · sₖ
7:   Scale αₖ, i.e. αₖ := αₖ + (αₖ − ᾱₖ)) · |hₖ|²
8:   if αₖ ≤ 0 then                 /* make Hessian positive definite*/
9:      λ̄ₖ := 2 · [λₖ − αₖ/|hₖ|²]
6:      ᾱₖ := −αₖ + λₖ · |hₖ|²
7:      λₖ := λ̄ₖ
8:   μₖ := −hₖᵀ · f'(θᵏ)            /* Calculate new step size */
9:   δₖ := μₖ/αₖ
10:  Measure the quality Δₖ of the approximation
11:  if δₖ ≥ 0 then                 /* successful reduction in error */
12:     success := true
13:     Compute new conjugate direction hₖ according to (12)
14:     λ̄ₖ := 0
14:     if Δₖ > 0.75 then           /* reduce scale parameter */
15:        λₖ := 0.25 · λₖ
16:  else                           /* reduction in error is not possible */
17:     λ̄ₖ := λₖ
18:     success := false
19:  if Δₖ < 0.25 then              /* increase scale parameter */
20:     λₖ := λₖ + (1 − Δₖ) · αₖ/|hₖ|²
21:  if converged terminate
22:  else go to step 3
```

Methodology

Data were generated from four Bayesian networks whose main characteristics are described in Table 2. From each target network, we generated a test set of 10000 data cases, and training sets of 100, 200, 500 and 1000 data cases with a fraction of 0, 0.1, 0.2 and 0.3 of values missing at random of the observed nodes. The values of latent nodes were never observed. For each training set, five different random initial sets of parameters were tried. To do so, we initialized each conditional probability table entry as a uniform random sample from $[0,1]$, and normalized the conditional probability tables afterwards. We ran each algorithm on each data set starting from each of the generated initial sets of parameters. We used a simple, typical stopping

Table 2. Description of the networks used in the experiments

Name	Description
Alarm	Well-known benchmark network for the ICU ventilator management. It consists of 37 nodes and 752 parameters [2]. No latent variables.
Insurance	Well-known benchmark network for estimating the expected claim costs for a car insurance policyholder. It consists of 27 nodes (12 latent) and over 1419 parameters [3].
3-1-3, 5-3-5	Two artificial networks with a feed-forward architecture known from neural networks [3]. There are three fully connected layers of $3 \times 2 \times 3$ (resp. $5 \times 3 \times 5$) nodes. Nodes of the first and third layers have 3 possible states, nodes of the second 2.

criterion. We stopped when a limit of 200 iterations was exceeded or a change in average log-likelihood per case was less than 10^{-5} from one iteration to the next. We chose this low threshold to see the behaviour of the algorithms when they are close to a solution. The EM algorithm can be very slow if high accuracy of the estimate is necessary, whereas (advanced) gradients work well under these conditions. Although discrete Bayesian network are said to be robust against small changes, there exist situations where small variations can lead to significant changes in computed queries [5].

All learned models were evaluated on the test set using a standard measure, the *normalized-loss*

$$\frac{1}{N} \sum_{l=1}^{N} (\log P^*(\mathbf{d}_l) - \log P(\mathbf{d}_l)) ,$$

where P^* is the probability of the generating distribution. Normalized-loss measures the additional penalty for using an approximation instead of the true model, and is also a cross-entropy estimate. The closer the normalized-loss is to zero, the better. We report on the average performance and running time. The latter is dominated by the Bayesian network inference, because except for the likelihood evaluations all computations are linear. Therefore, we consider *EM-equivalent* iterations, i.e. likelihood evaluations. Each iteration of the scaled conjugate gradient consists of two EM-equivalent iterations.

Results

The results on the normalized-loss measure are shown in Tables 3 and 4. Without setting priors, SCG clearly outperformed EM, as Table 3 shows. Therefore, we concentrate in the evaluation on the results with priors, cf. Table 4. We omit CG from further investigations because it did not terminated in reasonable time on **Insurance** and **Alarm**. No run was finished after one week on a Pentium III, 450 MHz. However, on **3-1-3** and **5-3-5**,

Table 3. Average normalized-loss of **Insurance, Alarm, 3-1-3,** and **5-3-5** on an independent test set consisting of 10000 data cases **without BDeu priors**. A bold term indicates the method which estimated the best model (CG not considered). When there is no value for conjugate gradient (CG) this indicates that one run of it took longer than one week on this domain

Domain / # hidden (in %)	Meth.	Number of data cases			
		100	200	500	1000
Insurance	EM	12.20 ± 2.03	9.18 ± 0.48	2.56 ± 0.41	0.72 ± 0.18
0.0	SCG	**8.40 ± 0.74**	**3.92 ± 0.76**	**1.33 ± 0.29**	**0.56 ± 0.19**
0.1	EM	20.98 ± 2.55	12.25 ± 2.15	3.03 ± 0.75	0.90 ± 0.15
	SCG	**8.10 ± 1.42**	**4.22 ± 1.19**	**1.76 ± 0.28**	**0.76 ± 0.04**
0.2	EM	22.94 ± 2.20	14.98 ± 3.01	4.60 ± 0.67	0.97 ± 0.19
	SCG	**8.85 ± 2.62**	**5.80 ± 0.29**	**2.64 ± 0.49**	1.06 ± 0.19
0.3	EM	28.78 ± 4.44	14.87 ± 1.63	6.11 ± 0.75	1.47 ± 0.16
	SCG	**12.26 ± 0.75**	**7.04 ± 1.17**	**3.62 ± 0.70**	1.43 ± 0.08
Alarm	EM	−0.66 ± 0.01	−0.44 ± 0.02	−0.15 ± 0.01	−0.03 ± 0.00
0.0	SCG	3.79 ± 0.18	2.41 ± 0.20	1.01 ± 0.05	0.53 ± 0.04
0.1	EM	13.34 ± 1.38	7.01 ± 0.87	3.30 ± 0.08	1.42 ± 0.07
	SCG	**5.94 ± 0.21**	**2.53 ± 0.13**	**1.04 ± 0.05**	**0.54 ± 0.02**
0.2	EM	18.47 ± 1.43	8.91 ± 0.76	2.75 ± 0.43	1.61 ± 0.06
	SCG	**6.06 ± 0.26**	**3.18 ± 0.09**	**1.09 ± 0.07**	**0.64 ± 0.03**
0.3	EM	16.63 ± 1.31	11.41 ± 0.79	3.98 ± 0.13	1.68 ± 0.05
	SCG	**6.76 ± 0.40**	**4.07 ± 0.27**	**1.70 ± 0.09**	**0.76 ± 0.04**
3-1-3	EM	**0.30 ± 0.26**	**0.12 ± 0.00**	**0.03 ± 0.00**	**0.02 ± 0.00**
0.0	SCG	0.57 ± 0.04	**0.12 ± 0.03**	**0.03 ± 0.00**	**0.02 ± 0.00**
	CG	0.56 ± 0.08	0.10 ± 0.00	0.03 ± 0.00	0.02 ± 0.00
0.1	EM	0.46 ± 0.35	0.18 ± 0.02	**0.06 ± 0.00**	**0.02 ± 0.00**
	SCG	**0.31 ± 0.05**	**0.17 ± 0.02**	**0.06 ± 0.00**	**0.02 ± 0.00**
	CG	0.31 ± 0.04	0.17 ± 0.02	0.06 ± 0.00	0.02 ± 0.00
0.2	EM	2.08 ± 0.90	0.28 ± 0.08	**0.07 ± 0.01**	**0.04 ± 0.00**
	SCG	**0.86 ± 0.31**	**0.20 ± 0.08**	0.11 ± 0.03	**0.04 ± 0.00**
	CG	0.79 ± 0.26	0.24 ± 0.09	0.11 ± 0.02	0.04 ± 0.00
0.3	EM	1.47 ± 0.18	0.90 ± 0.16	**0.06 ± 0.00**	0.06 ± 0.05
	SCG	**0.77 ± 0.28**	**0.68 ± 0.22**	**0.06 ± 0.00**	**0.02 ± 0.00**
	CG	0.79 ± 0.31	0.72 ± 0.16	0.06 ± 0.00	0.03 ± 0.00
5-3-5	EM	17.36 ± 0.92	8.38 ± 1.23	1.86 ± 0.41	0.43 ± 0.08
0.0	SCG	**6.00 ± 0.88**	**2.96 ± 1.32**	**0.94 ± 0.09**	**0.28 ± 0.02**
	CG	5.15 ± 0.84	3.28 ± 0.61	0.84 ± 0.12	0.29 ± 0.02
0.1	EM	19.37 ± 2.14	11.76 ± 2.05	3.03 ± 0.34	1.03 ± 0.22
	SCG	**7.52 ± 1.59**	**4.43 ± 0.66**	**1.47 ± 0.34**	**0.47 ± 0.05**
	CG	6.50 ± 0.65	4.08 ± 0.35	1.34 ± 0.07	0.44 ± 0.06
0.2	EM	24.81 ± 1.67	13.88 ± 2.74	4.09 ± 0.63	1.04 ± 0.16
	SCG	**7.64 ± 0.93**	**5.58 ± 0.66**	**1.56 ± 0.61**	**0.72 ± 0.07**
	CG	8.08 ± 1.00	4.50 ± 0.50	1.62 ± 0.24	0.72 ± 0.09
0.3	EM	26.53 ± 2.76	13.59 ± 2.25	4.88 ± 0.93	1.59 ± 0.30
	SCG	**8.32 ± 1.15**	**5.53 ± 0.70**	**2.68 ± 0.15**	**0.95 ± 0.19**
	CG	7.99 ± 0.72	4.45 ± 0.26	2.68 ± 0.45	0.92 ± 0.12

CG showed the expected behaviour. Considering only the normalized-loss, it reached normalized-losses slightly closer to zero than EM and SCG (mean difference around 0.001). The number of iterations on **3-1-3** was on average two times higher than for EM, on **5-3-5** they were on average twice lower. However, the number of EM-equivalent iterations was in all cases much higher than for EM and SCG. Furthermore, CG was sensitive to the initial set of parameters.

Table 4. Average normalized-loss of **Insurance, Alarm, 3-1-3**, and **5-3-5** on an independent test set consisting of 10000 data cases **with BDeu priors**. The mean and standard deviation of five restarts are reported. A bold term indicates the method which estimated the best model (CG not considered). When there is no value for conjugate gradient (CG) this indicates that one run of it took longer than one week on this domain

Domain / # hidden (in %)	Meth.	Number of data cases			
		100	200	500	1000
Insurance	EM	0.99 ± 0.04	**0.64 ± 0.03**	0.37 ± 0.02	0.20 ± 0.01
0.0	SCG	1.82 ± 1.76	0.62 ± 0.02	**0.33 ± 0.03**	**0.51 ± 0.60**
0.1	EM	1.08 ± 0.05	0.68 ± 0.04	0.37 ± 0.02	0.23 ± 0.01
	SCG	1.14 ± 0.35	0.69 ± 0.11	0.89 ± 0.67	0.30 ± 0.08
0.2	EM	1.07 ± 0.03	0.76 ± 0.03	0.45 ± 0.02	0.25 ± 0.02
	SCG	1.35 ± 0.48	**0.76 ± 0.06**	**0.44 ± 0.06**	**0.37 ± 0.24**
0.3	EM	1.35 ± 0.04	0.94 ± 0.02	**0.52 ± 0.03**	0.29 ± 0.01
	SCG	**1.59 ± 0.71**	**1.25 ± 0.39**	0.62 ± 0.13	**0.38 ± 0.09**
Alarm	EM	0.81 ± 0.00	**0.62 ± 0.00**	**0.29 ± 0.00**	0.16 ± 0.00
0.0	SCG	1.01 ± 0.25	0.94 ± 0.37	0.33 ± 0.02	**0.17 ± 0.00**
0.1	EM	1.28 ± 0.00	0.68 ± 0.00	0.31 ± 0.00	0.18 ± 0.00
	SCG	**1.81 ± 0.74**	**0.87 ± 0.27**	0.31 ± 0.00	0.19 ± 0.00
0.2	EM	1.10 ± 0.00	0.64 ± 0.00	0.31 ± 0.00	0.24 ± 0.00
	SCG	1.31 ± 0.19	**1.25 ± 0.80**	0.31 ± 0.00	0.24 ± 0.01
0.3	EM	1.22 ± 0.00	**0.65 ± 0.00**	**0.41 ± 0.00**	**0.21 ± 0.00**
	SCG	**1.28 ± 0.08**	0.66 ± 0.01	1.12 ± 0.96	0.22 ± 0.00
3-1-3	EM	**0.15 ± 0.01**	0.07 ± 0.00	**0.03 ± 0.00**	0.02 ± 0.00
0.0	SCG	0.16 ± 0.02	**0.07 ± 0.00**	0.03 ± 0.00	**0.02 ± 0.00**
	CG	0.17 ± 0.04	0.07 ± 0.00	0.04 ± 0.02	0.02 ± 0.00
0.1	EM	**0.13 ± 0.02**	0.13 ± 0.03	0.05 ± 0.00	**0.02 ± 0.00**
	SCG	**0.13 ± 0.01**	**0.11 ± 0.01**	0.06 ± 0.03	0.02 ± 0.00
	CG	0.13 ± 0.02	0.11 ± 0.02	0.05 ± 0.00	0.02 ± 0.01
0.2	EM	**0.20 ± 0.01**	**0.16 ± 0.04**	**0.05 ± 0.00**	**0.03 ± 0.00**
	SCG	0.25 ± 0.05	0.10 ± 0.01	0.05 ± 0.00	0.04 ± 0.03
	CG	0.20 ± 0.03	0.10 ± 0.01	0.06 ± 0.00	0.03 ± 0.00
0.3	EM	**0.15 ± 0.00**	**0.16 ± 0.01**	**0.03 ± 0.00**	0.05 ± 0.05
	SCG	0.17 ± 0.02	0.17 ± 0.01	0.04 ± 0.03	**0.02 ± 0.00**
	CG	0.16 ± 0.02	0.18 ± 0.04	0.03 ± 0.00	0.02 ± 0.00
5-3-5	EM	0.23 ± 0.01	0.22 ± 0.01	0.15 ± 0.01	0.11 ± 0.00
0.0	SCG	**0.19 ± 0.03**	**0.21 ± 0.01**	**0.13 ± 0.02**	**0.10 ± 0.01**
	CG	0.14 ± 0.07	0.13 ± 0.07	0.09 ± 0.05	0.06 ± 0.03
0.1	EM	0.19 ± 0.01	0.21 ± 0.01	0.18 ± 0.00	0.13 ± 0.01
	SCG	**0.14 ± 0.04**	**0.20 ± 0.01**	**0.15 ± 0.02**	**0.09 ± 0.02**
	CG	0.12 ± 0.06	0.13 ± 0.07	0.11 ± 0.06	0.08 ± 0.04
0.2	EM	0.16 ± 0.01	0.18 ± 0.01	0.16 ± 0.01	0.14 ± 0.01
	SCG	**0.12 ± 0.03**	**0.14 ± 0.03**	**0.14 ± 0.02**	**0.13 ± 0.01**
	CG	0.09 ± 0.05	0.10 ± 0.05	0.09 ± 0.05	0.08 ± 0.04
0.3	EM	0.25 ± 0.01	0.17 ± 0.00	0.25 ± 0.01	0.15 ± 0.00
	SCG	**0.20 ± 0.04**	**0.16 ± 0.01**	**0.21 ± 0.03**	**0.13 ± 0.02**
	CG	0.15 ± 0.08	0.10 ± 0.05	0.15 ± 0.08	0.09 ± 0.05

Quality: EM and SCG reached similar normalized-losses. As expected, a higher number of data cases led to a lower normalized-loss. Similarly, priors led to a lower normalized-loss. The absolute differences in averaged normalized-losses between using priors and not using priors are much smaller for SCG than for EM. Both were relatively insensitive to the initial set of parameters. The EM tended to have a slightly lower variation, but a *one-tailed, paired sampled t test* over all experiments (with priors) showed that

Table 5. Average EM equivalent iterations performed by EM and SCG. The mean and standard deviation of five restarts are reported (bracket term: without BDeu priors). An entry of 199 ± 0 for EM indicates that it exceeded the maximum number of iteration in all restarts

Domain / # hidden (in %)	Method	Number of data cases			
		100	200	500	1000
Insurance 0.0	EM	$100 \pm 28(144 \pm 34)$	$143 \pm 40(199 \pm 0)$	$188 \pm 17(199 \pm 0)$	$188 \pm 13(199 \pm 0)$
	SCG	$95 \pm 40(307 \pm 75)$	$132 \pm 24(272 \pm 67)$	$138 \pm 42(380 \pm 34)$	$156 \pm 82(335 \pm 130)$
0.1	EM	$129 \pm 37(128 \pm 32)$	$178 \pm 19(193 \pm 9)$	$168 \pm 25(199 \pm 0)$	$177 \pm 27(199 \pm 0)$
	SCG	$42 \pm 14(280 \pm 36)$	$136 \pm 58(302 \pm 74)$	$91 \pm 67(318 \pm 43)$	$89 \pm 54(355 \pm 42)$
0.2	EM	$171 \pm 26(153 \pm 29)$	$161 \pm 42(199 \pm 0)$	$196 \pm 6(199 \pm 0)$	$178 \pm 19(199 \pm 0)$
	SCG	$64 \pm 25(251 \pm 68)$	$141 \pm 47(357 \pm 67)$	$112 \pm 42(335 \pm 44)$	$106 \pm 71(377 \pm 29)$
0.3	EM	$132 \pm 38(180 \pm 20)$	$156 \pm 27(199 \pm 0)$	$187 \pm 13(199 \pm 0)$	$180 \pm 36(199 \pm 0)$
	SCG	$77 \pm 41(272 \pm 69)$	$79 \pm 43(328 \pm 49)$	$89 \pm 57(312 \pm 59)$	$88 \pm 51(312 \pm 50)$
Alarm 0.0	EM	$1 \pm 0(1 \pm 0)$	$1 \pm 0(1 \pm 0)$	$1 \pm 0(1 \pm 0)$	$1 \pm 0(1 \pm 0)$
	SCG	$66 \pm 18(97 \pm 12)$	$75 \pm 46(114 \pm 13)$	$104 \pm 20(148 \pm 15)$	$130 \pm 31(162 \pm 14)$
0.1	EM	$7 \pm 0(10 \pm 1)$	$8 \pm 0(13 \pm 3)$	$6 \pm 0(9 \pm 0)$	$6 \pm 0(8 \pm 0)$
	SCG	$83 \pm 36(130 \pm 8)$	$83 \pm 46(131 \pm 17)$	$140 \pm 18(147 \pm 12)$	$113 \pm 14(158 \pm 9)$
0.2	EM	$11 \pm 0(17 \pm 2)$	$11 \pm 0(16 \pm 3)$	$10 \pm 2(13 \pm 2)$	$9 \pm 0(12 \pm 1)$
	SCG	$84 \pm 22(167 \pm 26)$	$87 \pm 13(174 \pm 27)$	$100 \pm 7(170 \pm 32)$	$120 \pm 26(142 \pm 9)$
0.3	EM	$16 \pm 1(25 \pm 2)$	$14 \pm 0(23 \pm 2)$	$15 \pm 1(18 \pm 1)$	$13 \pm 1(17 \pm 1)$
	SCG	$83 \pm 27(163 \pm 22)$	$108 \pm 26(173 \pm 7)$	$69 \pm 45(164 \pm 6)$	$106 \pm 10(174 \pm 11)$
3-1-3 0.0	EM	$21 \pm 10(27 \pm 7)$	$21 \pm 2(22 \pm 2)$	$18 \pm 3(22 \pm 4)$	$16 \pm 1(18 \pm 1)$
	SCG	$49 \pm 24(58 \pm 15)$	$63 \pm 6(72 \pm 30)$	$66 \pm 15(83 \pm 6)$	$60 \pm 7(67 \pm 14)$
0.1	EM	$28 \pm 4(46 \pm 13)$	$26 \pm 7(35 \pm 17)$	$27 \pm 7(36 \pm 8)$	$23 \pm 7(27 \pm 8)$
	SCG	$54 \pm 11(76 \pm 11)$	$57 \pm 13(72 \pm 35)$	$44 \pm 8(87 \pm 15)$	$62 \pm 5(76 \pm 8)$
0.2	EM	$24 \pm 9(36 \pm 11)$	$52 \pm 36(43 \pm 14)$	$30 \pm 6(36 \pm 6)$	$27 \pm 6(28 \pm 7)$
	SCG	$54 \pm 23(58 \pm 15)$	$60 \pm 7(89 \pm 10)$	$51 \pm 10(61 \pm 20)$	$54 \pm 17(80 \pm 11)$
0.3	EM	$37 \pm 9(40 \pm 8)$	$46 \pm 8(62 \pm 18)$	$32 \pm 3(36 \pm 3)$	$35 \pm 14(40 \pm 17)$
	SCG	$55 \pm 36(70 \pm 10)$	$68 \pm 26(78 \pm 12)$	$57 \pm 23(80 \pm 12)$	$57 \pm 12(74 \pm 16)$
5-3-5 0.0	EM	$97 \pm 7(38 \pm 6)$	$92 \pm 40(52 \pm 13)$	$120 \pm 39(78 \pm 17)$	$113 \pm 24(85 \pm 7)$
	SCG	$75 \pm 37(151 \pm 60)$	$134 \pm 33(108 \pm 48)$	$151 \pm 50(166 \pm 7)$	$150 \pm 37(142 \pm 6)$
0.1	EM	$160 \pm 12(56 \pm 16)$	$116 \pm 19(67 \pm 14)$	$142 \pm 25(84 \pm 11)$	$162 \pm 14(126 \pm 20)$
	SCG	$80 \pm 48(207 \pm 45)$	$130 \pm 13(184 \pm 34)$	$112 \pm 22(141 \pm 21)$	$99 \pm 35(158 \pm 19)$
0.2	EM	$128 \pm 35(72 \pm 17)$	$190 \pm 19(78 \pm 16)$	$160 \pm 10(105 \pm 20)$	$141 \pm 10(100 \pm 19)$
	SCG	$99 \pm 62(180 \pm 38)$	$97 \pm 37(168 \pm 25)$	$150 \pm 58(140 \pm 37)$	$114 \pm 23(163 \pm 25)$
0.3	EM	$166 \pm 33(65 \pm 7)$	$99 \pm 15(78 \pm 16)$	$142 \pm 29(112 \pm 22)$	$157 \pm 23(131 \pm 17)$
	SCG	$72 \pm 42(170 \pm 21)$	$88 \pm 20(179 \pm 29)$	$121 \pm 28(151 \pm 9)$	$130 \pm 26(143 \pm 22)$

only a 0.045 difference in mean of EM and SCG ($p = 0.05$) was significant. EM reached normalized-losses slightly closer to zero in all domains but **5-3-5** where a 0.02 difference in mean was significant ($p = 0.05$) favoring SCG. However, the t test takes all restarts into account, but in practice, one usually selects for each method the best model out of the set of restarts. To compare the performance of the best models, we applied a *sign test*. The sign test yielded that the number of cases, where the best model of SCG outperformed EM's best model, was large, 43/21, but there were variations over the domains: **Insurance** 12/4, **Alarm** 9/7, **3-1-3** 6/10, and **5-3-5** 16/0.

Number of Iterations: Table 5 summarizes the number of EM-equivalent iterations. EM was significantly faster than SCG on **Alarm** . However, this did not carry over to all domains. A *one-tailed, paired sampled t test* (over all corresponding experiments with priors) showed that EM ran significantly faster than SCG on **Alarm**, $p = 0.00005$, and on **3-1-3**, $p = 0.0005$. The difference in mean decreased from 82 for **Alarm** to 24 for **3-1-3**. This changed for **5-3-5** and **Insurance**. Here, the same test showed that SCG ran significantly faster than EM on **5-3-5**, $p = 0.0005$, and on **Insurance**, $p = 0.00005$. The difference in mean increased from 23 for **5-3-5** to 62 for **Insurance**.

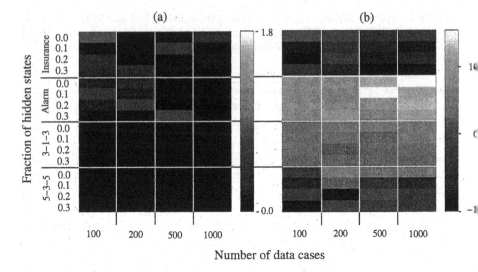

Fig. 2. Differences between SCG's and EM's behaviours. In both pictures, the mean of five restarts were subtracted. (a) Differences (SCG − EM) of average normalized-loss over **Insurance**, **Alarm**, **3-1-3**, and **5-3-5** on an independent test set consisting of 10000 data cases. (b) Difference (SCG − EM) of EM equivalent iterations

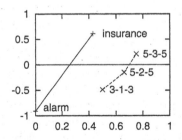

Fig. 3. Percentage of latent parameters vs. SCG's speed-up. The domains are grouped into network structures with (solid line) and without (dashed line) encoded prior information. Both curves confirm the theory that the more latent parameters in the data cases, the slower the convergence of the EM algorithm (when compared to scaled conjugate gradients)

Thus, an acceleration of conjugate gradients is possible. Next to the quality of the estimated parameters, SCG seems to have some EM-like characteristics. It is fast, and (relatively) insensitive to the initial set of parameters. This is remarkable given that gradients are usually considered to perform worse than EM (on discrete Bayesian networks). However, the behaviour of both SCG and EM varied over the domain.

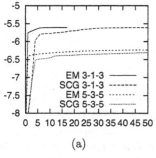
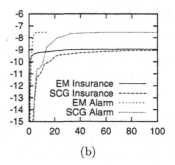

(a)   (b)

Fig. 4. Averaged learning curves. (a) Learning curves averaged over five restarts for 1000 data cases, no missing values. (b) Learning curves averaged over five restarts for 1000 data cases. Fraction of 0.3 missing values for **Alarm**, no missing values for **Insurance**

"EM-easy" vs. "EM-hard": As explained in Section 4, it is reasonable to classify domains into "EM-easy" and "EM-hard" ones. Figure 2 shows the differences in average normalized-loss and in EM equivalent iterations. **Alarm** and **Insurance** are networks where prior knowledge is encoded into the structure, whereas **3-1-3** and **5-3-5** have uninformative structures. Figure 2 (a) shows that on average EM tended to yield better normalized-loss on prior knowledge networks. However, within both types of network structures, the fraction of latent parameters seems to correlate to "EM-hard" domains (in particular with respect to the number of iterations). This was predicted by the theory, cf. Section 4. Figure 2 (b) shows that SCG converged faster on networks with a higher number of latent parameters, i.e. **Insurance** and **5-3-5**. Figure 3(a) shows the percentage of latent parameters vs. speed-up which is defined as the quotient of average EM-equivalent iterations (over corresponding restarts) of EM and SCG minus one. To test whether SCGs performs well on "EM-hard" domains, i.e. that these domains are "SCG-easy" ones, we investigated the network **5-2-5**. It is a variant of **5-3-5** with only two latent variables. The observable variables have two, the latent one three states. We tested **5-2-5** with the same experiments as before but restricted to 1000 data cases. SCG's speed-up on **5-2-5** was -0.15, i.e. between the speed-up of **3-1-3** and **5-3-5**. Although this is preliminary, the result is promising.

Finally, Figs. 4(b) and 4(c) show that the EM algorithm was usually faster when far away from the solution. Therefore, the EM seems to be the domain independent method of choice. But, the results show that SCG is not only competitive but that it can be superior to EM, and in contrast to CG, it learned **Insurance** and **Alarm**.

7   Related Work

Both, EM and gradients are well-known parameter estimation techniques. For a general introduction see [17,16]. Bishop [4] and Møller [18] show how to use (scaled) conjugate gradients to estimate parameters of neural networks. Lauritzen [15] introduced EM for Bayesian networks (see also Heckerman [10] for a nice tutorial). The original work on gradient-based approaches in the context of Bayesian networks was done by Binder *et al.* [3], some recent work by Jensen [13,14]. However, we are neither aware of any direct experimental comparison between the two approaches nor of research investigating accelerated conjugate gradients such as scaled ones. Binder *et al.* [3] used traditional conjugate gradients, and did not compare more advanced gradient-based and EM-based approaches. Bauer *et al.* [1] reported on experiments with different learning algorithms for Bayesian networks, but they focused on accelerations of the EM. They did not report on results of conjugate gradients. Jensen [13] derived a formula for the gradient which relates it to sensitivity analyses of Bayesian networks. Jensen did not compare it with the EM. Thiesson [26] discussed (traditional) conjugate gradient accelerations of the EM, but does not report on experiments. Ortiz and Kaelbling [23] conducted experiments with several learning algorithms (mainly accelerated EMs) for continuous models, namely density estimation with a mixture of Gaussian. They argued "that conjugate gradient by itself is not a good idea since it requires a very precise line search when it is far away from a solution". Our experiments do not support this for discrete Bayesian networks. Ortiz and Kaelbling also classified their domains into "easy" and "hard" ones. Like in our case, accelerated methods improved convergence on hard problems. However, they accelerated EM and not conjugate gradients.

8   Conclusions

In this case study, we experimentally compared EM and gradient-based algorithms for maximum likelihood parameter estimation on several Bayesian networks. To overcome the expensive line search of traditional conjugate gradients, we applied *scaled conjugate gradients*. They are novel in the context of Bayesian networks.

The experiments show that scaled conjugate gradients are competitive with EM in both quality of the estimate and convergence speed. To the best of our knowledge, this is the first time that gradient-based techniques are reported to be competitive with EM with respect to quality and speed for discrete Bayesian networks. In general EM seems to be the domain independent method of choice, but which technique is superior depends on properties of the domain.

We identified (in theory and in practice) "EM-easy" and "EM-hard" domains. Scaled conjugate gradients seem especially well suited in the presence of many latent parameters, i.e. "EM-hard" instance. Given that gradients are more flexible – as they allow us to learn non-multinomial parameterization (e.g. noisy-or) using the chain-rule of derivatives and conditional probability distributions using Bayes' rule – and that they are strongly related to methods in neural networks training, scaled conjugate gradients are a promising alternative to EM for parameter estimation of (discrete) Bayesian networks. Our results confirm that the number of latent parameters determines "EM-hard" instances.

A very promising line of future research seems to be accelerating EM using scaled conjugate gradients, and applying stochastic scaled conjugate gradients [19] to the problem of *online-learning* of Bayesian networks.

Acknowledgements The authors would like to thank Luc De Raedt and Wolfram Burgard for helpful discussions. The first author was supported by the European Union IST programme, IST-2001-33053 (Application of Probabilistic Inductive Logic Programming - APRIL).

References

1. E. Bauer, D. Koller, and Y. Singer. Update Rules for Parameter Estimation in Bayesian Networks. In D. Geiger and P. P. Shenoy, editors, *Proceedings of the Thirteenth Annual Conference on Uncertainty in Artificial Intelligence (UAI-97)*, pages 3–13, Providence, Rhode Island, USA, 1997. Morgan Kaufmann.
2. I. Beinlich, H. Suermondt, R. Chavez, and G. Cooper. The ALARM monitoring system: A case study with two probabilistic inference techniques for belief networks. In J. Hunter, editor, *Proceedings of the Second European Conference on Artificial Intelligence and Medicine (AIME-89)*, volume 38 of *Lecture Notes in Medical Informatics*, pages 247–256, City University, London, UK, 1989. Springer-Verlag.
3. J. Binder, D. Koller, S. Russell, and K. Kanazawa. Adaptive Probabilistic Networks with Hidden Variables. *Machine Learning*, 29(2/3):213–244, 1997.
4. C. M. Bishop. *Neural Networks for Pattern Recognition*. Oxford University Press, 1995.
5. H. Chan and A. Darwiche. When do numbers really matter? *Journal of Artificial Intelligence Research (JAIR)*, 17:265–287, 2002.
6. D. M. Chickering, D. Geiger, and D. Heckerman. Learning Bayesian networks is NP-hard. Technical Report MSR-TR-94-17, Microsoft Research, Advanced Technology Division, Microsoft Corporation, One Microsoft Way, Redmond, WA 98052, USA, 1994.
7. G. F. Cooper. The computational complexity of probabilistic inference using Bayesian belief networks. *Artificial Intelligence*, 42:393–405, 1990.
8. A. Dempster, N. Larid, and D. Rubin. Maximum likelihood from incomplete data via the EM algorithm. *Journal of the Royal Statistical Society*, 39:1–38, 1977.
9. P. E. Gill, W. Murray, and M. H. Wright. *Practical Optimization*. Acadamic Press, 1981.

10. D. Heckerman. A tutorial on learning with Bayesian networks. Technical Report MSR-TR-95-06, Microsoft Research, Advanced Technology Division, Microsoft Corporation, One Microsoft Way, Redmond, WA 98052, USA, 1995.

11. D. Heckerman, D. Geiger, and D. M. Chickering. Learning Bayesian networks: The combination of knowledge and statistical data. *Machine Learning*, 20(3):197–243, 1995.

12. Mathworks Inc.   Optimization   Toolbox   2.2   for   Matlab. http://www.mathworks.com.

13. F. V. Jensen. Gradient descent training of bayesian networks. In A. Hunter and S. Parsons, editors, *Proceedings of the Fifth European Conference on Symbolic and Quantitative Approaches to Reasoning with Uncertainty (ECSQARU-99)*, volume 1638 of *Lecture Notes in Computer Science*, pages 190–200. Springer, 1999.

14. F. V. Jensen. *Bayesian networks and decision graphs*. Springer-Verlag, 2001.

15. S. L. Lauritzen. The EM algorithm for graphical association models with missing data. *Computational Statistics and Data Analysis*, 19:191–201, 1995.

16. D. G. Luenberger. *Linear and Nonlinear Programming*. Addison-Wesley, 2. edition, 1984.

17. G. McLachlan and T. Krishnan. *The EM Algorithm and Extensions*. Wiley, New York, 1997.

18. M. Møller. A Scaled Conjugate Gradient Algoritm for Fast Supervised Learning. *Neural Networks*, 6:525–533, 1993.

19. M. Møller. Supervised Learning on Large Redundant Training sets. *International Journal Neural Systems*, 4(1):15–25, 1993.

20. M. Møller. *Efficient Training of Feed-Forward Neural Networks*. PhD thesis, Computer Science Department, Aarhus University, Denmark, 1997.

21. I. Nabney. *NETLAB: Algorithms for Pattern Recognition*. Advances in Pattern Recognition. Springer-Verlag, 2001. http://www.ncrg.aston.ac.uk/netlab/.

22. J. Nocedal. Theory of algorithms for unconstrained optimization. *Acta Numerica*, pages 199–242, 1992.

23. L. E. Ortiz and L. P. Kaelbling. Accelerating EM: An Empirical Study. In K. B. Laskey and H. Prade, editors, *Proceedings of the Fifteenth Annual Conference on Uncertainty in Articial Intelligence (UAI-99)*, pages 512–521, Stockholm, Sweden, 1999. Morgan Kaufmann.

24. J. Pearl. *Reasoning in Intelligent Systems: Networks of Plausible Inference*. Morgan Kaufmann, 2. edition, 1991.

25. W. H. Press, S. A. Teukolsky, W. T. Vetterling, and B. P. Flannery. *Numerical Recipes in C: The Art of Scientific Computing*. Cambridge University Press, 2. edition, 1993. http://www.nr.com.

26. B. Thiesson. Accelerated quantification of Bayesian networks with incomplete data. In U. M. Fayyad and R. Uthurusamy, editors, *Proceedings of First International Conference on Knowledge Discovery and Data Mining*, pages 306–311, Montreol, Canada, 1995. AAAI Press.

Learning Essential Graph Markov Models from Data

Robert Castelo[1] and Michael D. Perlman[2]

[1] Grup de Recerca en Informàtica Biomèdica, Departament de Ciències Experimentals i de la Salut, Universitat Pompeu Fabra, Psg. Marítim 37–49, 08003 Barcelona, Spain
robert.castelo@cexs.upf.es
[2] Department of Statistics, University of Washington, Box 354322, Seattle WA 98195–4322, USA
michael@ms.washington.edu

Abstract. In a model selection procedure where many models are to be compared, computational efficiency is critical. For acyclic digraph (ADG) Markov models (aka DAG models or Bayesian networks), each ADG Markov equivalence class can be represented by a unique chain graph, called an *essential graph (EG)*. This parsimonious representation might be used to facilitate selection among ADG models. Because EGs combine features of decomposable graphs and ADGs, a scoring metric can be developed for EGs with categorical (multinomial) data. This metric may permit the characterization of local computations directly for EGs, which in turn would yield a learning procedure that does not require transformation to representative ADGs at each step for scoring purposes, nor is the scoring metric constrained by Markov equivalence.

1 Introduction

Model selection, or statistical learning, among ADG models is constrained due to rapid growth in the number of ADGs as the number of nodes increases, so computational efficiency becomes critical. To this end, [10, eqn. (37)] show that the Bayes factor (the ratio of the two marginal likelihoods, cf. [9, pg. 250]) between two decomposable (DEC) graphical Markov models that differ in one single adjacency can be expressed in terms of at most four cliques from both junction trees, thereby facilitating selection among such models. Analogously, for the case of acyclic digraph Markov models, it suffices to compare the local scores for the variable whose parent set changes.

From a non-causal perspective we are interested in learning Markov equivalence classes of ADGs. [1] showed that each (possibly large) ADG Markov equivalence class can be represented by a *single* unique chain graph, called an *essential graph (EG)*, which combines features of decomposable graphs and ADGs. In this study we utilize this economical representation to facilitate selection among non-causal ADG models by directly scoring EG models. In particular, a scoring metric can be developed for EGs with categorical (multinomial) data. This metric may permit the characterization of local computations directly for EGs, which in turn would yield a learning procedure that does not require transformation to representative ADGs at each step for scoring purposes.

This characterization may depend, for instance, on a particular class of *local transformations* for EGs proposed by [19] that require at most two edge changes in the EG or an accompanying ADG, whose local scoring functions can be obtained. This yields a learning procedure that does not require transformation to representative ADGs at each step for scoring purposes. Furthermore, exploiting the recent works by [8] and [3] may lead to a characterization of local computations for our scoring metric that respects the graphical Markov model inclusion order - cf. [4, Section 3].

After reviewing terminology in Section 2, in Section 3 we present the factorization for pdfs that are Markov with respect to an EG G. The scoring metric for EGs with categorical data is given in Section 4, and related to those for decomposable graphs and ADGs in Section 5. The local transformations and corresponding local computations for EGs are discussed in Section 6, while concluding remarks are in Section 7.

2   Background Concepts

A random variable is denoted by an upper case letter indexed by a number or a lowercase letter, e.g. X_1 or X_i. A set or a vector of random variables is denoted by an upper case letter, e.g. X, which may be indexed by an uppercase letter, e.g. X_A. For notational convenience, we sometimes abbreviate X_i by i, $X_A \equiv \{X_i | i \in A\}$ by A, etc. All random variables in this article will be discrete.

A graph G is a pair (V, E) where V is a set of vertices and E is a set of edges. When every edge in E is undirected, G is an *undirected graph* (UG). When every edge in E is directed, G is a *directed graph*. If a directed graph G has no directed cycles, then G is an *acyclic digraph* (ADG).

Let G be a graph with vertex set V. The collection of random variables $X_V \equiv \{X_v | v \in V\}$ arises as a categorical dataset and follows a multinomial distribution P defined on a product space $\mathcal{X} = \times(\mathcal{X}_i | i \in V)$. We will use the term *level* to refer to a particular member $x \in \mathcal{X}$. In the present context, a graphical Markov model (GMM), denoted by $\mathbf{M}(G)$, is a family $\{P_\theta\}$ of multinomial distributions on $\mathcal{X}$ that satisfy the Markov properties determined by the graph G. In the Bayesian formulation, the parameter θ is itself random and follows a prior distribution, or *law*, usually denoted by π. It is common to assume that π is Dirichlet, the natural conjugate prior for the multinomial family, under which the posterior distribution of θ remains Dirichlet.

We shall use standard graph-theoretic terminology, such as *pa* for *parents*, *nd* for *nondescendants*, *bd* for *boundary*, and *nb* for *neighbors*. Refer to Lauritzen (1996) for standard GMM results and terminology.

A *chain graph* $G = (E, V)$ is a graph that may have both directed and undirected edges but is *adicyclic*, that is, has no fully or partially directed cycles. Let $\mathcal{T} \equiv \mathcal{T}(G)$ denote the set of *chain components* of G, i.e., the connected components of the graph obtained by removing all arrows from G. Let $\mathcal{D} \equiv \mathcal{D}(G)$ be the acyclic digraph with vertex set $\mathcal{T}$ and where, for $\tau_1, \tau_2 \in \mathcal{T}$, $\tau_1 \to \tau_2 \in \mathcal{D}$ iff $t_1 \to t_2$ for at least one pair $t_1 \in \tau_1$, $t_2 \in \tau_2$. (By adicyclicity, this is well defined.)

Let G be an undirected chordal (decomposable) graph with a set of cliques $\mathcal{C}$ and a set of separators $\mathcal{S}$. Note that some $S \in \mathcal{S}$ may occur in $\mathcal{S}$ more than once because it may repeat in the perfect numbering sequence of the cliques - cf. [10, pg. 1278, 1311]. [10, Theorem 2.6] showed that the unique Markov distribution over G having a specified consistent familiy $\{f_C \mid C \in \mathcal{C}\}$ of pdfs for its clique marginals, is given by the pdf

$$f(x) = \frac{\prod_{C \in \mathcal{C}} f_C(x_C)}{\prod_{S \in \mathcal{S}} f_S(x_S)}, \qquad (1)$$

where the separators set $\mathcal{S}$ incorporates $\nu(S)$ repetitions of $S \in \mathcal{S}$ that may occur in any given perfect numbering of the cliques $C \in \mathcal{C}$.

3   Factorization of a Multivariate Distribution According to an Essential Graph

Let D be an ADG with vertex set V and let $[D]$ denote its Markov equivalence class, that is, the set of all ADGs D' over V that determine the same Markov properties as D. [22] showed that $D' \in [D]$ if and only if D and D' have the same *skeleton* ($\equiv$ underlying undirect graph) and *immoralities* ($\equiv$ induced subgraphs of the form $a \to b \leftarrow c$). [1] showed that the entire Markov equivalence class $[D]$ is uniquely determined by the *essential graph* (EG) D^* defined as follows:

$$D^* = \cup\{D' \mid D' \in [D]\}.$$

Here the union is obtained by the rule that for any pair of vertices (a, b), the union of directed edges between a and b of like orientation is the common directed edge, while the union of directed edges between a and b with at least one opposite orientation is an undirected edge. Furthermore, they characterized those graphs that may occur as essential graphs by the following theorem:

Theorem 01 *A graph $G = (V, E)$ is the essential graph D^* for some ADG D with vertex set V iff G satisfies the following four conditions:*

(i) *G is a chain graph.*
(ii) *For each chain component $\tau \in \mathcal{T}(G)$, the UG G_τ is chordal.*
(iii) *G has no flags, i.e., no induced subgraphs of the form $a \to b - c$.*

(iv) *Each arrow $a \to b$ in G is "strongly protected", that is, occurs in at least one of the following four configurations as an induced subgraph of G:*

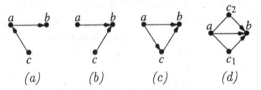

(a)   (b)   (c)   (d)

[2] provide the following definitions for the LWF and AMP block-recursive Markov properties for general chain graphs, hence in particular for essential graphs.

Definition 01 *A probability measure P on a product probability space $\mathcal{X} \equiv \times(\mathcal{X}_v | v \in V)$, where V indexes a vector of random variables $X = (X_1, \ldots, X_n)$, is said to be* LWF *(resp.,* AMP*) block-recursive G-Markovian if P satisfies the following conditions C1, C2, and C3 (resp., C1, C2, and C3*):*

(C1) $\forall \tau \in \mathcal{T} : X_\tau \perp\!\!\!\perp X_{(nd_\mathcal{D}(\tau) \setminus pa_\mathcal{D}(\tau))} \,|\, X_{pa_\mathcal{D}(\tau)}[P]$, *i.e., P is local $\equiv$ global $\mathcal{D}$-Markovian on $\mathcal{X}$.*

(C2) $\forall \tau \in \mathcal{T} :$ *the conditional distribution $P_{\tau | pa_\mathcal{D}(\tau)}$ is global G_τ-Markovian on $\mathcal{X}_\tau$.*

(C3) $\forall \tau \in \mathcal{T}, \forall \sigma \subseteq \tau : X_\sigma \perp\!\!\!\perp X_{(pa_\mathcal{D}(\tau) \setminus pa_G(\sigma))} \,|\, X_{pa_G(\sigma) \cup nb_G(\sigma)}[P]$.

(C3*) $\forall \tau \in \mathcal{T}, \forall \sigma \subseteq \tau : X_\sigma \perp\!\!\!\perp X_{(pa_\mathcal{D}(\tau) \setminus pa_G(\sigma))} \,|\, X_{pa_G(\sigma)}[P]$.

Note that if the chain graph G is in fact an essential graph, then by condition (iii) and Theorem 4 of [2], its AMP and LWF block-recursive Markov properties coincide. It follows from C1 and Lemma 4.1 of [2] that if a probability distribution P is LWF or AMP block-recursive G-Markovian for a chain graph G, then its pdf f admits the following recursive factorization:

$$f(x|G, \theta_G) = \prod_{\tau \in \mathcal{T}} f(x_\tau | x_{pa_G(\tau)}, \theta_\tau) \,. \tag{2}$$

(Here, since we shall consider P to be a member of a parametric statistical model, we include θ_τ to denote the parameters occurring in the conditional pdf and set $\theta_G = (\theta_\tau | \tau \in \mathcal{T})$.)

Note that $\theta_\tau = (\theta_{\tau, \rho} | \rho \in \mathcal{X}_{pa_G(\tau)})$, and therefore the parametrization in (2) involves only those parameters $\theta_{\tau, \rho}$ for which $\rho = x_{pa_G(\tau)}$:

$$f(x|G, \theta_G) = \prod_{\tau \in \mathcal{T}} f(x_\tau | x_{pa_G(\tau)}, \theta_{\tau, \rho}) \,. \tag{3}$$

(See the discussion in [9, Section 9.2].)

For the remainder of this paper, G shall denote an essential graph. By (ii), every chain component $\tau \in \mathcal{T}$ induces a chordal $\equiv$ decomposable graph

G_τ, so each term $f(x_\tau|x_{pa_G(\tau)}, \theta_{\tau,\rho})$ admits a further factorization according to the cliques and clique separators of G_τ (cf. (4) in Theorem 02 below) as shown by [10], Theorem 2.6. In fact, we now show that this further factorization is sufficient as well as necessary for a pdf to determine a G-Markovian distribution. (No positivity assumptions are needed - cf. [10, pg. 1279].)

Let $\mathcal{C}_\tau$ (resp., $\mathcal{S}_\tau$) denote the set of cliques (resp., clique separators) for the decomposable undirected graph G_τ, $\tau \in \mathcal{T} \equiv \mathcal{T}(G)$. Let P be a distribution on $\mathcal{X}$ that admits a pdf f.

Theorem 02 *Let G be an essential graph. The distribution P is AMP $\equiv$ LWF G-Markovian iff f factorizes as*

$$f(x|G, \theta_G) = \prod_{\tau \in \mathcal{T}} \left[\frac{\prod_{C \in \mathcal{C}_\tau} f(x_C|x_{pa_G(\tau)}, \theta_{C,\rho})}{\prod_{S \in \mathcal{S}_\tau} f(x_S|x_{pa_G(\tau)}, \theta_{S,\rho})} \right] . \tag{4}$$

PROOF. As already noted above, if P is AMP $\equiv$ LWF block-recursive G-Markovian then f satisfies (4). Conversely, if (4) holds it is straightforward to show that P satisfies C1 and C2 and that

$$f(x_\tau|x_{pa_G(\tau)}, \theta_{\tau,\rho}) = f(x_\tau|x_{pa_D(\tau)}, \theta_{\tau,\rho}) \tag{5}$$

for every $\tau \in \mathcal{T}$, so

$$X_\tau \perp\!\!\!\perp X_{(pa_D(\tau) \setminus pa_G(\tau))} \,|\, X_{pa_G(\tau)} \,[P] . \tag{6}$$

But if $\emptyset \neq \sigma \subseteq \tau \in \mathcal{T}$ then $pa_G(\sigma) = pa_G(\tau)$ because G is an essential graph and satisfies (iii), hence C3* holds.

Note that, in the previous theorem, $\theta_{C,\rho}$ and $\theta_{S,\rho}$ are subsets of $\theta_{\tau,\rho}$.

4   Bayesian Scoring Metric for Multinomial Data

Let $D = \{x^{(1)}, x^{(2)}, \ldots, x^{(N)}\}$ be a set of N exchangeable observations sampled from a multinomial distribution comprising the set of random variables $X = \{X_1, \ldots, X_n\}$. From a Bayesian perspective, we wish to choose a model, represented by an essential graph G, according to its posterior probability $p(G|D)$. By Bayes' theorem,

$$p(G|D) \propto f(D|G)p(G) , \tag{7}$$

where $f(D|G)$ is the integrated likelihood given by

$$f(D|G) = \int_{\theta_G} f(D|G, \theta_G)\pi(\theta_G)d\theta_G , \tag{8}$$

with respect to a prior distribution π. The reason for integrating out the parameters θ_G in (8) is that for the purpose of model comparison, we require the likelihood unconditional on any fixed parameter values. The Bayesian scoring metric for a model G given data D is the logarithm of the right-hand side of (7):

$$sc(G|D) = \log[f(D|G)p(G)] \,. \tag{9}$$

To calculate the integrated likelihood $f(D|G)$ from (8), we use Theorem 02 and make two independence assumptions about the parameters. First, the parameters $(\theta_\tau | \tau \in \mathcal{T})$ are *a priori* independent. This assumption is analogous to the one of global independence formulated by [21] for ADG Markov models. For that reason, we will refer to this assumption as the *global independence* assumption hereafter. Under this assumption we may write $\pi(\theta_G)$ as a product of the densities, where each of them involves only the parameters regarding one chain component $\tau \in \mathcal{T}$:

$$\pi(\theta_G) = \prod_{\tau \in \mathcal{T}} \pi(\theta_\tau) \,. \tag{10}$$

Second, for each $\tau \in \mathcal{T}$, the parameters $(\theta_{\tau,\rho} | \rho \in \mathcal{X}_{pa_G(\tau)})$ are also *a priori* independent. This assumption is known as *local independence* [21], and permits a factorization of $\pi(\theta_\tau)$ as

$$\pi(\theta_\tau) = \prod_{\rho \in \mathcal{X}_{pa_G(\tau)}} \pi(\theta_{\tau,\rho}) \,. \tag{11}$$

Because G_τ is decomposable for each τ, we can apply the results of [10]. We begin by assuming that $\pi(\theta_{\tau,\rho})$, the prior law for $\theta_{\tau,\rho}$, is *strong hyper Markov*, which allows us to factorize it as

$$\pi(\theta_{\tau,\rho}) = \frac{\prod_{C \in \mathcal{C}_\tau} \pi(\theta_{C,\rho})}{\prod_{S \in \mathcal{S}_\tau} \pi(\theta_{S,\rho})} \,. \tag{12}$$

Therefore, we just need to specify prior laws for each clique $C \in \mathcal{C}_\tau$. For $\pi(\theta_{\tau,\rho})$ being strong hyper Markov, we will consider a *hyper Dirichlet* law for each $\theta_{C,\rho}$, denoted by $\mathcal{D}(\theta_{C,\rho}; \vartheta_{C,\rho})$, which is specified by a hyperparameter set of positive numbers $\vartheta_{C,\rho} = (N'_{C1}, \ldots, N'_{Cq(C)})$ and a Dirichlet distribution with density

$$\pi(\theta_{C,\rho}) = \pi(\theta_{C,\rho} | \vartheta_{C,\rho}) = \frac{\Gamma(\sum_{k=1}^{q(C)} N'_{Ck})}{\prod_{k=1}^{q(C)} \Gamma(N'_{Ck})} \prod_{k=1}^{q(C)} \theta_C^{N'_{Ck} - 1} \,, \tag{13}$$

where $q(C)$ is the cardinality of the product space $\mathcal{X}_C$.

The above collection of hyper Dirichlet laws will determine a unique hyper Dirichlet law for $\theta_{\tau,\rho}$, which is strong hyper Markov, provided that the collection $\mathcal{D}(\theta_{C,\rho}; \vartheta_{C,\rho}), C \in \mathcal{C}$ is specified such that they are (pairwise) *hyperconsistent*. Two laws, $\mathcal{L}_A$ for θ_A and $\mathcal{L}_B$ for θ_B, are hyperconsistent if they both induce the same law for $A \cap B$ [10, pg. 1280]. In the case of hyper Dirichlet laws, they are hyperconsistent as long as for any two cliques C and

D such that $C \cap D \neq \emptyset$ and any given lth level $x_{\{C \cap D\}l} \in \mathcal{X}_{\{C \cap D\}}$, their corresponding hyperparameters ϑ_C and ϑ_D satisfy [10, pg. 1304]

$$\sum_{x_{Ck} \supset x_{\{C \cap D\}l}} N'_{Ck} = \sum_{x_{Dk} \supset x_{\{C \cap D\}l}} N'_{Dk} . \tag{14}$$

For a saturated multinomial model under a hyper Dirichlet prior law, the marginal probability distribution of a dataset $D_C = \{x_C^{(1)}, x_C^{(2)}, \ldots, x_C^{(N)}\}$, where $x_C^{(m)} \in \mathcal{X}_C$, $m = 1, \ldots, N$, and $\mathcal{X}_C = \times(\mathcal{X}_i | i \in C)$, is given by

$$f_C(D_C | \vartheta_C) = \frac{\Gamma(N'_C)}{\Gamma(N'_C + N)} \prod_{k=1}^{q(C)} \frac{\Gamma(N'_{Ck} + N_{Ck})}{\Gamma(N'_{Ck})} , \tag{15}$$

where $N'_C = \sum_{k=1}^{q(C)} N'_{Ck}$ and N_{Ck} are the counts of the contingency table associated with D_C. (Thus $N = \sum_k N_{Ck}$ is the size of the sample.)

The assumption of a strong hyper Markov prior law for θ_G allows us to use properties of such a prior law to proceed from expression (8). First, by (2), (10) and the exchangeability of the records of D, we can further factorize $f(D|G, \theta_G)$ and $\pi(\theta_G)$ to obtain

$$f(D|G) = \int_{\theta_\tau} \cdots \int \prod_{m=1}^{N} \prod_{\tau \in \mathcal{T}} f(x_\tau^{(m)} | x_{pa_G(\tau)}^{(m)}, \theta_\tau) \pi(\theta_\tau) d\theta_\tau . \tag{16}$$

By global independence, we can interchange the product and the integral, yielding

$$f(D|G) = \prod_{\tau \in \mathcal{T}} \int_{\theta_\tau} \prod_{m=1}^{N} f(x_\tau^{(m)} | x_{pa_G(\tau)}^{(m)}, \theta_\tau) \pi(\theta_\tau) d\theta_\tau \equiv$$

$$\equiv \prod_{\tau \in \mathcal{T}} f(D_\tau | D_{pa_G(\tau)}) . \tag{17}$$

Next, consider the following proposition.

Proposition 01 *[10, Prop. 5.6]*
For a decomposable undirected graph, if the prior law of θ is strong hyper Markov, then the marginal distribution of X is Markov.

Now apply this result to the conditional distribution determined by $f(D_\tau | D_{pa_G(\tau)})$. By the assumption of local independence, the integrated distribution $f(D_\tau | D_{pa_G(\tau)})$ of the data is Markov with respect to the graph G:

$$f(D|G) = \prod_{\tau \in \mathcal{T}} \left[\frac{\prod_{C \in \mathcal{C}_\tau} f_C(D_C | D_{pa_G(\tau)})}{\prod_{S \in \mathcal{S}_\tau} f_S(D_S | D_{pa_G(\tau)})} \right] . \tag{18}$$

Since $|\mathcal{C}_\tau| = |\mathcal{S}_\tau| + 1$, this is equivalent to

$$f(D|G) = \prod_{\tau \in \mathcal{T}} \left[\frac{1}{f_{pa_G(\tau)}(D_{pa_G(\tau)})} \cdot \frac{\prod_{C \in \mathcal{C}_\tau} f_{C,pa_G(\tau)}(D_{C,pa_G(\tau)})}{\prod_{S \in \mathcal{S}_\tau} f_{S,pa_G(\tau)}(D_{S,pa_G(\tau)})} \right], \quad (19)$$

where every $f_A(D_A)$, A being either $pa_G(\tau)$, $\{C, pa_G(\tau)\}$ or $\{S, pa_G(\tau)\}$, is replaced by the marginal probability distribution for a saturated multinomial model, i.e. (15), just as [10] do with their (41) and (6). Since any separator $S \in \mathcal{S}$ is, by definition, included in some clique $C \in \mathcal{C}$ the marginal probability $f_{S,pa(\tau)}$ corresponds to the marginalization of $f_{C,pa(\tau)}$ over $C \backslash S$. The same is true for $f_{pa(\tau)}$ which corresponds to the marginalization of $f_{C,pa(\tau)}$ over C. From this it follows that the term $\Gamma(N_C')/\Gamma(N_C' + N)$ in (15) will be the same in the marginals $f_{C,pa(\tau)}$, $f_{S,pa(\tau)}$ and $f_{pa(\tau)}$ such that they will cancel in (19). Therefore, we can write the integrated likelihood for an essential graph G as follows:

$$f(D|G) = \prod_{\tau \in \mathcal{T}(G)} \left[\left(\prod_{k=1}^{q(pa(\tau))} \frac{\Gamma(N_k')}{\Gamma(N_k' + N_k)} \right) \cdot \right.$$
$$\left. \cdot \frac{\prod_{C \in \mathcal{C}_\tau} \prod_{k=1}^{q(C \cup pa(\tau))} (\Gamma(N_k' + N_k)/\Gamma(N_k'))}{\prod_{S \in \mathcal{S}_\tau} \prod_{k=1}^{q(S \cup pa(\tau))} (\Gamma(N_k' + N_k)/\Gamma(N_k'))} \right], \quad (20)$$

where, for clarity, we have not specified the sets of variables associated with the N_k' and N_k, and they follow from the context of their running indexes, $q(pa(\tau))$, $q(C \cup pa(\tau))$ and $q(S \cup pa(\tau))$.

For the purpose of model comparison, it is necessary to specify a family of compatible prior laws that permit carrying out computations in a local manner. Such compatible families have been discussed in the context of decomposable models [10, Section 6.2] and ADG models [14,13,20]. The development of specific compatible priors for EG models requires an entire discussion on its own about the topic, but a straightforward uninformative approach is the assignment given by

$$N_{Ck}' = \frac{1}{|\mathcal{X}_{C,pa(\tau)}|} \cdot \quad (21)$$

Its compatibility follows from the equivalence of the EG factorization to the ADG factorization (see Section 5) and the results given by [14].

As part of our prior knowledge, the hyperparameters specified by expression (21) imply that we do not have any preference among the levels of each of the marginal contingency tables formed by the variables from the cliques in $\mathcal{C}_\tau$ and the parents $pa(\tau)$. The consequences of such a policy as prior knowledge for the parameters of the model are best understood by examining the expression of the variance of one of the parameters $\theta_i \in \theta_G$. The variance of

θ_i indicates how much the mean of θ_i may vary in the light of new data [11, Chapter 5, eqn. (7)]:

$$\text{Var}(\theta_i) = \frac{N_i'(N' - N_i')}{(N')^2(N' + 1)} \, , \tag{22}$$

where, recall, $N' = \sum_k N_k'$. This expression shows that the larger the values in ϑ are, the smaller the variance, as noted for instance in [5]. Since the assignment in (21) is the smallest positive hyperconsistent assignment one can give to the hyperparameters in ϑ, it follows immediately that we let the data determine as much as possible the shapes of the parameters. For this reason, this type of assignment is often also known as an *uninformative* assignment.

5   Equivalence with Respect to Other Factorizations

Every ADG or DEC model is Markov equivalent to an EG Markov model and every EG Markov model is Markov equivalent to a ADG model. From this fact it must follow that the factorization in Theorem 02 is, in fact, equivalent to those provided by any ADG or DEC graph in the equivalence class. We can see this by means of the following two cases:

- A given DEC graph U is in the Markov equivalence class represented by the EG G.
  This case is straightforward, as G will have one single chain component, $\mathcal{T} = \{\tau\}$ with $pa_G(\tau) = \emptyset$, which would be an undirected chordal graph. The factorization of the EG model G is identical to the factorization for U (1).
- A given ADG D is in the Markov equivalence class represented by the EG G.
  In this case we will see that for every clique C_τ in a chain component τ, the corresponding term $f(x_C|x_{pa_G(\tau)}, \theta_{C,\rho})$ can be further factorized and this will lead to transforming the whole factorization of G into a factorization for the Markov equivalent ADG D.
  The cliques, and separators, factorize in the following way

$$\prod_{C \in \mathcal{C}_\tau} f(x_C|x_{pa_G(\tau)}, \theta_{C,\rho}) = \prod_{C \in \mathcal{C}_\tau} \prod_{i=1}^{|C|} f(x_i|x_1, \dots, x_{i-1}, x_{pa_G(\tau)}, \theta_{C,\rho}).$$
$$\tag{23}$$

By performing this factorization according to any perfect numbering of the vertices - cf. [17, pg. 15] in G we will find repetitions of some terms $f(x_i|x_j, \dots, x_{i-1}, x_{pa_G(\tau)}, \theta_{C,\rho})$ because the intersections among the cliques in $\mathcal{C}_\tau$ may be non-empty. Since the set of separators S_τ corresponds to the intersections among the cliques, these repeated terms will cancel in (4). Because $|\mathcal{C}_\tau| = |\mathcal{S}_\tau| + 1$, exactly one of these repeated terms will remain, and therefore it follows that the factorization will consist of one term per random variable.

6   Local Computations and Bayes Factors

In order to apply the scoring criterion for EG model selection, we need a characterization of local computations for essential graph Markov models. This characterization will depend on the concept of *local transformation* that we use.

In decomposable models, a local transformation is the addition or removal of an undirected edge. In ADG models, a local transformation is the addition, removal or reversal of an arc.

For essential graphs, we can find different proposals for local transformations in [6][19][7][8][3]. We analyze first the transformations provided in [19], which we now sketch for completeness.

First note that if an arrow $a \rightarrow t \in \tau$ occurs in an EG G, then by Theorem 01(iii), for every other $t' \in \tau$ the arrow $a \rightarrow t'$ must occur in G as well. The collection $\{a \rightarrow t \mid t \in \tau\}$ is called an *arrow bundle* in G. The transformations are as follows:

Aα Remove an undirected edge, as long as the chain component remains chordal.

Aβ Add an undirected edge, as long as the chain component remains chordal.

Bα Remove an arrow bundle between two comparable chain components as long as strong protection is preserved.

Bβ Add an arrow bundle between two comparable chain components as long as strong protection is preserved.

Bγ Add an arrow bundle between two non-comparable chain components with different parent sets as long as strong protection is preserved.

Bδ Add an undirected edge between two vertices of two non-comparable chain components with the same parent sets as long as strong protection is preserved.

Cα Remove a collection of immoralities formed from two non-adjacent chain components with a single vertex and a third chain component which contains the collision vertices, as long as strong protection is preserved.

Cβ Add a collection of immoralities formed from two non-adjacent chain components with a single vertex and a third chain component which will contain the collision vertices, as long as strong protection is preserved.

For the transformation Aα, we may identify two cases: the case where the removal of an undirected edge leaves the chain component connected, and the case where the chain component is split into two chain components. In the former case, we only need to compute the Bayes factor provided by [10]. In the latter case, the set $\mathcal{T}(G)$ of chain components is enlarged by a new chain component. Therefore, the Bayes factor involves the comparison of the former larger chain component against the product of the two smaller new chain components.

For the transformation Aβ it suffices to use the Bayes factor provided by [10] since the transformation is within a single chain component which remains chordal.

In the transformations in Bα, Bβ and Bγ the set $\mathcal{T}(G)$ of chain components does not change, and only one arrow bundle involving two chain components will be added or removed. Therefore it suffices to compare the corresponding term in (20). The term should be entirely recomputed, though, due to the fact that the terms in (20) involve the current parent vertices which may change by any of these three operations.

In the transformation in Bδ, the set $\mathcal{T}(G)$ of chain components does change, by merging two chain components into a single one. In any case, if we have stored the computations made for the two merging components, still only computations for the new larger chain component will be necessary.

In the transformations in C, the set $\mathcal{T}(G)$ of chain components does not change, and it will suffice to compare the term in (20) that corresponds to the chain component containing the collision vertices.

6.1   Inclusion Friendly Local Computations

The graphical Markov inclusion order, or inclusion order for short, is a partial order among the graphs that belong to a common class of GMMs and have the same number of vertices. This partial order is defined as follows: a graph G precedes a different graph G', denoted $G \subseteq G'$, if and only if all the conditional independence statements that can be read off from G can be also read off from G'. For a more thorough description the reader should consult [4, Section 3]. A trivial example is that of the fully connected graph preceeding the fully disconnected graph. From the concept of inclusion order, follows the concept of inclusion boundary:

Definition 02 *[16]*
Let H, K, L be three GMMs. Let $H \prec L$ denote that $H \subset L$ and for no K, $H \subset K \subset L$. The inclusion boundary of the GMM G, denoted by $\mathit{IB}(G)$, is

$$\mathit{IB}(G) = \{H \,|\, H \prec G\} \cup \{L \,|\, G \prec L\}.$$

Several authors [16,4,8] have shown that learning algorithms for ADG models that perform local transformations which can reach any GMM in the inclusion boundary, perform better than those without this feature.

[18] conjectured a graphical characterization of the inclusion order for ADGs which basically tells us that the inclusion boundary for a given ADG is only one adjacency away. [15] proved Meek's conjecture for a the particular case where two ADGs differ in one single adjacency and recently, [8] has

proved Meek's conjecture in its general form. This implies we should strive to provide local transformations and local computations for EGs that may differ in one single adjacency.

The recent works by [8] and [3] provide such local transformations for EGs, combined with local computations for ADGs, because they use a scoring metric for ADGs. However, in both works the resulting graph, while being a chain graph, may not be a valid EG representation and therefore further transformations to obtain the corresponding EG may be necessary. The fact that we have (yet) no cheap graphical characterization of the inclusion boundary in terms of EGs, makes it difficult to find efficient local computations for a scoring metric specific for EGs, as the one presented in this paper. We expect, however, that this open problem can be solved in the future and therefore obtain an inclusion-friendly learning algorithm based exclusively upon the EG representation.

To illustrate our approach to this open problem, we shall consider the necessary local computations for a particular example. First, for each chain component $\tau \in \mathcal{T}$, consider a perfect ordering of the cliques - cf. [17, pg. 14] of G_τ as $C_1, \ldots, C_k$. Let $H_j = \bigcup_{i=1}^{j} C_i$, $S_{j+1} = C_{j+1} \cap H_j$, $R_{j+1} = C_{j+1} \backslash H_j$. The expression of the marginal likelihood of the data in (18) may be now written as

$$f(D|G) = \prod_{\tau \in \mathcal{T}} \left[f_{C_1|pa_G(\tau)}(D_{C_1}|D_{pa_G(\tau)}) \prod_{i=2}^{k} f_{R_i|S_i,pa_G(\tau)}(D_{R_i}|D_{S_i,pa_G(\tau)}) \right] .$$
$$(24)$$

Now consider the EGs in Figure 1.

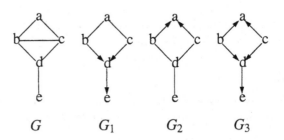

$$G \qquad G_1 \qquad G_2 \qquad G_3$$

Fig. 1. The EGs G_1, G_2 and G_3 are obtained from G by removing the undirected edge b–c

The EG G is formed by a single chain component. By removing the edge b–c we may obtain either G_1, G_2 or G_3 which have, respectively, 3, 2 and 5 chain components. Using expression (24), we are going to write down the different Bayes factors for comparing the EGs G_1, G_2, G_3 against the EG G. In G_1, the cliques in the chain component $\tau = \{a, b, c\}$ may follow the perfect ordering

$C_1 = \{a, b\}, C_2 = \{a, c\}$. Thus, we obtain $H_1 = C_1, S_2 = C_2 \cap H_1 = \{a\}, R_2 = \{C_2 \backslash H_1\} = \{c\}$. The other chain components, $\tau = \{d\}$ and $\tau = \{e\}$, are singletons and, therefore, each of them has just one clique containing one single vertex. By (24) the Bayes factor of the EG G_1 against G may be written as

$$\frac{f(D|G_1)}{f(D|G)} = \frac{f_{a,b}(D_{a,b})f_{c|a}(D_c|D_a)f_{d|b,c}(D_d|D_{b,c})f_{e|d}(D_e|D_d)}{f_{a,b,c}(D_{a,b,c})f_{d|b,c}(D_d|D_{b,c})f_{e|d}(D_e|D_d)},$$

which after some simplification becomes:

$$\frac{f(D|G_1)}{f(D|G)} = \frac{f_{a,b}(D_{a,b})f_{a,c}(D_{a,c})}{f_a(D_a)f_{a,b,c}(D_{a,b,c})}. \tag{25}$$

Similar to the case of Bayes factors in decomposable models [10, pg. 1300], expression (25) provides a formula to test the conditional independence $X_b \perp\!\!\!\perp X_c|X_a$.

The Bayes factor for G_2 against G is

$$\frac{f(D|G_2)}{f(D|G)} = \frac{f_{a|b,c}(D_a|D_{b,c})f_{b,d}(D_{b,d})f_{c|d}(D_c|D_d)f_{e|d}(D_e|D_d)}{f_{a,b,c}(D_{a,b,c})f_{d|b,c}(D_d|D_{b,c})f_{e|d}(D_e|D_d)} =$$

$$= \frac{f_{a,b,c}(D_{a,b,c})f_{b,c}(D_{b,c})f_{b,d}(D_{b,d})f_{c,d}(D_{c,d})}{f_{a,b,c}(D_{a,b,c})f_{b,c,d}(D_{b,c,d})f_{b,c}(D_{b,c})f_d(D_d)} = \frac{f_{b,d}(D_{b,d})f_{c,d}(D_{c,d})}{f_{b,c,d}(D_{b,c,d})f_d(D_d)},$$

which corresponds to a formula to test whether $X_b \perp\!\!\!\perp X_c|X_d$. Finally, the Bayes factor for G_3 against G is

$$\frac{f(D|G_3)}{f(D|G)} = \frac{f_{a|b,c}(D_a|D_{b,c})f_b(D_b)f_c(D_c)f_{d|b,c}(D_d|D_{b,c})f_{e|d}(D_e|D_d)}{f_{a,b,c}(D_{a,b,c})f_{d|b,c}(D_d|D_{b,c})f_{e|d}(D_e|D_d)} =$$

$$= \frac{f_b(D_b)f_c(D_c)}{f_{b,c}(D_{b,c})},$$

which corresponds to a formula to test whether $X_b \perp\!\!\!\perp X_c|\emptyset$.

7 Concluding Remarks

We have presented a scoring metric for EGs with discrete multinomial data. Such a scoring metric opens the way to devise a learning, or selection, procedure that relies only on the EG representation. Another advantage is that the EG scoring metric automatically assigns the same score to equivalent ADG models and bypasses the need to impose strong constraints on the prior distribution of both graphs and parameters as discussed by [1, Section 7.2].

We have investigated possible ways of performing local computations with this scoring metric. We have briefly analized the question of finding the corresponding local computations for local transformations on single adjacencies, which would respect the inclusion order. In that analysis, we can see how this scoring metric in fact provides formulae for investigating the conditional independence restrictions being removed (or added) while performing model comparison. This feature is important if we want to monitor, and understand, each step in the learning process.

References

1. Andersson, S., Madigan, D., and Perlman, M. (1997). A characterization of Markov equivalence classes for acyclic digraphs. Annals of Statistics, 25:505–541.
2. Andersson, S., Madigan, D., and Perlman, M. (2001). Alternative Markov properties for chain graphs. Scandinavian Journal of Statistics, 28:33–85.
3. Auvray, V. and Wehenkel L. (2002). On the construction of the inclusion boundary neighborhood for Markov equivalence classes of Bayesian network structures. In Proc. of the Eighteenth Conf. on Uncertainty in Art. Int.. Morgan Kaufmann.
4. Castelo, R. and Kočka, T. (2003). On an inclusion driven learning of Bayesian networks. Journal of Machine Learning Research, 4:527-574.
5. Castillo, E., Hadi, A., and Solares, C. (1997). Learning and updating of uncertainty in Dirichlet models. Machine Learning, 26:43–63.
6. Chickering, D. (1996). Learning equivalence classes of Bayesian network structures. In Proc. of the Twelfth Conf. on Uncertainty in Art. Int., pg. 150–157. Morgan Kaufmann.
7. Chickering, D. (2002a). Learning equivalence classes of Bayesian-network structures. Journal of Machine Learning Research, 2:445-498.
8. Chickering, D. (2002b). Optimal Structure Identification with Greedy Search. Journal of Machine Learning Research, 3:507–554
9. Cowell, R., Dawid, A., Lauritzen, S., and Spiegelhalter, D. (1999). Probabilistic Networks and Expert Systems. Springer-Verlag, New York.
10. Dawid, P. and Lauritzen, S. (1993). Hyper-Markov laws in the statistical analysis of decomposable graphical models. Annals of Statistics, 21(3):1272–1317.
11. DeGroot, M. H. (1970). Optimal Statistical Decisions. McGraw-Hill.
12. Frydenberg, M. (1990). The chain graph Markov property. Scandinavian Journal of Statistics, 17:333–353.
13. Geiger, D. and Heckerman, D. (1998). Parameter priors for directed acyclic graphical models and the characterization of several probability distributions. Tech. Rep. MSR-TR-98-67, Oct. 1998, Microsoft Research.
14. Heckerman, D., Geiger, D., and Chickering, D. (1995). Learning Bayesian networks: The combination of knowledge and statistical data. Machine Learning, 20:194–243.
15. Kočka, T., Bouckaert, R., and Studený, M. (2001). On characterizing inclusion of Bayesian networks. In Proc. of the Seventeenth Conf. on Uncertainty in Art. Int., pg. 261–268. Morgan Kaufmann.

16. Kočka, T. and Castelo, R. (2001). Improved learning of Bayesian networks. In Proc. of the Seventeenth Conf. on Uncertainty in Art. Int., pg. 269–276. Morgan Kaufmann.

17. Lauritzen, S. (1996). Graphical Models. Oxford University Press, Oxford.

18. Meek, C. (1997). Graphical models, selecting causal and statistical models. PhD Thesis, Carnegie Mellon University.

19. Perlman, M. (2001). Graphical model search via essential graphs. In *Algebraic Methods in Statistics and Probability*, V. 287, American Math. Soc, Providence, Rhode Island.

20. Roverato, A. and Consonni, G. (2001). Compatible Prior Distributions for DAG models. Tech. Rep. 134, Sept. 2001, University of Pavia.

21. Spiegelhalter, D. and Lauritzen, S. (1990). Sequential updating of conditional probabilities on directed graphical structures. Networks, 20:579–605.

22. Verma, T. and Pearl, J. (1990). Equivalence and synthesis of causal models. In Proc. of the Sixth Conf. on Uncertainty in Art. Int., pg. 255–268. Morgan Kaufmann.

23. Wilks, S.S. (1962). Mathematical Statistics. Wiley, New York.

Fast Propagation Algorithms for Singly Connected Networks and their Applications to Information Retrieval

Luis M. de Campos, Juan M. Fernández-Luna, and Juan F. Huete

Departamento de Ciencias de la Computación e Inteligencia Artificial.
Escuela Técnica Superior de Ingeniería Informática. Universidad de Granada.
18071 - Granada, Spain
{lci,jmfluna,jhg}@decsai.ugr.es

Abstract. There are some problems in which the time required for evaluating a Bayesian network must not be long. This means that not all the propagation algorithms could be applied in these situations, and therefore an appropriate method must be chosen. A type of evaluation that fulfills this requirement is known generically as "anytime algorithm". They are able to return an approximate output at any moment, solution that could be enough for some problems. In this paper two methods framed in this approach are presented. These are modifications of the Pearl's exact propagation algorithm for singly connected networks, designed with the final aim of reducing the propagation effort in domain problems where the number of variables is very large and the domain knowledge could be represented using that topology. As a very suitable test bed for measuring their quality, the field of Information Retrieval was selected. Results of a detailed experimentation are presented as well, showing a good performance.

1 Introduction

Bayesian networks (BNs) are one of the most useful tools to represent problems pervaded with uncertainty, because, among other features, they provide an efficient codification of joint probability distributions. Also, their inference mechanisms, i.e., propagation algorithms, allow to reason under this situation. But, in general, inference in Bayesian networks is an NP-hard problem [3]. When dealing with BNs in which the number of variables is very large and the stored probability tables are also of a very big size, the inference process may be a problem. In terms of running time, it may be very difficult to obtain the required results in an acceptable time. That is the reason why researchers have been looking for exact and approximate methods to reduce propagation time, preparing them to run in real-time situations.

One of the problematic fields is the *Information Retrieval* (IR) area, in which when a user submits a query to an interactive Information Retrieval System (IRS), that allows the storage and access to millions of documents in a library or in the Internet, it must return the set of documents which are considered relevant for that query using a short response time. Considering this type of system implemented by means of BNs, where the number of variables that contain is very high (in most of the cases, huge), if it is not able to cope with this temporal restriction, it will not be very useful, no matter how effective it is in terms of retrieval success (ability to retrieve only those documents that satisfy the user's information need).

As previously mentioned, one attempt to solve this problem in practical situations is to run the inference by means of approximate propagation algorithms. Faster than the exact ones, they can be used with any kind of topologies and the loss of precision in the result is almost noticeable. Nevertheless, even this approach may be excessively time consuming when dealing with IR problems [5]. Other alternative is running "anytime evaluation" algorithms [7,11,9], returning estimates of posterior probabilities in any stage of the inference process. Also, we may reduce the complexity of the topology of the graph, for example by using, instead of general graphs, *polytrees* (*singly connected graphs*). For these networks there are exact and efficient propagation methods that run in a time polynomial with the number of nodes. Obviously, the use of exact propagation algorithms in a simplified model is another form of approximation. However, exact propagation in polytrees is also time consuming when using huge networks, hence we look for a method to accelerate the response of the system.

Let us imagine a large network where we want to propagate the impact of a relatively small number of evidences. It could be sensible to think that those nodes in the network further from the evidences will not change a lot their probability, because the "influence" of the evidences has been weakened more and more through the propagation process[1]. Therefore, perhaps we could save time by restricting the propagation process within some neighbourhood of the instantiated variables. Alternatively, we could selectively prune the propagation when we detect that we are not getting any profit (posterior probabilities of the nodes will remain almost unchanged). In this case, we could use any kind of propagation algorithm, even exact, reducing the running time and obtaining results very similar to those obtained by the original algorithm. The key is just to look for a way of driving the propagation.

Exploiting these ideas, in this paper we present two different methods to approximate the results of an exact propagation algorithm, concretely Pearl's propagation algorithm for polytrees [13]. We have applied them to large networks containing knowledge extracted from collections of documents, in the IR context. Therefore, in Section 2, we briefly introduce some of IR basic concepts. In Section 3 we outline our BN-based retrieval model, with which we have carried out the experiments. The two proposed propagation methods are described in Section 4. Section 5 contains the description and the results of the experiments. Finally, some concluding remarks are presented in Section 7.

---

[1] Although quite frequent, this is not always the case. For instance, if there are deterministic relationships.

2   Preliminaries: Information Retrieval

A good definition of Information Retrieval (IR) is given by Salton and McGill[20] who point out that this area deals with representation, storage, organization and access of information items. The software that puts into practice these tasks is known as an *Information Retrieval System* (IRS). Mainly, it matches user *queries* (formal statements of information needs) to documents stored in a database (the *document collection*). In our case, the *documents* will always be the textual representations of any data objects.

An *IR model* is a specification about how to represent documents and queries, and how to compare them. Many IR models, as the Vector Space model or Probabilistic models [1], do not use the documents themselves but a kind of document surrogates, usually in the form of weighted vectors of *terms* or *keywords*, which try to characterise the document's information content. Queries are also represented in the same way. When a user formulates a query, this is compared with each document from the collection and a score that represents its relevance (matching degree) is computed. Later, the documents are sorted in decreasing order of relevance and returned to the user.

Probabilistic models [1,2] compute the probability of relevance given a document and a query (the probability that a document satisfies a query) and are based on the 'Probability ranking principle'. This principle states that the best overall retrieval effectiveness will be achieved when documents are ranked in decreasing order of their probability of relevance [19].

Also founded on probabilistic methods, *Bayesian networks*[13] have been proved to be a good model to manage uncertainty, even in the IR environment, where they have already been successfully applied as an extension of probabilistic IR models [21,16,18,10].

The most commonly used measures to evaluate IR systems, in terms of retrieval effectiveness, are *recall (R)* (the proportion of relevant documents retrieved), and *precision (P)* (the proportion of retrieved documents that are relevant, for a given query). The relevance or irrelevance of a document is based, for test collections, on the *relevance judgments* expressed by experts for a fixed set of queries [17]. By computing the precision for a number of values of recall we obtain a recall-precision plot. If a single measure of performance is desired, the average precision for all the recall values considered may be used. Finally, if we are processing together a set of queries, the usual approach is to report mean values of the selected performance measure(s).

3   The Bayesian Network Retrieval Model

Our retrieval model is based on a BN[2] composed of two different sets of nodes: $\mathcal{T}$, containing binary random variables representing the terms in the collection, and $\mathcal{D}$, corresponding also to binary random variables, but in this

---

[2] There exist other retrieval models also based on BNs, see [21,16].

case related to the documents which belong to the collection. A variable T_i associated to a term takes its values from the set $\{\bar{t}_i, t_i\}$, where $\bar{t}_i$ stands for 'the term T_i is not relevant', and t_i represents 'the term T_i is relevant'. A variable referring to a document D_j, has its domain in the set $\{\bar{d}_j, d_j\}$, where $\bar{d}_j$ and d_j respectively mean 'the document D_j is not relevant', and 'the document D_j is relevant'.

With respect to the topology of the network [8,6], there are arcs going from term nodes to those document nodes where these terms appear, and there are not arcs connecting document nodes. To give our model the ability to represent dependences between terms, we use an automatic learning algorithm that takes the set of documents in a collection as the input and generates a polytree of terms as the output (a graph in which there is no more than one undirected path connecting each pair of nodes). For polytrees there are efficient specific algorithms for learning and performing exact inference. The learning algorithm, specifically designed for the IR context, is described in [4]. Figure 1 displays an example of BN for IR.

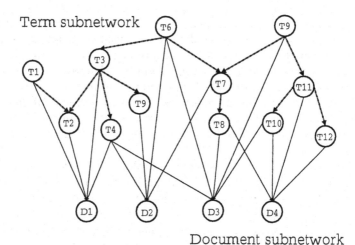

Fig. 1. The BN Retrieval Model.

The next step is to estimate the probability distributions stored in each node of the network. All the root nodes will store marginal distributions. For each root term node we use the following estimator for all nodes: $p(t_i) = \frac{1}{M}$ and $p(\bar{t}_i) = \frac{M-1}{M}$ (M being the number of terms in the collection).

The nodes with parents (term and document nodes) will store a set of conditional probability distributions, one for each of the possible configurations of the parent nodes. Term nodes will store the conditional probabilities $p(T_i|\pi(T_i))$, where $\pi(T_i)$ is a configuration associated to the set of parents of T_i, $\Pi(T_i)$. The estimator used in this case is based on the Jaccard coefficient [17]: If we define $n(\bar{t}_i)$, $n(\pi(T_i))$ and $n(\bar{t}_i, \pi(T_i))$ as the number of documents where the term variables T_i and $\Pi(T_i)$ take the values $\bar{t}_i$, $\pi(T_i)$ and $(\bar{t}_i, \pi(T_i))$, respectively, then

$$p(\bar{t}_i \mid \pi(T_i)) = \frac{n(\bar{t}_i, \pi(T_i))}{n(\bar{t}_i) + n(\pi(T_i)) - n(\bar{t}_i, \pi(T_i))} \tag{1}$$
$$p(t_i \mid \pi(T_i)) = 1 - p(\bar{t}_i \mid \pi(T_i))$$

Finally, we have to estimate the conditional probabilities for the document nodes, $p(D_j|\pi(D_j))$. The problem is that a document node may have a high number of parents, so that the number of conditional probabilities that we need to estimate and store may be huge. Therefore, instead of explicitly computing these probabilities, we use a *probability function*, which returns a probability value when it is called during the inference stage, each time that a conditional probability is required. For any configuration $\pi(D_j)$ of $\Pi(D_j)$ (i.e., any assignment of values to all the term variables in D_j), we define the conditional probability of relevance of D_j as follows:

$$p(d_j \mid \pi(D_j)) \propto \frac{\sum_{T_i \in D_j, T_i = t_i} tf_{ij} idf_i^2}{\sqrt{\sum_{T_i \in D_j} tf_{ij}^2 idf_i^2}}, \tag{2}$$

where tf_{ij} is the frequency of the ith term in the jth document, and idf_i is the inverse document frequency of such a term in the collection.

Once the network has been built, it can be used to obtain a relevance value for each document given a query Q. Each term q in the query Q is considered as an evidence for the propagation process, and therefore instantiated to "the term is relevant". Then, the propagation process is run, thus obtaining the posterior probability of relevance of each document, $p(d_j \mid Q)$. Later, the documents are sorted according to their corresponding probability and shown to the user.

Taking into account the large number of nodes in the BN and the fact that it contains cycles and nodes with a great number of parents, general purpose inference algorithms cannot be applied due to efficiency considerations, even for small document collections. To solve this problem, we have designed a specific inference method that takes advantage of both the topology of the network and the kind of probability function used for document nodes. This exact method [8] is called *propagation + evaluation*: (1) An exact propagation in the term subnetwork (using Pearl's algorithm for polytrees); the results of this first stage are $p(t_i|Q), \forall T_i$; (2) An evaluation of the probability function (2), using the information obtained in the previous propagation. Therefore, the relevance of each document in the collection is computed as follows:

$$p(d_j \mid Q) \propto \frac{\sum_{T_i \in D_j} tf_{ij} idf_i^2 \cdot p(t_i \mid Q)}{\sqrt{\sum_{T_i \in D_j} tf_{ij}^2 idf_i^2}} \tag{3}$$

In [8] it is shown how, by using probability functions of the type of (2), our method returns the same values as an exact propagation would in the whole network.

4 Proposals for Reducing the Propagation Time

Let us start with an example that will illustrate the problem that we have detected in exact propagations in relatively large polytrees. The MEDLARS and CACM test collections contains 7170 and 7562 terms, respectively (the polytrees underlying the term subnetwork will have these numbers of nodes). Using our model (for different queries), we realized that only 2112 and 1865 terms, on the average, had changed their probabilities with respect to the prior distributions. This fact means that 71% and 75% of the terms do not receive any influence from the evidences (the terms in the queries) but they are involved in the propagation process, and the propagation wastes time passing messages from and to these nodes.

We believe that for those nodes which are so far from any evidence, the information that arrives at them is so weakened that does not provoke almost any change in their original beliefs. However, those terms nearer to the evidences will be updated more substantially. Therefore, if we were able to introduce some control in the propagation, we could save a great amount of time, being the posterior probabilities of the terms very similar to those obtained using the full propagation.

To test this hypothesis we have used, as mentioned before, Pearl's algorithm for polytrees, because we opted for this topology as the basis for our IR model. This algorithm makes use of the fact that, in a polytree, each node T_i divides it in two independent components: The polytree composed of T_i's descendants, and the one containing T_i's ancestors. Therefore, the propagation process computes $p(T_i \mid Q)$ by combining the information that arrives from both polytrees by means of message passing from one side to the other, and vice versa. The posterior probability is obtained by means of

$$p(T_i \mid Q) = \frac{\lambda_i(T_i) \cdot \rho_i(T_i)}{k}, \tag{4}$$

where $\lambda_i(T_i)$ is the information, obtained from λ messages, coming from T_i's children and $\rho_i(T_i)$ is the information, in this case a combination of ρ messages, originated in its parents. k is a normalizing constant.

In our implementation of the algorithm, the propagation of several evidences is done in turn. Therefore, for each evidence, a node will receive only one λ or ρ message, and then it will compute the λ_s and ρ_s messages that will be sent to its parents and children, respectively. This process is repeated for each instantiated variable.

We have modified the propagation algorithm in two ways, designing two different techniques to control the propagation to the terms: radial and linear propagation. In both cases the idea is to avoid to send messages to all the nodes in the network, only those nodes where a significant change in their posterior probabilities is expected will be processed.

4.1   Radial Propagation

The main hypothesis on which this method is based is that the nodes closer to the evidences will receive more influence from them, changing considerably their probabilities. However, those which are further will receive "weakened" messages that will cause no (or a very small) change. So, the method is very simple: Starting from each evidence q, to propagate just to those nodes which are placed in the graph to a distance from q^3 lower than or equal to a fixed radius, r. If a node is at radius r, then it will not send any message to any other node. An example is displayed in Fig. 2, where the term T_4 has been instantiated and $r = 2$. The shadowed area corresponds to those nodes that are visited during the propagation.

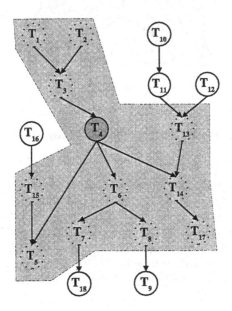

Fig. 2. Example of radial propagation.

4.2   Linear Propagation

In this case, when a node T_i sends a message to a second node T_j, the impact of this message on the probability of the second node is measured. If it is great

---

[3] The number of links that are in the path that connects two nodes.

enough (greater than a threshold), then T_j will also send its corresponding messages to its parents and children, continuing the propagation. Otherwise, the propagation, for that path, will be stopped at this node T_j. In this way, this second technique is "linear" instead of "radial".

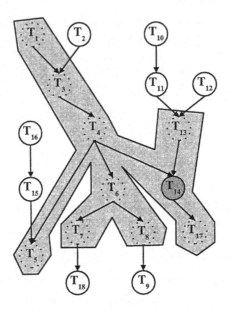

Fig. 3. Example of linear propagation.

How do we measure the impact of a λ or ρ message sent by a node, T_i, on a second node, T_j? We compare the posterior probability of T_j given the evidences propagated up till now (without combining the message that is being sent at this moment), $p(T_j|Q')$, and the posterior probability of T_j, once the current message has been taken into account, $p(T_j|Q'')$. The comparison is carried out using the Kullback-Leibler's Cross Entropy, CE, that measures the distance between two probability distributions in the following way:

$$CE(p(T_j|Q'), p(T_j|Q'')) = \sum_{\bar{t}_j, t_j} p(\mathbf{t_j}|Q') \log \frac{p(\mathbf{t_j}|Q')}{p(\mathbf{t_j}|Q'')} \tag{5}$$

If $CE(p(T_j \mid Q'), p(T_j \mid Q'')) > u$, u being a threshold, then the propagation will go on, and T_j will send to its neighbours the corresponding λ and ρ messages. Otherwise, the current path will be pruned. Figure 3 shows an example where the instantiated term is T_{14}. The shadowed area corresponds to the nodes visited during the propagation.

5   Experiments and Results

With the aim of comparing the behaviour of the two previous approximate methods with respect to exact propagation, we have carried out a battery of experiments with five term subnetworks, corresponding to five standard medium-size test collections that are commonly used in IR experiments: ADI, CACM, CISI, CRANFIELD, and MEDLARS. Their number of documents, terms and queries, respectively, are displayed in Table 1.

Table 1. Main features of the five test collections.

Collection	Documents	Terms	Queries
ADI	82	828	35
CACM	3204	7562	64
CISI	1460	4985	76
CRANFIELD	1398	3857	225
MEDLARS	1033	7170	30

For each collection, and using exact propagation in the term subnetwork for all the queries, we have measured the average running time (AT), in seconds, and the average number of nodes involved in the propagation process (ANN). Finally, running the evaluation stage in the document subnetwork, the average precision at the eleven standard recall points (AP11) is also computed. The results are displayed in Table 2.

Table 2. Results using exact propagation for all the test collections.

	ADI	CACM	CISI	CRANFIELD	MEDLARS
AT	0.0134	0.0589	0.1612	0.0159	0.0577
ANN	2570	8476	21110	2535	8370
AP11	0.4130	0.3760	0.2005	0.4314	0.6180

With respect to Radial Propagation, we have carried out experiments for each of the following radii: $0, 1, 2, 5, 10, 15, 20$. For Linear Propagation, the experiments have been performed according to the following thresholds for the cross entropy: $0.001, 0.005, 0.01, 0.05, 0.1, 0.5$. In all the cases, we have computed AT, ANN and AP11 measures. In order to measure the possible loss of precision of the approximate methods, we have also computed the average square error (AE) between the exact and the approximate posterior distributions per term[4] and query. Tables 3 and 4 show the results of the experiments.

Table 3. Results of the experimentation with the Radial Propagation method.

	ADI	CACM	CISI	CRANFIELD	MEDLARS
r=0					
AE	0.014598	0.001843	0.002341	0.001021	0.001064
AT	0.0	0.0	0.0	0.0	0.0
ANN	6	10	27	9	10
AP11	0.4699	0.3572	0.2201	0.4319	0.5548
r=1					
AE	0.011838	0.000694	0.001413	0.000343	0.000452
AT	0.000285	0.002968	0.001607	0.000977	0.000999
ANN	18	109	162	59	72
AP11	0.4500	0.3813	0.2111	0.4301	0.6090
r=2					
AE	0.006623	0.000183	0.001145	0.000086	0.000152
AT	0.000468	0.009062	0.004642	0.002088	0.002333
ANN	43	526	550	202	255
AP11	0.4365	0.3761	0.2134	0.4311	0.6152
r=5					
AE	0.000893	0.0000008	0.000762	0.0000006	0.000003
AT	0.002285	0.0387499	0.034910	0.008311	0.016666
ANN	247	4721	4659	1259	2240
AP11	0.4159	0.3760	0.2134	0.4314	0.6151
r=10					
AE	0.000003	1.3e-12	0.000038	1.9	1.1e-10
AT	0.003428	0.058281	0.116339	0.014488	0.047999
ANN	919	8260	15675	2355	6967
AP11	0.4130	0.3760	0.2129	0.4314	0.6150
r=15					
AE	1.8e-11	0.0	1.8e-7	0.0	0.0
AT	0.007428	0.059843	0.155178	0.015599	0.058333
ANN	1691	8473	20499	2519	8282
AP11	0.4130	0.3760	0.2129	0.4314	0.6150
r=20					
AE	2.6e-11	0.0	4.2e-9	0.0	0.0
AT	0.009428	0.058906	0.160089	0.015511	0.055999
ANN	2227	8476	21064	2535	8366
AP11	0.4130	0.3760	0.2129	0.4314	0.6150

---

[4] Considering only terms whose posterior probability is modified when using exact propagation.

Table 4. Results of the experimentation with the Linear Propagation method.

	ADI	CACM	CISI	CRANFIELD	MEDLARS
u=0.001					
AE	0.000393	0.000199	0.001192	0.000102	0.000066
AT	0.003714	0.035781	0.023660	0.008799	0.016333
ANN	162	282	313	134	311
AP11RP	0.4133	0.3762	0.2133	0.4311	0.6214
u=0.005					
AE	0.000917	0.000383	0.001287	0.000180	0.000141
AT	0.002285	0.025312	0.017053	0.006177	0.012666
NVN	3969	8893	20673	17093	4955
ANN	113	139	184	76	165
AP11	0.4131	0.3752	0.2122	0.4312	0.6209
u=0.01					
AE	0.001234	0.000505	0.001357	0.000232	0.000198
AT	0.004571	0.021249	0.014642	0.005822	0.009666
ANN	97	101	150	60	120
AP11	0.4111	0.3763	0.2128	0.4305	0.6203
u=0.05					
AE	0.002945	0.001071	0.001912	0.000576	0.000401
AT	0.002285	0.013124	0.007321	0.003422	0.005999
ANN	56	31	51	23	58
AP11	0.4162	0.3792	0.2131	0.4291	0.6151
u=0.01					
AE	0.004466	0.001358	0.002070	0.000749	0.000684
AT	0.000571	0.009374	0.006517	0.002488	0.003999
ANN	37	18	38	15	26
AP11	0.4132	0.3738	0.2084	0.4286	0.6064
u=0.5					
AE	0.008970	0.001713	0.002227	0.000955	0.000974
AT	0.000571	0.004843	0.004374	0.001733	0.002666
ANN	15	11	30	10	11
AP11	0.4526	0.3825	0.2045	0.4215	0.5655

To facilitate the analysis of the previous results, data presented in tables 3 and 4 have been plotted in graphs. Figures 4, 6 and 8 (for radial propagation) and 5, 7 and 9 (for linear propagation), represent the percentages of change of the performance measures for the approximate methods with respect to the exact one. Figures 10 and 11 shows the graphs for average square errors.

In Figs. 8 and 9 we can observe that, in terms of retrieval success, the approximate propagation methods do not lose effectiveness, except for very small radii or very high thresholds in some cases (the retrieval success is even increased in some other cases). With respect to precision, we can see in Figs.

10 and 11 that the mean square errors are quite low, even for small radii and high thresholds. Therefore, at least for the type of networks being considered, the approximate posterior probabilities are quite acceptable.

From the point of view of the efficiency, the number of visited nodes (Figs. 6 and 7) is reduced dramatically when we use linear propagation (more than 93% in all the cases), being more moderate for radial propagation (although it is also substantial, more than 44%, for values of the radius up to 5). A similar pattern can be observed for the running times using radial propagation (Fig. 4): In general, this method spends the same time as exact propagation would for high radii (15 or 20) but it becomes more useful for small radii (for example, for $r = 5$, the percentages of change fluctuate between 34% and 83%). The running times for linear propagation (Fig. 5) are much smaller (39% of reduction in the worst case and 97% in the best case). However, the ratio between the average time and the number of nodes considered, $\frac{AT}{ANN}$, is much greater for linear propagation than for radial propagation, because in the first case we have to compute a cross entropy measure each time we visit a node.

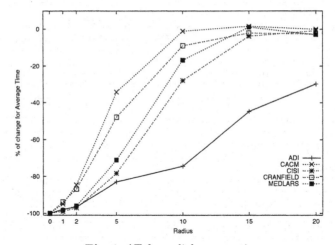

Fig. 4. AT for radial propagation.

6   Related works

In [9] the reader may find a good survey on real-time inference algorithms for BNs. Among them, some of the existing anytime algorithms are reviewed. In this section, we shall briefly describe some relevant papers related to our approach.

An example of resolution of the problem that we face in this paper is presented in [7]. The authors propose an anytime algorithm (LPE), which

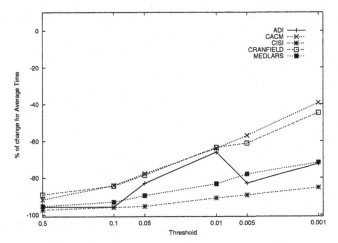

Fig. 5. AT for linear propagation.

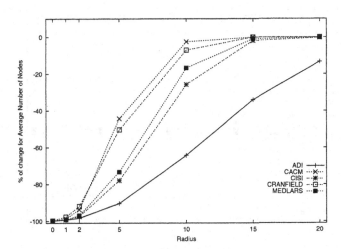

Fig. 6. ANN for radial propagation.

computes interval bounds on the probability, i.e., they look for intervals in which the exact probability is included for what they call a query node, i.e., the objective node. The algorithms begins with an "active subset", composed of a query node originally, and include new neighbours while a stopping condition is not fulfilled. The differences between our proposal and theirs are the following: Our algorithms return probabilities for each node in the network, while Draper and Hanks' returns probability intervals just for the query nodes. Also, we begin propagation with evidences; on the other hand, they do with query nodes. LPE carries out a complete propagation in the active subset each time that the algorithm have to add new neighbours to

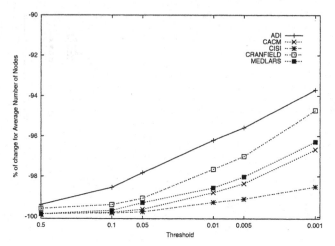

Fig. 7. ANN for linear propagation.

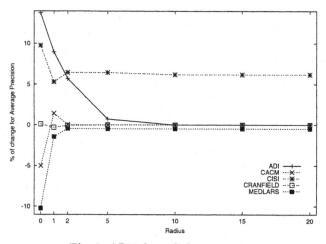

Fig. 8. AP11 for radial propagation.

it, being in some cases very inefficient, while our proposals follow the normal flow of propagation, being more efficient.

A different approach is presented in [12] to design an anytime inference method for BNs. In this case, the authors play with the precision of the state space of variables. The granularity is modulated progressively, giving as a consequence an estimation of the exact results that answer a query.

A third paper that proposes a solution for this problem, more closely related to ours, is [11]. In this work, a general BN is converted into a polytree, whose root is the query node. Since that moment, the posterior probability of that query node is computed incorporating, each time that the algorithm is

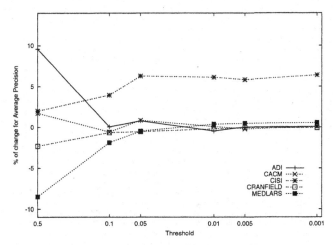

Fig. 9. AP11 for linear propagation.

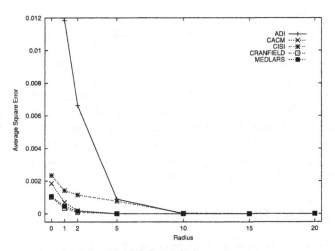

Fig. 10. AE for radial propagation.

iterated, a new descendant, which sends messages to its parents. Again, there are several main differences with respect to our approach: These authors only are interested in the posterior probability of a node; the algorithm starts the propagation with the query node, and each time that a new node is visited, it has to recompute all the probabilities of its ancestors.

Iterative Structured Variable Elimination (ISVE) [14] is an approach based on recursive probability models, that iteratively constructs larger BN fragments and solves them in order to answer a query. Therefore, in each iteration of the algorithm, it could present an approximated probability distribution.

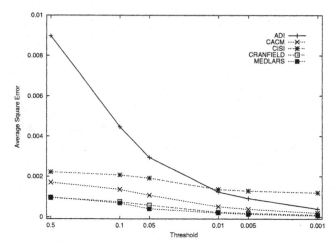

Fig. 11. AE for lineal propagation.

We could say that this method is dynamic in terms of the network structure, as an opposite approach of ours, which works with a stable graph.

A final and recent paper is [15] that deals with BN propagation in environments in which resources as memory and time are restricted, as in embedded systems. In this case, the network is transformed using the variable elimination algorithm in several sub-networks. In each one of them, a different inference method will be applied depending on the constrained resources. The authors' system are able to produce results in any moment only deciding the appropriate algorithm for each different case. This is a different approach with respect to ours because the authors do not modify any algorithm, except the one which divides the original network.

7   Concluding Remarks

In this paper we have presented two modifications of Pearl's exact propagation for polytrees, designed with the aim of reducing the propagation effort. The underlying idea was to avoid to propagate in the whole network, because the nodes situated farthest from the evidences do not almost change their posterior probabilities. The first approach, the Radial Propagation, only propagates the evidences to those nodes which are situated in the network to a given distance of the evidences. The Linear Propagation maintains the message passing to the nodes in a path, while the cross entropy of the posterior probabilities (before and after combining the messages) is greater than a given threshold.

Our experiments show that these methods can be successfully applied to large polytrees, as those found in the IR context, without decreasing the retrieval performance. Therefore, we expect to be able to manage document collections of a more realistic size, like TREC, where the number of terms is much larger (more than 500000).

In the IR context, an additional important saving of time due to application of these methods is the evaluation of a lesser number of documents using (3) because more documents do not change their probabilities, what it means that they are not affected by the query (any of the terms that they contain do not change their probabilities), and consequently do not have to be evaluated, saving time in the document evaluation stage. Therefore, these two propagation methods presented in this paper show important time reductions in the retrieval process: the methods themselves are faster, and as a direct consequence of the way in which they work, less documents are evaluated.

Taking into account the number of variables involved in a collection, with this methodology, as future works, we are planning to use more complex graphs than a polytree to represent term relationships, improving the effectiveness of the model, without being a very important problem the propagation in that structure. To put in practice this idea, a proper solution would be, therefore, to modify the exact propagation algorithm corresponding to the used topology, trying to save time by means of not doing useless propagations. Thus, our next objectives are:

- To learn more complex graphs for the term subnetwork,
- to study the possible propagation algorithms for those topologies,
- and improve them with the aim of saving time and not losing too much precision in the results.

In this working framework, Bayesian networks will leave to be judged as "intractable" in the peculiar IR field, because they will be able to represent huge collections, and answer users' queries in a very short time.

Acknowledgments: This work has been supported by the Spanish MCYT and FIS, under Projects TIC2000-1351 and PI021147, respectively.

References

1. Baeza-Yates, R., Ribeiro-Neto, B. (1999) Modern Information Retrieval. Addison-Wesley, Essex.
2. Crestani, F., Lalmas, M., van Rijsbergen, C. J, Campbell, L. (1991) Is this Document Relevant?... Probably. A Survey of Probabilistic Models in Information Retrieval. ACM Computing Survey. **30:4**, 528–552
3. Cooper, G. F. (1990) The Computational Complexity of Probabilistic Inference Using Bayesian Belief Networks. Artificial Intelligence. **42**, 393–405
4. de Campos, L. M., Fernández-Luna, J. M., Huete. J. F. (1998) Query Expansion in Information Retrieval Systems Using a Bayesian Network-based Thesaurus. Proceedings of the 14th Uncertainty in Artificial Intelligence Conference. 53–60.

5. de Campos, L. M., Fernández-Luna, J.M., Huete, J.F. (2000) Building Bayesian Network-based Information Retrieval Systems. Proceedings of the 11th International Workshop on Database and Expert Systems Applications. 543–550

6. de Campos, L. M., Fernández-Luna, J.M., Huete, J.F. (2003) The BNR Model: Fundations and Performance of a Bayesian Network-based Retrieval Model. To appear in the International Journal of Approximate Reasoning.

7. Draper, L. D., Hanks, S. (1994) Localized Partial Evaluation of Belief Networks. Proceedings of the 10th Uncertainty in Artificial Intelligence Conference. 170–177

8. Fernández-Luna, J. M. (2001) Modelos de Recuperación de Información Basados en Redes de Creencia (in Spanish). Ph.D.thesis, E.T.S. Ingeniería Informática. Universidad de Granada

9. Guo H., Hsu, W. (2002) A Survey of Algorithms for Real-Time Bayesian Network Inference. Proceedings of the AAAI/KDD/UAI-2002 Joint Workshop on Real-Time Decision Support and Diagnosis Systems. 1–12

10. Ghazfan, D., Indrawan, M., Srinivasan, B. (1996) Toward Meaningful Bayesian Networks for Information Retrieval Systems. Proceedings of the IPMU'96 Conference. 841–846

11. Jitnah, N., Nicholson, A. E. (1997) treeNets: A Framework for Anytime Evaluation of Belief Networks. Proceedings of the ECSQUARU'97 Conference. 350–364

12. Liu, C. L, Wellman, M. P. (1995) On State-Space Abstraction for Anytime Evaluation of Bayesian Networks. Proceedings of the IJCAI Workshop on Anytime Algorithms and Deliberation Schedulling. 91–98.

13. Pearl, J. (1988) Probabilistic Reasoning in Intelligent Systems: Networks of Plausible Inference. Morgan and Kaufmann, San Mateo, California.

14. Pfeffer, A., Koller, D. (2000) Semantics and Inference for Recursive Probability Models. Proceedings of the 7th National Conference on Artificial Intelligence. 538–544

15. Ramos, F. T., Cozman, F. G., Ide, S. I. (2002) Embedded Bayesian Networks: Anyspace, Anytime Probabilistic Inference. Proceedings of the AAAI/KDD/UAI Joint Workshop on Real Time Decision Support and Diagnosis Systems. 12 – 19.

16. Reis Silva, I. (2000) Bayesian Networks for Information Retrieval Systems. Ph.D. thesis, Universidad Federal de Minas Gerais.

17. van Rijsbergen, C. J. (1979) Information Retrieval. Second Edition. Butter Worths, London

18. Ribeiro–Neto, B. A., Muntz, R. R. (1996) A Belief Network Model for IR. Proceedings of the 19th ACM–SIGIR Conference on Research and Development in Information Retrieval (SIGIR'96). 253–260

19. Robertson, S. E. (1977) The probability ranking principle in IR. Journal of Documentation. **33:4**, 294–304

20. Salton G., McGill, M. J. (1983) Introduction to Modern Information Retrieval, McGraw-Hill.

21. Turtle, H. R. (1990) Inference Networks for Document Retrieval. Ph.D. thesis, University of Massachusetts.

Continuous Speech Recognition Using Dynamic Bayesian Networks: A Fast Decoding Algorithm

Murat Deviren and Khalid Daoudi

INRIA-LORIA, Speech Group, B.P. 101 - 54602 Villers lès Nancy, France

Abstract. State-of-the-art automatic speech recognition systems are based on probabilistic modeling of the speech signal using Hidden Markov Models (HMMs). Recent work has focused on the use of dynamic Bayesian networks (DBNs) framework to construct new acoustic models to overcome the limitations of HMM based systems. In this line of research we proposed a methodology to learn the conditional independence assertions of acoustic models based on structural learning of DBNs. In previous work, we evaluated this approach for simple isolated and connected digit recognition tasks. In this paper we evaluate our approach for a more complex task: continuous phoneme recognition. For this purpose, we propose a new decoding algorithm based on dynamic programming. The proposed algorithm decreases the computational complexity of decoding and hence enables the application of the approach to complex speech recognition tasks.

1 Introduction

One of the basic building blocks of a speech recognition system is the acoustic modeling of sub-speech units (i.e. words, phones, bi-phones, etc.). The accuracy of acoustic modeling is a key point in the performance of a recognition system. The most popular probabilistic models used for this task are hidden Markov models (HMMs). Recently, there has been an increasing interest in a more general class of probabilistic models which include HMM only as a special case: *dynamic Bayesian networks* (DBNs). An important and attractive property of the DBNs formalism is the ability to encode complex dependency relations. In the last few years there has been several approaches to exploit this property in speech recognition systems [1], [4], [3], [2], [13], [14]. These provide several useful capabilities that are not readily achievable by HMM based systems because of restrictive Conditional Independence (CI) assertions of HMMs.

In [6], we proposed a methodology to learn the CI assertions of acoustic models based on structural learning of DBNs. As in HMMs, we consider that the observations are governed by a hidden process. On the other hand, we do not make any *a priori* dependency assumption between the hidden and observed processes. Rather, we give data a relative freedom to dictate the appropriate dependencies. In other words, we learn the dependencies between (hidden and observable) variables *from data*. The principle of this methodology is to search over all possible "realistic" dependencies, and to choose the ones which best explain the data. We use a structure learning algorithm based on minimum description length (MDL) scoring metric. This approach has the advantage to *guarantee* that the resulting model represents speech with higher fidelity than HMMs. Moreover, a *control* is given to the user to specify the maximal dependency structure and hence to make a trade-off between modeling accuracy and model complexity. The proposed approach is also technically very attractive because all the computational effort is performed in training phase.

We presented the application of this approach to isolated speech recognition in [6]. In [7], we extended our approach for continuous speech recognition and presented preliminary results on a connected digit recognition task. However the applicability of the decoding algorithm used in [7] is limited because of its computational complexity. In this paper, we provide in particular a much faster decoding algorithm which handles different model structures for different acoustic units. The algorithm is based on a dynamical programming procedure and leads to a substantial gain in complexity as compared to the algorithm in [7].

In our previous contributions, we evaluated our methodology on simple databases and tasks. In this paper, we evaluate our system on a phoneme recognition task using the TIMIT database. This database has been specifically designed for the development and evaluation of automatic speech recognition systems using phoneme models [11]. We provide a comparison of our system with an HMM based system using the HTK toolkit [1]. The results show that our methodology of learning the model dependencies from data still leads to significant improvement with respect to standard HMM modeling.

In the next section, we introduce the DBNs terminology. In section 3, we define the class of DBNs we use in our setting. We then briefly summarize the structural learning algorithm and our training strategy. In section 5, we describe a fast decoding algorithm for continuous speech recognition using the set of DBNs we consider. Finally, we illustrate the performance of our approach on a phoneme recognition task using the TIMIT database.

2   Dynamic Bayesian Networks

Our approach is based on the framework of *dynamic Bayesian networks* (DBNs). DBN theory is a generalization of Bayesian network (BN) theory

---

[1] The Hidden Markov Model Toolkit (HTK) is a portable toolkit for building and manipulating hidden Markov models. It is widely used in speech community and generally considered as a reference system for HMM based speech recognition. See http://htk.eng.cam.ac.uk/ for details on the toolkit.

to dynamic processes. Briefly, the BN formalism consists of associating a directed acyclic graph to the joint probability distribution (JPD) $P(\mathbf{X})$ of a set of random variables $\mathbf{X} = \{X_1, ..., X_n\}$. The nodes of this graph represent the random variables while the arrows encode the conditional independencies (CI) which (are supposed to) exist in the JPD. The set of all CI relations, which are implied by the separation properties of the graph, are termed the *Markov properties*. Formally, a Bayesian Network is defined as a pair $B = (S, \Theta)$, where S is a directed acyclic graph, Θ is a parameterization of a set $\{P(X_1|\mathbf{\Pi}_1), \ldots, P(X_n|\mathbf{\Pi}_n)\}$ of conditional probability distributions (CPDs), one for each variable, and $\mathbf{\Pi}_i$ is the set of parents of node X_i in S. The set of CPDs defines the associated JPD as:

$$P(\mathbf{X}) = \prod_{i=1}^{n} P(X_i|\mathbf{\Pi}_i) \ . \tag{1}$$

A dynamic Bayesian network extends this factored JPD representation to dynamic processes to encode the joint probability distribution of a time evolving set $\mathbf{X}[t] = \{X_1[t], \ldots, X_n[t]\}$ of variables. Following a similar notation as the static BN representation, the JPD for the dynamic process $\mathbf{X}[t]$ on a finite time interval $[1, T]$ is represented in factored form as:

$$P(\mathbf{X}[1], \ldots, \mathbf{X}[T]) = \prod_{t=1}^{T} \prod_{i=1}^{n} P(X_i[t]|\mathbf{\Pi}_i[t]) \tag{2}$$

where $\mathbf{\Pi}_i[t]$ denotes the parents of $X_i[t]$ in the graphical structure of the DBN. The graphical structure of a DBN can be viewed as a concatenation of several static BNs linked with temporal arcs. We call each of these static networks a time slice[2] of the DBN. In the most general case, if no assumptions are imposed on the underlying dynamic process, the graph structure and the numerical parameterization of a DBN can be different for each time slice. In this case the encoding of the JPD can be extremely complex.

In the literature the DBN representation is most often used for stationary first order Markov processes. In this case, a simplified DBN definition in terms of two static BNs is used [8]. The DBN is decomposed into an initial network B_I and a transition network $B_\rightarrow$. The principle of the definition relies on the stationarity assumption which implies that the structure and the parameters of the DBN are identical for each time slice. The repeating time slice is defined by the transition network $B_\rightarrow$ that encodes the time-invariant transition probability distribution over two consecutive time slices, $P(\mathbf{X}[t]|\mathbf{X}[t-1])$ for all $t > 1$. Due to *first* order Markov assumption it suffices to define the transition network over the variables of two successive time slices. The initial network B_I defines the structure in the boundary $t = 1$

---

[2] A time slice is defined as the collection of the set of variables $\mathbf{X}[t]$ in a single time instant t and their associated parents $\mathbf{\Pi}[t]$ in the graph structure.

and encodes the distribution over the variables $\mathbf{X}[1]$. The JPD for a finite time interval is obtained by *unrolling* the transition network for a sufficient number of time slices. The unrolling mechanism consists of replicating the structure and parameters of the transition network for each time t. Using this representation, the JPD is factored as :

$$P(\mathbf{X}[1],\ldots,\mathbf{X}[T]) = P_{B_I}(\mathbf{X}[1]) \prod_{t=2}^{T-1} P_{B_\rightarrow}(\mathbf{X}[t]|\mathbf{X}[t-1]) , \qquad (3)$$

where $P_{B_I}(.)$ and $P_{B_\rightarrow}(.)$ are the probability densities encoded by the initial and transition networks, respectively.

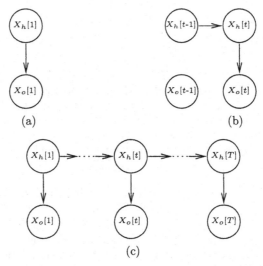

Fig. 1. HMM represented as a DBN : (a) initial network (B_I), (b) transition network $(B_\rightarrow)$ and (c) unrolled DBN where $X_h[t]$ denotes the hidden state variable at time t and $X_o[t]$ denotes the observation at time t

As an example, Fig. 1 shows the representation of a classical hidden Markov model as a DBN using this interpretation. The corresponding initial and transition networks are given in Fig. 1 (a) and (b) respectively. Each node represents a random variable $X_h[t]$ or $X_o[t]$, whose value specifies the state or the observation at time t. The Markov properties (dependency semantics) of an HMM, are completely defined by these networks, i.e., the hidden process $X_h[t]$ is first order Markov and each observation $X_o[t]$ depends only on the current state of the hidden process. The HMM for a finite time interval $[1, T]$ is given in Fig. 1 (c) which is obtained by unrolling the transition network in Fig. 1 (b).

In this paper, we extend the above representation to non-Markov stationary processes. Namely, we consider stationary processes $\mathbf{X}[t]$ with the following conditional independence property :

$$P(\mathbf{X}[t]|\mathbf{X}[1],\ldots,\mathbf{X}[t+\tau_f]) = P(\mathbf{X}[t]|\mathbf{X}[t-\tau_p],\ldots,\mathbf{X}[t+\tau_f]) \qquad (4)$$

for some positive integers τ_p and τ_f. The interval $[t-\tau_p, t+\tau_f]$ specifies the dependency window for the variables $\mathbf{X}[t]$. Graphically, (4) implies that a variable at time t can have parents in the interval $[t-\tau_p, t+\tau_f]$.

As in [8] we represent the DBN in terms of several static BNs defined on individual time slices with different independence assertions. Using stationarity assumption, we define a transition network over the variables in the interval $[t-\tau_p, t+\tau_f]$. The transition network encodes time-invariant transition probability defined in (4) on the repeating structure. However, the CI property in (4) is not valid for the first τ_p and the last τ_f time slices of the DBN. In these boundary time slices, the dependency time window vary with t for each variable $X_i[t] \in \mathbf{X}[t]$. Assuming $T \geq \tau_p + \tau_f + 1$, the CI relations for these time slices can be stated as follows :

- for the first τ_p time slices, $t \in [1, \tau_p]$,

$$P(\mathbf{X}[t]|\mathbf{X}[1],\ldots,\mathbf{X}[t+\tau_f]) = P(\mathbf{X}[t]|\mathbf{X}[t],\ldots,\mathbf{X}[t+\tau_f]) \ . \qquad (5)$$

- for the last τ_f time slices, $t \in [T-\tau_f+1, T]$,

$$P(\mathbf{X}[t]|\mathbf{X}[1],\ldots,\mathbf{X}[T]) = P(\mathbf{X}[t]|\mathbf{X}[t-\tau_p],\ldots,\mathbf{X}[T]) \ . \qquad (6)$$

Hence, the overall DBN structure is decomposed with respect to three time intervals : In $[1, \tau_p]$ and $[T-\tau_f+1, T]$ we have τ_p and τ_f different boundary structures, respectively. For $t \in [\tau_p+1, T-\tau_f]$ we have the repeating transition structure. With this consideration, the DBN is represented with $(\tau_p + \tau_f + 1)$ static BNs: τ_p initial networks $(B_{I_1}, \ldots, B_{I_{\tau_p}})$, τ_f final networks $(B_{F_{T-\tau_f+1}}, \ldots, B_{F_T})$ and a transition network $(B_\rightarrow)$. The initial and final networks encode the non-stationary dynamics of the process on the boundaries. The transition network encode the stationary dynamics. The DBN for a finite time interval $[1, T]$ is again constructed by unrolling the transition network for a required number of time slices. However care must be taken to properly concatenate the initial and final networks.

Using this representation the factorization of the JPD can be reinterpreted in terms of the probability densities of each static BN as:

$$P(\mathbf{X}[1],\ldots,\mathbf{X}[T]) = \prod_{t=1}^{\tau_p} P_{B_{I_t}}(\mathbf{X}[t]|\mathbf{\Pi}[t]) \prod_{t=\tau_p+1}^{T-\tau_f} P_{B_\rightarrow}(\mathbf{X}[t]|\mathbf{\Pi}[t])$$

$$\prod_{t=T-\tau_f+1}^{T} P_{B_{F_t}}(\mathbf{X}[t]|\mathbf{\Pi}[t]) \ , \qquad (7)$$

where $\mathbf{\Pi}[t]$ denotes the set of parents for the variables $\mathbf{X}[t]$ in the corresponding initial, transition and final networks. The parameterization of the associated conditional probabilities for each of these networks, are defined individually.

As an example consider the non-Markov process $\mathbf{X}[t] = \{X_h[t], X_o[t]\}$ with the following CI assumptions :

$$P(X_h[t]|\mathbf{X}[1], \ldots, \mathbf{X}[t+1]) = P(X_h[t]|X_h[t-2], X_h[t-1]) \qquad (8)$$

$$P(X_o[t]|\mathbf{X}[1], \ldots, \mathbf{X}[t+1]) = P(X_o[t]|X_h[t-1], X_h[t], X_h[t+1]) \quad (9)$$

The unrolled DBN for $T = 5$ time slices is shown in Fig. 2. The maximum past and future dependence parameters are, $\tau_p = 2$ and $\tau_f = 1$. Therefore, the DBN can be represented with 2 initial, 1 final and a transition static networks. These are given in Figs. 3(a), (b) (c) and (d).

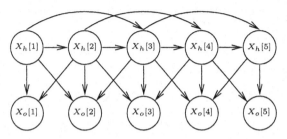

Fig. 2. DBN structure unrolled for $T = 5$ time slices

3   Structure Search Class

In acoustic modeling we consider that speech observations are governed by a discrete hidden process. Our goal is to learn the DBN structures (and their corresponding parameters) which "best" explain data, i.e., the "appropriate" dependencies between the hidden and observed variables of the speech process. This requires a search over a class of structures. Searching over all possible DBN structures would be computationally infeasible. Therefore, we restrict ourselves to a small but rich set of structures that represents only *realistic* dependencies, in a physical and computational sense.

In specifying the class of DBNs we first define the set of plausible dependencies among the hidden and observed processes of speech. First, we do not consider direct dependencies between observed variables. We believe this is a reasonable assumption because, short-time correlations in speech are indirectly considered by first and second order time derivatives of observation vectors. Second, we do not allow future dependencies between the hidden variables. This is because past dependencies must be considered given causality

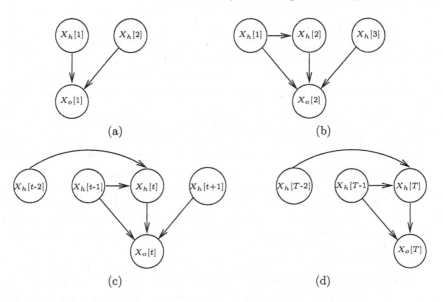

Fig. 3. (a), (b) Initial networks (in $t = 1, 2$), (c) transition network, and (d) final network (in $t = T$)

of speech. Thus, given that the graph has to be acyclic, we cannot have both past and future dependencies between hidden variables. Finally, we do not allow a hidden variable to have an observable one as a parent. This is because there exists no exact algorithm to infer BNs where continuous variables have discrete children. Therefore, the remaining authorized dependencies in the class of structures we are considering are the following.

Let $\mathbf{X}[t] = \{X_h[t], X_o[t]\}$ be the set of hidden and observed variables at time t. The allowed dependencies are:

- The hidden variable at time t is independent of $\mathbf{X}_1^{t-\kappa-1} = \{\mathbf{X}[1], \ldots, \mathbf{X}[t - \kappa - 1]\}$ given the last κ hidden variables, for $t > \kappa$,

$$P(X_h[t]|\mathbf{X}_1^{t-1}) = P(X_h[t]|X_h[t - \kappa], \ldots, X_h[t - 1]) . \tag{10}$$

- The observation variable at time t is independent of all other variables given the hidden variables in the time window $[t - \tau_p, t + \tau_f]$, for some positive integers τ_p and τ_f,

$$P(X_o[t]|\mathbf{X}_1^T \setminus \{X_o[t]\}) = P(X_o[t]|X_h[t - \tau_p], \ldots, X_h[t + \tau_f]) . \tag{11}$$

Equation (10) states that the hidden process is Markovian of order κ. Equation (11) states that the observation at time t depends on states of the hidden process in the interval $[t - \tau_p, t + \tau_f]$. Notice that using these independence properties we are able to take into account various phenomena in the speech production process. For instance, the non-Markov property on the observations may take into account the well known anticipation phenomena.

Based on these CI assumptions and stationarity we define a unique DBN structure using the triple (κ, τ_p, τ_f). When $(\kappa, \tau_p, \tau_f) = (1, 0, 0)$, the model reduces to the standard first-order HMM (Fig. 1). Using this structure specification we define a class of structures by specifying a lower and upper limit on the parameters κ, τ_p and τ_f. We set the lower limit as the first order HMM structure, i.e. $(\kappa, \tau_p, \tau_f) = (1, 0, 0)$. The upper limit $(\kappa_{max}, \tau_{p_{max}}, \tau_{f_{max}})$ provides a control mechanism to restrict the size of the search class and therefore reduce the computational complexity of the modeling. In this paper we use $(\kappa_{max}, \tau_{p_{max}}, \tau_{f_{max}}) = (2, 1, 1)$ as the upper bound on the structure search class. The structure of this network unrolled for $T = 5$ time slices is shown in Fig. 2.

Now that we have specified our class of structures, in the following we define the parameterization of the conditional probability densities defined by these structures. Consider a DBN with structure $S = (\kappa, \tau_p, \tau_f)$. If each discrete hidden variable $X_h[t]$ takes its values in the set of ordered labels $I = \{1 \ldots N\}$, and each observable variable $X_o[t]$ has a conditional mixture of Gaussians density, the numerical parameterization for each time slice is as follows:

$$P(X_h[t] = j | \Pi_h[t] = \mathbf{i}) = a_{\mathbf{i}j}[t], \quad for \quad j = 1 \ldots N \ .$$
$$P(X_o[t] | \Pi_o[t] = \mathbf{i}) \sim \sum_k c_k[t] \mathcal{N}(\mu_{\mathbf{i}_k}[t], \Sigma_{\mathbf{i}_k}[t]) \tag{12}$$

The (possibly) vector-index $\mathbf{i}$ is over all possible values of the variable's parents, which depend on the structure of the corresponding time slice. Given the stationarity assumption, the parameters of these conditional probabilities for the transition network are constant over time, when the structure is self-repeating, i.e.,

$$a_{\mathbf{i}j}[t] = a_{\mathbf{i}j} \ , \quad for \quad t \in [\kappa + 1, T]$$
$$c_k[t] = c_k \ , \mu_{\mathbf{i}_k}[t] = \mu_{\mathbf{i}_k} \ , \Sigma_{\mathbf{i}_k}[t] = \Sigma_{\mathbf{i}_k} \ , \quad for \quad t \in [\tau_p + 1, T - \tau_f]$$

For $S = (\kappa, \tau_p, \tau_f)$ we have $\max(\kappa, \tau_p)$ initial and τ_f final networks. The CPDs of each of these networks are specified on the variables of the corresponding time slice as in (12).

4   Learning Algorithm

In the previous section, we defined a set of plausible DBN structures for speech recognition. In this section we use this set as the structure search class for the learning algorithm. The learning problem for Bayesian networks is to find the optimal structure and parameters of the network given a set of training data. The general approach is to define a scoring metric that evaluates the optimality of a given BN and use a search algorithm that traces the space of allowed structures to maximize the scoring metric. In [8], the structure scoring rules for static BNs is extended to dynamic case. Using the representation of a DBN for a first order Markov process in terms of an initial and a transition network the authors show that the DBN learning problem can be reformulated as learning static BNs. The same reasoning can be applied for DBNs of non-Markov processes represented with several static BNs, i.e., initial, transition and final networks as described in Sect. 2.

Our goal to find the "optimal" DBN structure and the associated parameters for each acoustic unit v in the given set V. The optimality will be considered in the sense that the likelihood of observations is maximum and the structural complexity is minimum. We use the minimum description length (MDL) metric (or equivalently Bayesian Information Criteria (BIC)) to evaluate the complexity of a given network. Thus, the scoring of a network is given by:

$$Score_{MDL} = \log P(D|\Theta, S) - \frac{\log L}{2} \sum_{i=1}^{n} ||X_i, \Pi_i|| \tag{13}$$

where D is the observation set, L is the number of examples (realizations) in D and $||X, Y||$ is defined as the number of parameters required to encode the conditional probability, $P(X|Y)$. We use the set of structures defined in Sect. 3 as the search class for the learning algorithm

The principle of the learning algorithm is to start with some initial structure and parameters and to run a generalized version of the Expectation Maximization (EM) algorithm[9], called structural EM (SEM) which allows the update of structure and parameters iteratively. The block diagram of this algorithm is given in Fig. 4. At each step, first a parametric EM algorithm is performed to find the optimal parameters for the current structure. Second, a structural search is performed to find a structure that yields a higher score. The computation of the scoring metric for a candidate structure generally requires inference within the previous network. This algorithm guarantees an increase in the MDL score at each iteration, and converges to a local maximum. The details of the algorithm are given in [6,7] and the references within.

In the following, we describe our training strategy for continuous speech recognition based on context independent DBN models. We use phonemes as the basic acoustic unit for our experimental recognition system. Nevertheless the presented algorithms are applicable to any acoustic unit, i.e. syllable, word, etc..

In our setting, we initialize using the HMM structure $(\kappa, \tau_p, \tau_f) = (1, 0, 0)$ which is the lower bound of our search space. This initialization guarantees that the resulting DBN will model speech with a higher (or equal) fidelity as

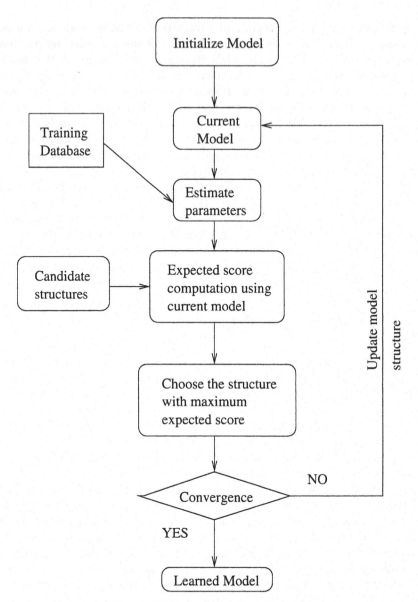

Fig. 4. *SEM algorithm block diagram*

compared to HMMs. The trade off between the complexity of the learning algorithm and the fidelity of the resulting model is controlled by the upper bound on the search space.

The learning algorithm requires a set of isolated observations for each DBN. However, in continuous speech recognition the training database consists of continuous utterances. Each utterance is labelled with the underlying acoustic unit sequence but in general the boundaries dividing the acoustic unit segments are either unavailable or unreliable. Therefore an embedded training strategy is necessary, that is, we make use of a reference HMM system to obtain the best segmentation of each training sentence using a forced alignment procedure with this reference system. We consider these segments as isolated observations for each acoustic unit and run the structural learning algorithm to obtain our DBN models. By doing so we rely on the segmentation provided by HMMs and improve the modeling accuracy within each segment.

5   Decoding Algorithm

Let us assume that we are given a set V of $|V|$ acoustic units, and a DBN model for each acoustic unit $v \in V$, with structure $(\kappa^v, \tau_p^v, \tau_f^v)$. The decoding problem is to identify the most likely sequence of these units, given a speaker utterance. In [7] we present a decoding algorithm based on the use of Dawid's algorithm [5] on a state-augmented DBN. Dawid's algorithm is a message propagation algorithm that allows the identification of the most likely sequence of hidden states in an arbitrary (discrete) Bayesian network given the observations [5]. Using our augmentation methodology we were able to represent all acoustic units' models in a single DBN. Hence, using Dawid's algorithm on this augmented DBN, one obtains the most likely sequence of acoustic units given the observations. The algorithm presented in [7], however, is rather a "brute force" approach which introduces substantial unnecessary computations. In the following we present an equivalent[3] but much faster algorithm which takes advantage of the particular DBNs we are dealing with.

As in [7], the first step is to construct a "maximal" DBN for each acoustic unit. The goal is to represent all acoustic models with the same graphical structure without violating the learned Conditional Independence (CI) relations. The maximal network structure $(\kappa^m, \tau_p^m, \tau_f^m)$ is chosen such that all the learned CI relations can be encoded using this maximal structure. Precisely,

$$\kappa^m = \max_v(\kappa^v), \quad \tau_p^m = \max_v(\tau_p^v), \quad \tau_f^m = \max_v(\tau_f^v) .$$

Then, the numerical parameterization of each maximal network is obtained from the corresponding DBN, without violating its (learned) Markov prop-

---

[3] in the sense that it is an exact inference algorithm and yields the same decoding results as in [7]

erties. This is achieved by setting several conditional probabilities of the maximal DBN to be equal to those given by the learned DBN. Precisely,

$$P_m(X_i[t]|\Pi_{it}^m = (\mathbf{i}, \mathbf{j})) = P(X_i[t]|\Pi_{it} = \mathbf{i})$$
$$\text{for all } \mathbf{i} \in \Pi_{it}, \mathbf{j} \in \Pi_{it}^m \setminus \Pi_{it} \tag{14}$$

where Π_{it}^m is the set of parents of $X_i[t]$ in the maximal DBN, and Π_{it} is the set of parents of $X_i[t]$ in the learned DBN.

Once the maximal network of each acoustic unit is constructed, the second step (which is the heart of this paper) consists, basically speaking, in operating a dynamical programming procedure to perform a "parallel" inference algorithm on "local clique potentials" (see [10] for details).

The principle of the algorithm is similar to Viterbi decoding algorithm that is widely used in HMM based speech recognition systems [12]. The main difference is that Viterbi algorithm is used for first order hidden Markov models whereas the proposed algorithm is derived for the set of DBN structures we propose in [6]. Indeed, if the maximal structure is $(1, 0, 0)$, i.e. HMM structure, our algorithm is identical to Viterbi decoding algorithm. We proceed now to describe this algorithm.

In the experiments we carry out later, the maximal network structure we obtain is $(\kappa^m, \tau_p^m, \tau_f^m) = (2, 1, 1)$. Given the technicality of the algorithm and for the sake of clarity and simplicity of the paper, we describe our decoding algorithm for this particular case. We emphasize however that this algorithm can be readily generalized to any kind of maximal network.

Let us denote the hidden and observed variables $X_h[t], X_o[t]$ as H_t and O_t respectively. Let H_t take its values in $\{1_v, \ldots, N_v, \forall v \in V\}$ to denote the hidden states for all acoustic unit models. As described in Sect. 2, the conditional probabilities of a DBN are defined according to the initial, transition and final networks. For the maximal structure $(2,1,1)$ there are 2 initial, 1 transition and 1 final networks. The conditional probabilities of these networks are denoted as follows (where $i, j, k \in \{1_v, \ldots, N_v, \forall v \in V\}$).

Initial networks :

$$a_j^v = P(H_1 = j) \,,$$
$$a_{jk}^v = P(H_2 = k|H_1 = j) \,,$$
$$b_{jk}^{Iv}(O_1) = P(O_1|H_1 = j, H_2 = k) \,.$$

Transition network :

$$a_{ijk}^v = P(H_{t+1} = k|H_t = j, H_{t-1} = i) \,,$$
$$b_{ijk}^v(O_t) = P(O_t|H_{t-1} = i, H_t = j, H_{t+1} = k) \,.$$

Final network :

$$b_{ij}^{Fv}(O_t) = P(O_t|H_{t-1} = i, H_t = j) \,.$$

Our aim is to find the most likely sequence of hidden states $H_1^T = \{H_1, \ldots, H_T\}$ given the observation sequence $O_1^T = \{O_1, \ldots, O_T\}$, and consequently the likelihood of observations along this sequence:

$$P_{max}(O_1^T) = \max_{H_1 \ldots H_T} P(O_1^T, H_1^T). \tag{15}$$

First note that the state transitions are restricted to a left-to-right topology and a model transition from model r to v is only allowed from the last state of r (N_r) to the first state of v (1_v). In order to provide a recursive formulation for (15) we define two intermediate quantities:

$$\delta_t^v(i, j, k) = \max_{H_1 \ldots H_{t-2}} P(O_1^t, H_1^{t-2}, H_{t-1} = i, H_t = j, H_{t+1} = k) \,,$$

$$\gamma_t^v(i, j) = \max_{H_1 \ldots H_{t-2}} P(O_1^t, H_1^{t-2}, H_{t-1} = i, H_t = j) \,.$$

$\delta_t^v(i, j, k)$ is the likelihood of the first t observations along the most likely state sequence up to H_{t-2} and for $H_{t-1} = i, H_t = j, H_{t+1} = k$. Similarly $\gamma_t^v(i, j)$ is the likelihood of the first t observations along the most likely state sequence up to H_{t-2} and for $H_{t-1} = i, H_t = j$ assuming that O_t is the last observation from model v. We need this second quantity to compute the likelihood when a model transition occurs at $t + 1$.

For each model v we initialize the recursion with $\delta_1^v(i, j, k) = a_j^v a_{jk}^v b_{jk}^{I_v}(O_1)$ which is the the likelihood of O_1 emitted from the initial network of v. Then for each t, the new evidence O_t can be emitted from either of the initial, transition and final networks of v. We know that $\delta_{t-1}^v(n, i, j)$ is the maximum likelihood for the observation sequence O_1^{t-1} for $H_{t-2} = n, H_{t-1} = i, H_t = j$.

If the emission is from the final network then O_t does not depend on H_{t+1} and the new likelihood is computed by maximizing $\delta_{t-1}^v(n, i, j)$ over n and incorporating the emission probability from the final network. We defined this term as $\gamma_t^v(i, j)$:

$$\gamma_t^v(i, j) = b_{ij}^{Fv}(O_t) \max_n [\delta_{t-1}^v(n, i, j)] \,. \tag{16}$$

This term is the maximum likelihood of the observation sequence O_1^t for $H_{t-1} = i, H_t = j$ for $i, j \in \{1_v, \ldots, N_v\}$. Considering the fact that for the full observation sequence O_1^T the last observation is emitted from the final network of some $v \in V$, the maximum likelihood for the full observation sequence is obtained from the following maximization over v, i, j:

$$P_{max}(O_1^T) = \max_{i,j,v} [\gamma_T^v(i, j)] \,. \tag{17}$$

To complete our formulation we continue with the derivation of the recursion formula for $\delta_t^v(i, j, k)$. If the emission is from initial or transition networks of v the incorporation of the new evidence O_t depends on the value of H_{t-2} such that:

- if $H_{t-2} \in \{1_v, \ldots, N_v\}$ then O_{t-1} was emitted from v and O_t will be emitted from the transition network of v. The new likelihood is computed by maximizing $\delta^v_{t-1}(n, i, j)$ over n and incorporating the emission probability from the transition network of v.

$$\delta^v_t(i, j, k) = a^v_{ijk} b^v_{ijk}(O_t) \max_n [\delta^v_{t-1}(n, i, j)] \tag{18}$$

- if $H_{t-2} \in \{N_r, \forall r \in V\}$ then O_{t-1} was emitted from the final network of r and a model transition occurs at t. Hence O_t will be emitted from the initial network of v with $H_t = 1_v$. In this case we need a maximization over n and r to specify the model that transits with the maximum likelihood. Therefore the likelihood term is computed as :

$$\delta^v_t(i, 1_v, k) = a^v_{1_v k} b^{I_v}_{1_v k}(O_t) \max_{n, r} [\gamma^r_{t-1}(n, N_r) P(v|r)] . \tag{19}$$

$P(v|r)$ is the transition probability from model r to v and it is given by the language model obtained from frequency counts of acoustic unit pair occurrences in the training corpus.

Now in order to specify the value[4] of H_{t-2} that maximizes the overall likelihood for O^t_1 we maximize among these two cases :

$$\delta^v_t(i, j, k) = \max \left\{ a^v_{ijk} b^v_{ijk}(O_t) \max_n [\delta^v_{t-1}(n, i, j)] , \right.$$
$$\left. a^v_{jk} b^{I_v}_{jk}(O_t) \max_{n, r} [\gamma^r_{t-1}(n, N_r) P(v|r)] \right\} . \tag{20}$$

The case that yields the maximum likelihood specifies the value of H_{t-2} in the state sequence.

The complete algorithm is given as follows where we introduce $\psi_t(i, j)$ for back tracking the maximization arguments. Once the maximum likelihood is computed, the most likely state sequence is obtained by back tracking the maximization arguments.

Initialization :

$$\delta^v_1(j, k) = a^v_j a^v_{jk} b^{I_v}_{jk}(O_1) ,$$
$$\gamma^v_1(j) = 0 ,$$
$$\psi_1(j) = 0 .$$

Recursion for $1 < t < T$:
    For $2_v \leq j \leq N_v$,

$$\delta^v_t(i, j, k) = a^v_{ijk} b^v_{ijk}(O_t) \max_n [\delta^v_{t-1}(n, i, j)] ,$$
$$\psi_t(i, j) = \arg \max_{n, v} [\delta^v_{t-1}(n, i, j)] .$$

---

[4] Notice that, we allowed H_{t-2} to take values only in $\{1_v, \ldots, N_v, N_r, \forall r \in V\}$. For all other cases no state transitions are allowed and the likelihood is zero.

For $j = 1_v$,

$$\delta_t^v(i,j,k) = \max \left\{ a_{ijk}^v b_{ijk}^v(O_t) \max_n [\delta_{t-1}^v(n,i,j)] \, , \right.$$

$$\left. a_{jk}^v b_{jk}^{Iv}(O_t) \max_{n,r} [\gamma_{t-1}^r(n, N_r) P(v|r)] \right\} \, ,$$

$$\psi_t(i,j) = \begin{cases} \arg\max_{n,v} [\delta_{t-1}^v(n,i,j)] \, . \\ \arg\max_{n,v} [\gamma_{t-1}^r(n, N_r) P(v|r)] \, . \end{cases}$$

For $j = N_v$,

$$\gamma_t^v(i,j) = b_{ij}^{Fv}(O_t) \max_n [\delta_{t-1}^v(n,i,j)] \, .$$

Termination :

$$\gamma_T^v(i,j) = b_{ij}^{Fv}(O_T) \max_n [\delta_{T-1}^v(n,i,j)] \, ,$$

$$\psi_T(i,j) = \arg\max_{n,v} [\delta_{T-1}^v(n,i,j)] \, ,$$

$$P_{max}(O_1^T) = \max_{i,j,v} [\gamma_T^v(i,j)] \, ,$$

$$(H_{T-1}, H_T) = \arg\max_{i,j} [\delta_T^v(i,j)] \, .$$

Back Tracking :

$$H_t = \psi_{t+2}(H_{t+1}, H_{t+2}), \; t = T - 2, \ldots, 1 \, .$$

5.1 Complexity Analysis

Before describing the computational complexity of the proposed algorithm, we shall refer to our initial decoding algorithm. As briefly described before, the decoding is performed using Dawid's algorithm on a state augmented model [7]. The complexity of Dawid's algorithm depends on the size of each clique on the junction tree and the number of values of each variable in the clique [5]. In our setting, the cliques[5] of the maximal network with structure $(\kappa^m, \tau_p^m, \tau_f^m) = (2, 1, 1)$ contains three consecutive hidden state variables (H_{t-1}, H_t, H_{t+1}). If we assume that $N_v = N \; \forall v \in V$, then each augmented variable takes $N|V|$ values. The number of distinct realizations for each clique is therefore $N^3|V|^3$. The computational complexity for an observation sequence of length T is $O(N^3|V|^3 T)$.

For the algorithm proposed in this paper, there are three major items in the recursive computation. These are computed for all $t = 2, \ldots, T - 1$ and $v \in V$. We analyze each of them separately for each t, v and then derive the asymptotic complexity accordingly.

---

[5] A constrained triangulation scheme results in a repeating clique tree for the transition network. The cliques for the initial and final networks deviate from this repeating form. These are not considered for asymptotic complexity analysis.

- $a_{ijk}^v b_{ijk}^v(O_t) \max_n [\delta_{t-1}^v(n, i, j)]$

  This term is computed for $1 \leq j \leq N$. The complexity depends on the number of distinct values of the (i, j, k) triple. This would be N^3 without any topological constraints yielding $2N^3$ multiplications and N^3 operations in maximization. Exploiting the left-to-right topology for the state sequence, i.e. if $H_{t-1} = i$, then $H_t = j \in \{i, i+1\}$, the number of distinct values for (i, j, k) reduces to $4N + 4$. The total number of operations is therefore $(4N+4 + 2(4N+4))$. Hence the complexity introduced by this term is $O(N)$ rather than $O(N^3)$.

- $a_{jk}^v b_{jk}^{Iv}(o_t) \max_{n,r} [\gamma_{t-1}^r(n, N) P(v|r)]$

  This term is computed only for $j = 1$. The computation requires $N + 1$ multiplications for each k and $N|V|$ multiplications for each n, r. The maximization also introduces $N|V|$ operations. The asymptotic complexity is $O(N|V|)$. The topological constraint does not reduce the complexity because it only affects the range of values for k to reduce the $N + 1$ multiplications to 3.

- $b_{ij}^{Fv}(O_t) \max_n [\delta_{t-1}^v(n, i, j)]$

  This term is computed only for $j = N$ with N multiplications and N^2 operations in maximization. The left-to-right topology limits the values of i to $\{N, N-1\}$. So the complexity reduces to $O(N)$ instead of $O(N^2)$.

Now considering the computation for all t and v, the overall asymptotic complexity "without" the topological constraints is $O(N^3|V|T + N|V|^2T + N^2|V|T)$. The improvement is significant as compared to $O(N^3|V|^3T)$. This improvement is due to 1) the parallel nature of the proposed algorithm and 2) exploitation of the structural properties of the DBNs we use. Further improvement is achieved by imposing the a left-to-right transition topology for the hidden states. In this case the dominant term in the asymptotic complexity is $O(N|V|^2T)$.

6   Experiments

In this section, we compare the performances of DBN models to standard HMMs. In previous work we evaluated our approach and presented promising results on isolated and connected digit recognition tasks [6,7]. In this paper, we evaluate our approach on a more complicated task, i.e, continuous phoneme recognition. Our experiments are carried out on the TIMIT database that has been designed to provide speech data for the acquisition of acoustic-phonetic knowledge and for the development and evaluation of automatic speech recognition systems [11]. The training corpus consists of 3695 sentences. The tests are performed on 192 sentences in the core test set. We train 48 phoneme models and a silence model. Each phoneme and the silence are modeled with 3-states for HMMs and DBNs, i.e $H_t = \{1, 2, 3\}$. The output probability density of each observed variable O_t is represented with a mixture of Gaussians with diagonal covariance matrices. In tests, we use a phoneme based bi-gram language model learned from the training corpus. In accuracy computations we neglect the confusions within 7 phoneme groups to yield 39 effective phonemes as in [11]. The acoustic parameterization is based on standard Mel frequency cepstral coefficients (MFCC). These are computed from 26 Mel scaled logarithmic filter-bank energies of 25ms frames. The frame rate is 10ms. The resulting parameter vector consists of 12 static MFCC coefficients and energy concatenated with first and second order time derivatives. Front-end parameterization and HMM modeling are performed using HTK toolkit.

As a first step we construct an HMM based system. This system is used as a reference to compare DBN based systems. We initialize the models using the phonetic transcriptions provided in the database [11]. Given the segment boundaries of each phoneme, we consider isolated observations for each model. We use a uniform segmentation for the state sequence and obtain initial estimates for the means and variances of the Gaussians. Next, we apply the Viterbi alignment procedure to compute the maximum likelihood state sequence for each phoneme segment and reestimate the models' parameters iteratively. These initial models are then refined by increasing the number of Gaussian mixtures and embedded parameter reestimation iterations. The phoneme recognition accuracies obtained by this system with different number of Gaussian mixtures are given in the first row of Table 1.

Table 1. Phoneme recognition accuracy of HMM and DBN based systems using different number of Gaussian mixture components (%).

nb mixtures	1	2	4	8	16	32	64
HMM	-	57.48	62.25	65.32	65.77	67.82	69.00
DBN-1	59.83	61.79	65.50	67.50	69.23	70.22	-
DBN-2	59.83	62.38	65.84	67.97	69.64	70.82	-

In the construction of DBN based systems we use an initialization step that relies on the HMM system defined above. Rather than using the phonetic segmentation of the database, we start with a more reliable segmentation obtained by a forced alignment procedure. We use HMMs with 8 mixtures of Gaussians for this task. The recognition accuracy obtained with these models on the core test set is **65.32** %. Using this segmentation we initialize single Gaussian DBN models by running the SEM algorithm. The upper bound on the structure search space is set to be $(2, 1, 1)$. The learned structure for each phoneme is given in Table 2. It is interesting to observe that for most of the phonemes the learned structure turns out to be non-Markov, i.e., with future dependencies. This result supports our specification of the class of allowed

Table 2. Results of structural learning algorithm.

(κ, τ_p, τ_f)	phonemes
(1,0,1)	b, uh, th, epi, zh
(1,1,0)	g, en, oy, d
(1,1,1)	dh, ah, ng, el, ow, w, ch, dx, q, aa, y, k, f, m, t, ey, ay, ae, eh, v, hh, jh, aw
(2,0,1)	ax
(2,1,1)	vcl, z, p, l, ix, s, cl, ih, n, sh, ao, r, iy, er, uw, s#

structures for acoustic modeling. Using this class of structures we are able to capture the non-Markov phenomena in speech.

We use two different schemes to further refine single Gaussian DBN models. In the first scheme (DBN-1), we fix the model structures obtained in the first run of the SEM algorithm and increase the number of mixtures of each model. This technique allows a finer tuning of model parameters with fixed dependency assumptions. In the second scheme (DBN-2), we reestimate the structures each time the number of mixtures is increased. In this case the dependency assumptions are updated as the model accuracy is increased which allows a better exploration of dependency relations in data.

Recognition is performed using the decoding algorithm presented in Sect. 5. The maximal structure for the set of learned structures is $(2,1,1)$. Recognition accuracies for different number of Gaussian mixtures are given in the second and third rows of Table 1 for DBN-1 and DBN-2, respectively. Our first observation from these tests is that DBN-2 systems perform better than DBN-1 systems which implies that updating model structure as the model is refined yields improved modeling.

From Table 1, it is clear that both DBN based systems perform better than HMM based systems with same number of Gaussian mixtures. However to make fair comparisons, we should compare a DBN system to an HMM, with an equivalent number of parameters. To do so, we compare our system to an HMM with the same number of states, but using an observation probability density with two times more mixture of Gaussians. These comparisons are provided as a bars-plot in Fig. 5. On the x-axis we indicate the number of mixture components for the DBN based systems. The equivalent HMM system at the same abscissa has two times more mixture components. At each abscissa we compare equivalent HMM, DBN-1 and DBN-2 systems presented in the same order from left to right. In all cases the DBN system yields a higher recognition accuracy than equivalent HMM.

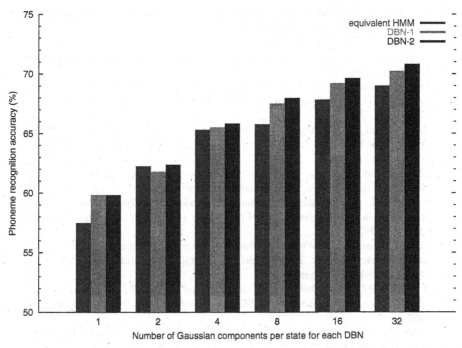

Fig. 5. Recognition accuracy comparisons of DBN based systems with equivalent HMM based systems

These results show that substantial gain in recognition accuracy can be obtained when model dependencies are learned from data. This supports our previous results on isolated and connected digit recognition experiments [6,7]. In addition, the computational complexity of decoding is greatly reduced using the proposed algorithm which enables the application of the methodology for complex speech recognition tasks.

References

1. J. A. Bilmes. Dynamic Bayesian multinets. In *16th Conference on Uncertainty in Artificial Intelligence*, 2000.
2. O. Cetin, H. Nock, K. Kirchhoff, J. Bilmes, and M. Ostendorf. The 2001 GMTK-based SPINE ASR system. In *International Conference on Spoken Language Processing*, 2002.
3. K. Daoudi, D.Fohr, and C. Antoine. Continuous multi-band speech recognition using Bayesian networks. In *Automatic Speech Recognition and Understanding Workshop*, 2001.
4. K. Daoudi, D.Fohr, and C. Antoine. Dynamic Bayesian networks for multi-band automatic speech recognition. *to appear in Computer Speech and Language*, 2001.

5. A.P. Dawid. Application of a general propagation algorithm for probabilistic expert systems. *Statistics and Computing*, (2):25–36, 1992.
6. M. Deviren and K. Daoudi. Structural learning of dynamic Bayesian networks in speech recognition. In *7th European Conference on Speech Communication and Technology*, 2001.
7. M. Deviren and K. Daoudi. Continuous speech recognition using structural learning of dynamic Bayesian networks. In *11th European Signal Processing Conference*, 2002.
8. N. Friedman, K. Murphy, and S. Russell. Learning the structure of dynamic probabilistic networks. In *14th Conference on Uncertainty in Artificial Intelligence*, 1998.
9. Nir Friedman. Learning belief networks in the presence of missing values and hidden variables. In *Int. Conf. Machine Learning*, 1997.
10. U. Kjaerulf. A computational scheme for reasoning in dynamic probabilistic networks. In *8th Conference on Uncertainty in Artificial Intelligence*, pages 121–129, 1992.
11. Kai-Fu Lee and Hsiao-Wuen Hon. Speaker-independent phone recognition using hidden Markov models. *IEEE Trans. Acoustics, Speech and Signal Processing*, 37(11):1641–1648, November 1989.
12. A. J. Viterbi. Error bounds for convolutional codes and an asymptotically optimal decoding algorithm. *IEEE Trans. Information Theory*, 13:260–269, 1967.
13. G. Zweig. *Speech Recognition with Dynamic Bayesian Networks*. PhD thesis, U.C., Berkeley, 1998.
14. G. Zweig, J. Bilmes, T. Richardson, K. Filali, K. Livescu, P. Xu, K. Jackson, Y. Brandman, E. Sandness, E. Holtz, J. Torres, and B. Byrne. Structurally discriminative graphical models for automatic speech recognition - results from the 2001 John Hopkins summer workshop. In *International Conference on Acoustics Speech and Signal Processing*, 2002.

Applications of Bayesian Networks in Meteorology

Rafael Cano[1], Carmen Sordo[2], and José M. Gutiérrez[2,3]

[1] Instituto Nacional de Meteorología
CMT/CAS, Santander, Spain.
[2] Dpto. Matemática Aplicada y C.C. (grupo de IA en Meteorología),
Universidad de Cantabria, 39005, Santander, Spain.
[3] (corresponding author): gutierjm@unican.es, http://grupos.unican.es/ai/meteo

Abstract. In this paper we present some applications of Bayesian networks in Meteorology from a data mining point of view. We work with a database of observations (daily rainfall and maximum wind speed) in a network of 100 stations in the Iberian peninsula and with the corresponding gridded atmospheric patterns generated by a numerical circulation model. As a first step, we analyze the efficiency of standard learning algorithms to obtain directed acyclic graphs representing the spatial dependencies among the variables included in the database; we also present a new local learning algorithm which takes advantage of the spatial character of the problem. The resulting graphical models are applied to different meteorological problems including weather forecast and stochastic weather generation. Some promising results are reported.

1 Introduction

The increasing availability of climate data during the last decades (observational records, radar and satellite maps, proxy data, etc.) led to the development of efficient statistical [1,2] and data mining [3] techniques in different areas of the atmospheric sciences, including climate variability and local weather forecast. Among this huge amount of data there are two important sources of information for statistical applications:

- *Climatological databases*, which collect local information (e.g., historical observations of precipitation or wind speed) from networks of stations within specific geographical regions.
- *Reanalysis databases*, which gather outputs of numerical atmospheric circulation models for long periods of time and different lead times (forecast steps). These outputs are gridded values of temperature, humidity, etc., on appropriate 3D grids covering the world, or an area of interest.

Climatological databases contain the statistical properties of the different local climatology existing in an area of study, whereas reanalysis databases contain the evolution of the atmosphere simulated by numerical models. Thus, this information can be used to analyze problems related with the atmosphere dynamics and with the possible impacts it may cause on the climatology of local regions. Some examples of application of statistical techniques in these problems include linear regression methods [4] for weather forecast, clustering methods [5] and principal component analysis [6] for identification of representative atmospheric patterns, etc. However, these techniques deal with special problems and with particular sets of data and they both fragment the information and assume ad-hoc spatial independencies in order to simplify the resulting problem-driven models.

In this paper we introduce probabilistic graphical models (Bayesian networks) in Meteorology as a data mining technique [7,8]. Bayesian networks automatically capture probabilistic information from data using directed acyclic graphs and factorized probability functions. The graph intuitively represents the relevant spatial dependencies and the resulting factorized probability function leads to efficient inference algorithms for updating probabilities when new evidence is available. Thus, Bayesian networks offer a sound and practical methodology for discovering probabilistic knowledge in databases and for building intuitive and tractable probabilistic models which can be easily used to solve a wide variety of problems.

We deal both with the automatic construction of Bayesian networks from meteorological data and with the application of the resulting models to different interesting problems. We assume that the audience is not familiar with meteorological concepts, so the prerequisites have been kept to a minimum and special care has been made to give all the necessary meteorological background for a full understanding of the examples.

The paper is organized as follows. We start in Sec. 2 describing the sources of data used in this paper. Then, we introduce some interesting meteorological problems and show their relationship with the data (Sec. 3). These problems are used throughout the paper to illustrate the application of Bayesian networks. In Section 4 we briefly describe Bayesian networks and analyze different methods to learn these models from data; Sec. 4.4 describes probabilistic inference in these models. Finally, in Sec. 5 several applications of Bayesian networks are described.

2   Area of Study and Available Data

In this work we consider the Iberian Peninsula as the geographical area of interest, and use daily data (precipitation and maximum wind speed) from a 100 stations' network (see Fig. 1(a)) provided by the Spanish weather service (Instituto Nacional de Meteorología, INM). The data covers the period from 1959 to 2000 and is representative of the local climatology of this area. The variables are considered continuous for some applications, but are also quantized for other applications. Precipitation (*Precip*) is quantized into four different states (1= "dry or very light rain",2= "light rain", 3= "moderate rain" and 4= "heavy rain"), according to the thresholds 0, 2, 10 and 20

mm/day, respectively (for instance, the event "heavy rain" corresponds to $Precip > 20mm$). Similarly, maximum wind speed ($Wind$) is quantized into three different states corresponding to the thresholds 0, 40 and 120 Km/h, respectively.

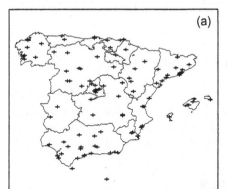 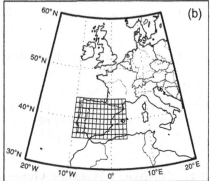

Fig. 1. (a) Geographical area of study and the location of the 100 stations considered in this work; (b) Fragment of the $1° \times 1°$ grid used by the ECMWF operative atmospheric model covering the area of interest.

On the other hand, we consider data which characterizes the state and evolution of the atmospheric variables. In particular, we use daily atmospheric circulation gridded fields provided by the European Center for Medium Weather Forecast (ECMWF) reanalysis project ERA-15 [9]. The fields are obtained integrating a numerical Atmospheric Circulation Model (ACM) on a global grid during the period from 1979 to 1993. To run the model for a particular date, the available observations are assimilated to obtain an initial condition for the atmosphere state (this is called the analysis); then, the integration of the model returns the predicted fields for different lead times (we shall only use the forecast corresponding to the state of the atmosphere 24 hours later). In this paper each circulation pattern is defined on the $1° \times 1°$ latitude and longitude grid covering the area of interest shown in Fig. 1(b). The patterns (data vectors) are formed considering the values of five meteorological variables (temperature, geopotential, U and V wind components, and humidity) at six vertical pressure levels (300, 500, 700, 850, 925, and 1000 mb) —the lower level corresponds to surface, whereas the higher level corresponds to approximately 10Km—. Collecting all this information, we have the following atmospheric pattern:

$$\mathbf{X_t} = (T_t^{1000}, \dots, T_t^{300}, H_t^{1000}, \dots, H_t^{300}, \dots, V_t^{1000}, \dots, V_t^{300}), \qquad (1)$$

where X_i^j denotes the j-pressure level field of variable X for date t. We get a total of 9×15 (grid) $\times 6$ (pressure levels) $\times 5$ (variables) = 4050 dimensions.

Note that the enormous size of these patterns is somehow misleading, since there exists a great correlation on all variables and space scales which makes the effective dimension much lower. In this paper, in order to simplify the treatment of atmospheric variables, we use a quantized variable S for the "state of the atmosphere" obtained by clustering the patterns (1) into 25 different classes using the k-means algorithm (further information and details about this process are given in [3,5]).

3   Some Common Problems in Meteorology

In this section we introduce some interesting problems from Meteorology which are suitable for a probabilistic treatment. We shall use later these problems to illustrate the applications of Bayesian networks, presenting some promising results and describing other interesting applications.

3.1   Statistical Weather Generators

Weather generators have been used extensively in agricultural and hydrological applications where high-spatial resolution and/or long sequential series of records are required to solve common problems [10]. Simply stated, these methods are statistical simulation techniques adapted for producing samples of meteorological variables preserving some observed statistical properties and resembling the local climatology (annual or seasonal means, variances, frequencies of events, etc.). Thus, statistical weather generators provide an alternative for filling in missing data and for producing indefinitely long series from finite observation records of a local station stored in a climatological database.

One of the most important variables for practical applications is precipitation. In this case we are interested both in the absence or presence of the event (discrete variable) and its amount or intensity (continuous). In this paper we focus on the occurrence of daily precipitation (the application of Bayesian networks to the intensity process is related with the problem presented in the following section). Markov chains were the first models applied to simulate precipitation binary series (absence/occurrence) preserving the temporal correlation and the basic statistics of a local station under study. It was shown that a first-order model could capture the persistent nature of daily precipitation (i.e., the basic correlation of the series) [11]. This led to simple and efficient weather generators for simulating series of precipitation occurrence for a single station [12]. However, when simultaneous series for a set of stations are required, the above models are inappropriate and more general frameworks are needed [13]. The simulation of spatial coherent series from multisite data is still a challenging problem.

Another challenging problem is the forecast of meteorological variables (e.g., daily precipitation) at a given location (e.g., Santander city).

3.2   Local Weather Forecast

The main tools available for weather forecast are the outputs of numerical ACMs, which provide future predicted values of atmospheric variables on a grid. However, the resolution of the global grids used in these models and the physical parametrization of sub-grid scale processes –such as cloud formation, orography, etc.– limit the skill of these models to capture variabilities of the weather on a local station (e.g., precipitation or wind speed in the city of Santander); i.e., the output fields are smoothed. This problem motivated the development of different statistical techniques to model and forecast local climatological time series: auto-regressive models [14], embedding techniques [15], neural networks [16], etc. These techniques capture the dynamics underlying the time series fitting the parameters to the historical observed data. However, it has been shown that the knowledge of statistical techniques alone is not sufficient to obtain skillful forecasts.

An alternative solution for this problem is combining the information in both climatological and reanalysis databases. In other words, statistical techniques can be applied to relate model outputs to local observations, leading to forecast models for adapting a gridded forecast to local climates in an straightforward way. For instance, given a database of atmospheric circulation patterns $\mathbf{X}_t$, used as predictors, and simultaneous historical records of a local variable Y_t (predictand), standard and modern statistical methods such as regression analysis and neural networks can be applied to obtain a model $\hat{y}_t = f(\mathbf{x}_t) + \epsilon_t$ for performing local forecast [17]. However, these models assume stationary atmosphere dynamics during the period of study (note that the global model is trained with all the available data), and this is by no means guaranteed.

Local techniques, such as the method of analogs [18] or nearest neighbors, provide a simple solution for this problem, since models are trained locally. In this case, it is assumed that similar circulation patterns lead to similar meteorological events. Thus, local predictions are derived from $\mathbf{X}_t$ by first looking for the set of its Nearest Neighbors (NNs), or analog ensemble, in a circulation reanalysis database using some desirable metric. Then, the local observations registered at the analog ensemble dates are used to obtain a forecast. Recently, some methods have substituted the analog ensemble by a discrete atmospheric circulation state variable S obtained by applying a clustering algorithm (hierarchical, k-means, etc.) to a reanalysis database (see [19,20] for details). The resulting states of the variable correspond to the different clusters ($S = c$ refers to the subset of the reanalysis database associated with cluster c). In this situation, an input pattern is assigned to the closest cluster c_t, and a local probabilistic prediction $P(Y_t = i)$ for the occurrence $Y = i$ at date t is simply obtained as the conditional probability of the phenomenon, given the cluster observations (i.e., given the value of the state variable):

$$P(Y_t = i) = P(Y = i | S = c_t), \ i = 1, \ldots, m, \tag{2}$$

where m are the possible states of variable y (four for the case of precipitation). As pointed out in [21] these models are simple naive Bayes classifiers which do not take into account the spatial dependencies of the variables.

Therefore, modeling spatial dependency among local stations is an important unsolved topic for the two problems above. Bayesian networks provide a simple and sound framework for this task.

4   Bayesian Networks. Learning from Data

In this section we briefly introduce Bayesian networks and describe the available learning algorithms to obtain these models from meteorological data. We also discuss the performance of these methods considering the spatial character of the variables involved.

4.1   Bayesian networks

Bayesian networks are known for providing a compact and simple representation of probabilistic information, allowing the creation of models associating a large number of variables [7,8]. For instance, let us consider the network of climatic stations shown in the graph in Fig. 1(a); for most of the problems described in Sec. 3, we would like to obtain a probabilistic model representing the whole uncertainty about precipitation (or wind speed) in the network. The basic idea of probabilistic networks is encoding the dependencies among a set of variables using a graphical representation (a directed acyclic graph) which is easy to understand and interpret. There is a node for each domain variable (rainfall at each of the stations), and edges connect attributes that are directly dependent on each other. The graph defines a decomposition of the high-dimensional probability density function into low-dimensional local distributions (marginal or conditional).

For instance, the graph shown in Fig. 2 represents a directed acyclic graph where the variables are represented pictorially by a set of nodes; one node for each variable (in this paper we consider the set of nodes $\{Y_1, \ldots, Y_n\}$, where the subindex refers to the station's number). These nodes are connected by directed links, which represent a dependency relationship: If there is an arrow from node Y_j to node Y_k, we say that Y_j is a parent of Y_k, or equivalently, Y_k is a child of Y_j. The set of parents of a node Y_i is denoted as Π_i. For instance, in the inset of Figure 2 each node has a single parent ($\Pi_A = \Pi_B = \{C\}$, $\Pi_C = \{D\}$, and so on). Directed graphs provide a simple definition of independence (d-separation) based on the existence or not of certain paths between the variables (see [8] for a detailed introduction to Bayesian networks).

The dependency/independency structure displayed by the acyclic directed graph can be translated to the joint Probability Density Function (PDF) of

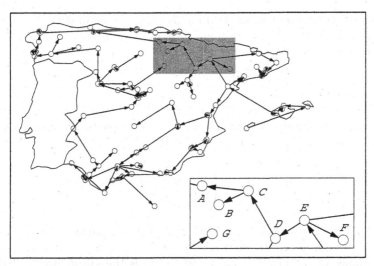

Fig. 2. Example of directed acyclic graph relating the stations shown in Fig. 1(a). The inset shows a magnification of the shaded area. The nodes shown in the inset correspond to the following cities: *A=Vitoria, B=Logroño, C=Pamplona, D=Zaragoza, E=Huesca, F=Lérida, G=Soria.*

the variables by means of a factorization as a product of conditional/marginal distributions as follows:

$$P(y_1, y_2, \ldots, y_n) = \prod_{i=1}^{n} P(y_i | \pi_i). \tag{3}$$

Therefore, the independencies from the graph are easily translated to the probabilistic model in a sound form. Depending on the discrete or continuous character of the variables, the conditional probabilities in (3) are specific parametric families. In this paper we consider two important types of Bayesian networks:

- *Multinomial Bayesian networks.* In a multinomial Bayesian network we assume that all variables are discrete, that is, each variable has a finite set of possible values (in this paper we shall consider four states for precipitation and three states for wind speed, as described in Sec. 2). We also assume that the conditional probability of each variable given its parents is multinomial and, hence, it is specified by a probability table given by the probabilities associated with the different combinations of values of the variables involved. Note that the number of parameters required to specify the full-dependence PDF in the left hand side term of (3) is $m^n = 4^{100}$, whereas the BN involves the specification of 100 conditional probability tables, one for each variable conditioned to its unitary parents' set.

- *Gaussian Bayesian networks.* In a Gaussian Bayesian network, the variables are assumed to have a multivariate normal distribution, $N(\mu, \Sigma)$, whose joint density function is given by

$$f(x) = (2\pi)^{-n/2} |\Sigma|^{-1/2} \exp\left\{-1/2(x - \mu)^T \Sigma^{-1}(x - \mu)\right\}, \qquad (4)$$

where μ is the n-dimensional mean vector, Σ is the $n \times n$ covariance matrix, $|\Sigma|$ is the determinant of Σ, and μ^T denotes the transpose of μ. In a Gaussian Bayesian network this PDF is specified as in (3) by:

$$f(x_i | \pi_i) \sim N\left(\mu_i + \sum_{j=1}^{i-1} \beta_{ij}(x_j - \mu_j), \ v_i\right), \qquad (5)$$

where β_{ij} is the regression coefficient of X_j in the regression of X_i on the parents of X_i, Π_i, and $v_i = \Sigma_i - \Sigma_{i\Pi_i}\Sigma_{\Pi_i}^{-1}\Sigma_{i\Pi_i}^T$ is the conditional variance of X_i, given $\Pi_i = \pi_i$, where Σ_i is the unconditional variance of X_i, $\Sigma_{i\Pi_i}$ is the vector of covariances between X_i and the variables in Π_i, and Σ_{Π_i} is the covariance matrix of Π_i. Note that β_{ij} measures the strength of the relationship between X_i and X_j. If $\beta_{ij} = 0$, then X_j is not a parent of X_i. Thus, the Gaussian Bayesian network is given by a collection of parameters $\{\mu_1, \ldots, \mu_n\}$, $\{v_1, \ldots, v_n\}$, and $\{\beta_{ij} \mid j < i\}$, as shown in (5).

4.2   Learning Algorithms

From a practical point of view, the application of Bayes Networks (BNs) in real-world problems depends on the availability of automatic learning procedures to infer both the graphical structure and the parameters of the conditional probabilities (the probability tables for the multinomial case, or the means, variances and regression coefficients for Gaussian networks). Several methods have been recently introduced for learning the graphical structure (structure learning) and estimating probabilities (parametric learning) from data (see [22] and references therein). Among these methods the so-called "score and search" consist of two parts:

1. A *quality measure* for computing the quality of the graphical structure and the estimated parameters for the candidate BNs.
2. A *search algorithm* to efficiently search the space of possible BNs to find the one with highest quality. Note that the number of all possible networks is huge even for a small number of variables, and so it is the search space (the search process has shown to be NP-hard in the general case).

Among the different quality measures proposed in the literature, Bayesian quality measures have a firm theoretical foundation. Each Bayesian net $B = (M, \theta)$, with network structure M and the corresponding estimated probabilities θ, is assigned a function of its posterior probability distribution given the available data D. This posterior probability distribution $p(B|D)$ is calculated as follows:

$$p(B|D) = p(M,\theta|D) = \frac{p(M,\theta,D)}{p(D)} \propto p(M)p(\theta|M)p(D|M,\theta), \qquad (6)$$

In the case of multinomial networks, assuming certain hypothesis about the prior distributions of the parameters and a uniform initial probability for all models, then the following quality measure is obtained [23]:

$$p(B|D) \propto \sum_{i=1}^{n} \left[\sum_{k=1}^{s_i} \left[\log \frac{\Gamma(\eta_{ik})}{\Gamma(\eta_{ik} + N_{ik})} + \sum_{j=0}^{r_i} \log \frac{\Gamma(\eta_{ijk} + N_{ijk})}{\Gamma(\eta_{ijk})} \right] \right], \qquad (7)$$

where n is the number of variables, r_i is the cardinal of the i-th variable, s_i the number of realizations of the parent's set Π_i, η_{ijk} are the "a priori" Dirichlet hyper-parameters for the conditional distribution of node i, N_{ijk} is the number of realizations in the database consistent with $y_i = j$ and $\pi_i = k$, and N_{ik} is the number of those consistent with $\pi_i = k$. Note that all the parameters in (7) can be easily computed from data.

In the case of Gaussian networks, a similar quality measure can be obtained assuming normal-Wishart distribution for the parameters (see [24] for a detailed description). Other quality measures have been introduced in the literature for multinomial and Gaussian networks but, for the sake of simplicity, in this paper we shall restrict to the above introduced measures.

Among the proposed search strategies, the $K2$ algorithm is a simple greedy search algorithm for finding a high quality Bayesian network in a reasonable time [25]. This algorithm takes advantage of the network's decomposability (which allow computing the separate contribution of each variable Y_i to the quality of the network (7)) and the non-unique representation of dependency models using graphs (equivalent graphs can be obtained reversing some of their links). The $K2$ iterative algorithm assumes that the nodes are ordered (to avoid cycles) and starts with a network with no links. For each variable Y_i, the algorithm adds to its parent set Π_i the node that is lower numbered than Y_i and leads to a maximum increment in the quality measure. The process is repeated until either adding new links does not increase the quality or a complete network is attained. A maximum number of parents for each node can also be enforced during the search. For instance, the graph in Fig. 2 was obtained applying the $K2$ learning algorithm to the quantized precipitation data, considering at most one single parent for each node (the database of precipitation records covers the period 1979-1993 for the network of stations shown in Fig. 1(a)). This Bayesian network do not provide an efficient model of the problem, but gives us a benchmark for comparing the performance of more realistic Bayesian networks trained with the data. For instance, Figure 3(a) shows a directed graph obtained increasing the threshold of the algorithm to two parents.

K2 Precipitation, score= -4.7697e+005

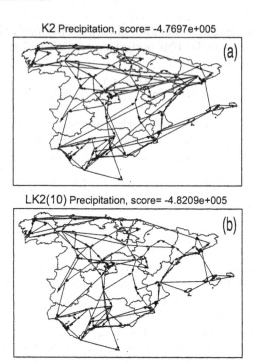

LK2(10) Precipitation, score= -4.8209e+005

Fig. 3. Directed acyclic graphs for precipitation obtained applying (a) the $K2$ algorithm and (b) the local $K2$ LK2 with 10 neighbors as candidate parents. In both cases, a maximum of two parents for each node are considered.

The $K2$ learning algorithm is simple and useful. However, further research has to be done in order to develop efficient learning methods for dealing with a large number of variables (stations) and huge amounts of data. Some preliminary works have recently appeared introducing promising ideas such as modularity, or data partitioning (see [26]). The spatial character of the meteorological problem presented in this paper can be taken into account for modifying the $K2$ algorithm to increase the speed and the meteorological significance of the obtained graphs. We call the resulting method the *local K2 learning algorithm* or simply *LK2*.

4.3   The Local K2 Learning Algorithm

In order to increase the efficiency of the $K2$ search strategy we modify the set of candidate parents of node Y_i, $\{1, \ldots, i-1\}$, to include only those nodes with similar climatology to the climatology of Y_i. This is done by computing the correlation of the observed records in different stations and obtaining the k-nearest neighbors for each station (the larger the parameter k, the more similar the algorithm to the original $K2$). This modification reduces the complexity of the search process, since now the set of candidate parents is of constant size. This fact is illustrated in Fig. 4 which shows the CPU time of the learning process versus the number of nodes for three different learning algorithms (the standard $K2$ and local $K2$ with five and ten neighbors, respectively). The savings in terms of computational time are clearly shown in this figure. Figure 3(b) shows the directed graph obtained using $Lk2$ algorithm with ten neighbors ($LK2(10)$) for precipitation. The score of the local graph is similar to the one obtained using the $K2$ algorithm (Fig. 3(a)). Moreover, from visual inspection, the meteorological significance of the $LK2$ graph is higher.

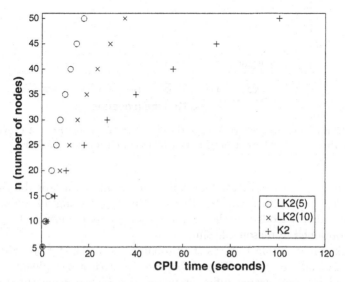

Fig. 4. Problem size (number of nodes) versus CPU time for three different learning algorithms: the standard $K2$ and local $K2$ with five $LK2(5)$ and ten $LK2(10)$ neighbors, respectively (Pentium IV 1.6 GHz).

On the other hand, Figure 5 compares the performance of the standard $K2$ algorithm versus $LK2$ with different number of neighbors $LK2(k)$ when applied to the precipitation data. From this figure we can see the exponential convergence of the score as a function of k, increasing from the threshold given by an empty graph, to the threshold given by the $K2$ algorithm. Thus, substantial saving in computing time can be achieved with a minor loss of score using a small set of neighbors as candidate parents.

4.4 Probabilistic Inference

When some evidence becomes available (e.g., we know that the event "heavy rain" is occurring in Madrid and Coruña $e = \{\text{Madrid}=4, \text{Coruña}=4\}$),

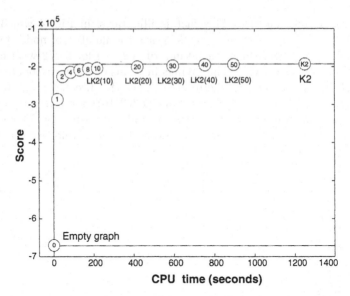

Fig. 5. Comparing the performance of the standard $K2$ algorithm with several local $K2$ algorithms with different neighbors $LK2(k)$ (Pentium IV 1.6 GHz).

Bayesian networks can be used for computing conditional probabilities of the form $P(y_i|e)$. This task can be efficiently done by taking advantage of the independencies encoded in the graph, so probabilistic answers to problems can be obtained in reasonable time.

For instance, Table 1 shows the probabilities of the stations *Santander*, *Zaragoza* and *Soria* (see Fig. 2 for their relative geographical positions) computed by propagating different pieces of evidence using three different graphs: the one obtained using the standard $K2$ algorithm (Fig. 3(a)), the given by $LK2(10)$ with a maximum of one parent per node (Fig. 2) and the one corresponding to $LK2(10)$ with a maximum of two parents per node (Fig. 3(b)). Table 1(a) shows the initial or *a priori* marginal probabilities for each station. From this table we can see that all the graphs provide similar marginal probabilities (all of them have been learnt from the same data). However, this is not the case for the probabilities conditioned to different pieces of evidence. From Tables (b)-(d) we observe a clear difference between the one-parent $LK2(10)$ sparse graph and the other two more dense Bayesian networks with closer scores shown in Fig. 3 (the mean difference between the probabilities provided by these two graphs is less than 3%).

5   Applications of Bayesian Networks

In this section we present some illustrative applications of Bayes networks to some topics related with the problems presented in Sec. 3.

Table 1. (a) Marginal probabilities for three nodes: Santander, Soria and Zaragoza, and (b)-(d) conditional probabilities given different pieces of evidence: $e_1 = \{Madrid = 4\}$, $e_2 = \{Coruña = 4\}$, and $e_3 = \{Madrid = 4, Coruña = 4\}$. Three different Bayesian networks have been considered to propagate the evidence: $K2$ refers to the standard $K2$ algorithm, whereas $LK2_1$ and $LK2_2$ refer to the local $K2$ algorithm with 10 neighboring nodes and a maximum of 1 and 2 parents per node, respectively.

(a) Initial probability, $P(y_k)$

State	Santander			Soria			Zaragoza		
	$K2$	$Lk2_1$	$Lk2_2$	$K2$	$Lk2_1$	$Lk2_2$	$K2$	$Lk2_1$	$Lk2_2$
0	0.502	0.502	0.502	0.613	0.613	0.616	0.741	0.736	0.741
1	0.211	0.211	0.211	0.223	0.215	0.216	0.163	0.158	0.162
2	0.250	0.250	0.250	0.155	0.162	0.158	0.091	0.099	0.092
3	0.037	0.037	0.037	0.009	0.010	0.010	0.005	0.006	0.005

(b) Conditional probability, $P(y_k|Madrid = 4)$

State	Santander			Soria			Zaragoza		
	$K2$	$Lk2_1$	$Lk2_2$	$K2$	$Lk2_1$	$Lk2_2$	$K2$	$Lk2_1$	$Lk2_2$
0	0.357	0.478	0.285	0.156	0.147	0.156	0.331	0.570	0.322
1	0.220	0.211	0.224	0.277	0.296	0.267	0.330	0.235	0.329
2	0.357	0.269	0.409	0.518	0.505	0.523	0.316	0.182	0.324
3	0.065	0.041	0.081	0.049	0.051	0.054	0.023	0.013	0.024

(c)(Conditional probability, $P(y_k|Coruña = 4)$

State	Santander			Soria			Zaragoza		
	$K2$	$Lk2_1$	$Lk2_2$	$K2$	$Lk2_1$	$Lk2_2$	$K2$	$Lk2_1$	$Lk2_2$
0	0.152	0.264	0.150	0.339	0.412	0.291	0.597	0.673	0.563
1	0.183	0.264	0.181	0.331	0.295	0.330	0.237	0.188	0.251
2	0.502	0.425	0.506	0.311	0.275	0.353	0.157	0.130	0.176
3	0.162	0.106	0.163	0.019	0.017	0.026	0.009	0.009	0.010

(d) Conditional probability, $P(y_k|Madrid = 4, Coruña = 4)$

State	Santander			Soria			Zaragoza		
	$K2$	$Lk2_1$	$Lk2_2$	$K2$	$Lk2_1$	$Lk2_2$	$K2$	$Lk2_1$	$Lk2_2$
0	0.127	0.264	0.095	0.077	0.068	0.064	0.285	0.535	0.266
1	0.170	0.205	0.161	0.256	0.279	0.235	0.341	0.250	0.341
2	0.520	0.425	0.540	0.603	0.591	0.621	0.348	0.200	0.364
3	0.183	0.106	0.204	0.064	0.062	0.079	0.026	0.014	0.028

5.1   Spatial Consistency in Stochastic Weather Generators

In the previous section, exact inference techniques have been applied to obtain the conditional probabilities of nodes in a Bayesian network given some evidence. There are also alternative methods for computing the probabilities approximately. The basic idea behind these methods is generating a sample of realizations from the joint PDF of the variables, and computing approximate values for the probabilities from the sample (as frequencies observed in the sample). The simulation is facilitated by the factorization given by the Bayesian network, since each node can be simulated independently (according to the values of the parents), following an appropriate ancestral ordering (see [8] for details). One of the most intuitive and simple simulation methods operates in a forward manner generating the instantiations one variable at a time; a variable is sampled only after all its parents have been sampled [27]. If the obtained realization match the evidence (if any), then it is included in the sample; otherwise it is rejected. The algorithm ends when the required number of acceptable instantiations is obtained.

The above algorithm provide us a simple method for generating stochastic weather from a Bayesian network in an spatially consistent manner. Instead of simulating values independently for each variable (as in standard weather generator methods) we simulate spatial realizations taking into account the constraints imposed by the dependencies in the graph. For the sake of simplicity, we illustrate the application of simulation algorithms in the case of discrete variables (precipitation), though a similar scheme is applicable to the continuous case. Figure 6 shows a sample of rain values (1, 2, 3, or 4) generated for 200 consecutive days for three different stations. We can see that the generated rain values are spatially consistent.

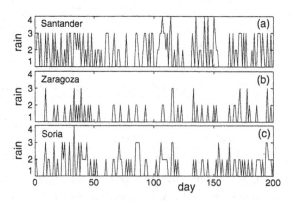

Fig. 6. Simulating a spatial time series of precipitation for 200 days applying the acceptance-rejection method to the marginal distribution (the simulation order is the one used for the $K2$ algorithm): (a) Santander, (b) Zaragoza, (c) Soria.

We can also simulate weather consistent with different evidences (wet or rainy episodes). For instance, Figs. 7 show precipitation simulations obtained when wet and rainy events are assumed to be occurring simultaneously at the staions *Madrid* and *Coruña* (note that this flexibility is not proper of standard weather generators [10]). These figures also exhibit spatial consistency, but now the simulated weather is adapted to the observed evidence.

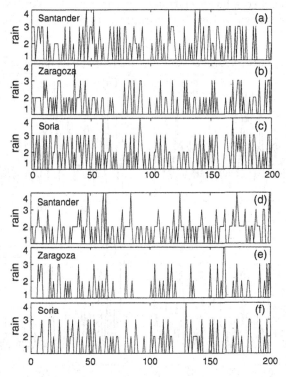

Fig. 7. Simulating a spatial time series of precipitation for 200 days considering the evidence *Madrid=4, Coruña=4* (figures (a)-(c)), and *Madrid=1, Coruña=1* (d)-(f).

5.2   Filling Missing Data.

Another interesting application of Bayesian networks is filling missing data. Missing values are usually present in observation records and some applications require complete data. Missing value estimation is a particular type of forecast problem which is extremely difficult for locations with high spatial variance. Missing data must be recovered preserving the main characteristics of the original data (distribution, mean, extremes, etc.). Standard techniques for this problem use regression models associated with neighboring stations with complete data archives. A Gaussian Bayesian network provide a global and robust alternative for this problem. An advantage of Bayesian networks is that any piece of information can be plugged in as evidence, obtaining an estimation of the missing values. For instance, Fig. 8 shows the values of maximum wind speed (Wind) estimated from a Gaussian Bayesian network considering as evidence the real values occurred at Madrid and Coruña (left column) and Madrid, Coruña, Ranón and Sondika stations (right column). The Gaussian network has been trained considering the Wind values observed in the stations shown in Fig. 1(a) during the period 1959-2000.

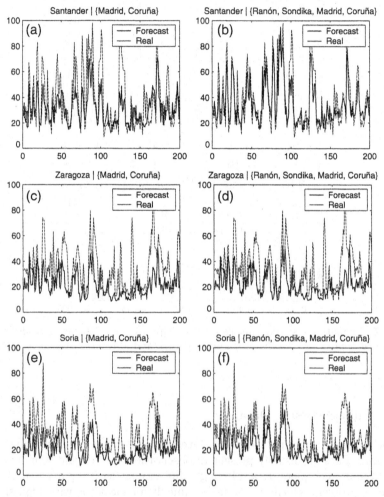

Fig. 8. Conditional mean of the normal distributions of Wind for Santander (a),(b), Zaragoza (c),(d), and Soria (e),(f). Evidences are the values observed at: Madrid and Coruña (left column), Ranón, Sondika, Madrid and Coruña (right column).

The estimated values are closer to the real ones when the evidence is informative for the corresponding stations (from Figs. 8(a) and (b) we see that Sondika and Ranón provide valuable information for Wind at Santander).

This technique can also be used to remove incoherent data from the database (the one leading to probabilities inconsistent with the observations of neighboring stations), and even for short-time automatic forecast (nowcasting) using the real-time observations available at the forecast time [28].

5.3   Local Weather Forecast.

As mentioned in Section 3.2, analog probabilistic forecast methods can be considered a particular type of naive classifiers, where each predictand variable is only conditioned to the atmospheric state S. Given the state of the atmosphere provided by a numerical ACM and considering four states for precipitation (dry, light rain, moderate rain, heavy rain), the naive Bayesian classifier computes the probability for each state at a local station as shown in (2). However, once again, this model does not guarantee the spatial consistency of the results (see [21] for details). This problem can be solved by using the Bayesian network to represent the spatial dependencies among the variables (instead of considering conditional independence of the stations, given S). Figure 9(left column) shows the observed values and the predicted probabilities for two months (January and February 1991) for the event "dry" in Santander, Zaragoza, and Soria using the Bayesian network shown in Fig.3(a).

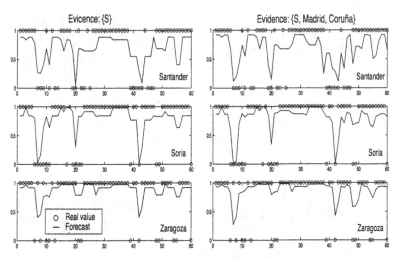

Fig. 9. Observations of the event "dry" in Santander, Zaragoza, and Soria (circles indicate the absence "0" or the occurrence "1"). Forecasts obtained with evidences: S (left column); S and the values observed in Madrid and Coruña (right column).

On the other hand, if we consider a continuous variable (such as Wind), we can condition each of the continuous nodes of the Gaussian network to the discrete atmospheric state, obtaining a conditioned Gaussian network [30]. In this case, the local forecast can be obtained as the conditional mean of the variables, given the state of the atmosphere predicted by a numerical ACM. Figure 10 shows the resulting values when considering 25 (left column) and 50 (right column) states for the atmospheric variable. This figure shows the sensitive dependence of the forecast on the number of states chosen for the atmospheric variable.

The above examples are illustrative and can not be considered competitive models in Meteorology. Further research is still needed to investigate the operative performance of these models.

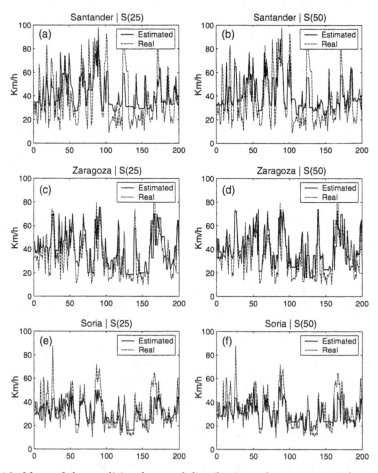

Fig. 10. Mean of the conditional normal distributions of maximum wind speed for Santander (a),(b), Zaragoza (c),(d), and Soria (e),(f).

Acknowledgments

The authors are grateful to the Instituto Nacional de Meteorología (INM) for providing us with the necessary data for this work. The authors are also grateful to the University of Cantabria, CSIC and the Comisión Interministerial de Ciencia y Tecnología (CICYT) (Project REN2000-1572) for partial support of this work. Finally, we would also like to thank K. Murphy for his work on the Matlab Bayes nets Toolbox [31], which helped us to obtain the results of this paper.

References

1. von Storch H., Navarra A. (1995). Analysis of Climate Variability. Applications of Statistical Techniques, second edition. Springer Verlag
2. von Storch H., Zwiers F.W. (1999). Statistical Analysis in Climate Research. Cambridge University Press
3. Cofiño A.S., Gutiérrez J.M., Jakubiak B., Melonek M. (2003) Implementation of data mining techniques for meteorological applications. In Realizing Teracomputing (W. Zwieflhofer and N. Kreitz editors), World Scientific Publishing, 256-271
4. Wilby R.L., Wigley T.M.L. (1997) Downscaling general circulation model output. A review of methods and limitations. Progress in Physical Geography **21**, 530-548
5. Enke W., Spekat A. (1997) Downscaling climate model outputs into local and regional weather elements by classification and regression. Climate Research **8**, 195-207
6. Kutzbach J. (1967) Empirical eigenvectors of sea-level pressure, surface temperature and precipitation complexes over North America. Journal of Applied Meteorology **6**, 791-802
7. Pearl J. (1988). Probabilistic Reasoning in Intelligent Systems: Networks of Plausible Inference. Morgan Kaufmann
8. Castillo E., Gutiérrez J. M., Hadi A. S. (1997). Expert Systems and Probabilistic Network Models. Springer-Verlag, New York (Free Spanish version http://personales.unican.es/gutierjm)
9. ECMWF ERA-15 Reanalysis project
   http://www.ecmwf.int/research/era/ERA-15/Project/
10. Wilby R.L., Wilks D.S. (1999) The weather generation game. A review of stochastic weather models. Progress in Physical Geography **23**, 329-357
11. Gabriel K.R., Neumann J. (1962) A Markov chain model for daily rainfall occurrence at Tel Aviv. Quarterly Journal of the Royal Meteorological Society **88**, 90-95
12. Richardson C.W. (1981) Stochastic simulation of daily precipitation, temperature, and solar radiation. Water Resources Research **17**, 182-190
13. Wilks D.S. (1999) Multisite downscaling of daily precipitation with a stochastic weather generator. Climate Research **11**, 125-136

14. Zwiers F., von Storch, H. (1990) Regime dependent auto-regressive time series modeling of the Southern Oscillation. Journal of Climate **3**, 1347–1360

15. Pérez-Muñuzuri V., Gelpi I.R. (2000) Application of nonlinear forecasting techniques for meteorological modeling. Annales Geohysicae **18** 1349–1359

16. Gardner M.W., Dorling S.R. (1998) Artificial neural networks (the multilayer perceptron). A review of applications in the atmospheric sciences. Journal of Applied Meteorology **39** 147–159

17. Klein W.H., Glahn, H.R. (1974) Forecasting local weather by means of model output statistics. Bulletin of the American Meteorological Society **55**, 1217–1227

18. Lorenz E.N. (1969) Atmospheric Predictability as Revealed by Naturally Occurring Analogues. Journal of the Atmospheric sciences **26**, 636–646

19. Cano R., López F.J., Cofiño A.S., Gutiérrez J.M., and Rodríguez M.A. (2001) Aplicación de métodos de clasificación al downscaling estadístico. In Actas del V Simposio Nacional de Predicción. Instituto Nacional de Meteorología, 235–240

20. Gutiérrez J.M., Cofiño A.S, Cano R., Rodríguez M.A. (2003) Clustering methods for statistical downscaling in short-range weather forecast. Submitted.

21. Cano R., Cofiño A.S., Gutiérrez J.M., Sordo C. (2002) Probabilistic networks for statistical downscaling and spatialisation of meteorological data Geophysical Research Abstracts **4**, 194

22. Jordan M.I. (1998). Learning in Graphical Models. MIT Press.

23. Geiger D., Heckerman, D. (1995) A characterization of the Dirichlet distribution with application to learning Bayesian networks. Proceedings of the Eleventh Conference on Uncertainty in Artificial Intelligence. Morgan Kaufmann Publishers, 196–207

24. Geiger D., Heckerman, D. (1994) Learning gaussian networks. In Proceedings of the Tenth Conference on Uncertainty in Artificial Intelligence. Morgan Kaufmann Publishers, 235–243

25. Cooper G.F., Herskovitz E. (1992) A Bayesian method for the induction of probabilistic networks from data. Machine Learning **9**, 309–347

26. Neil M., Fenton N., and Nielson L. (2000) Building large-scale Bayesian networks. The Knowledge Engineering Review **15**, 257–284

27. Henrion M. (1988) Propagation of uncertainty by logic sampling in Bayes' networks. In Uncertainty in Artificial Intelligence 2 (Lemmer, J. F. and Kanal, L. N., editors). North Holland, 149–164

28. Michaelides S.C., Pattichis C.S., Kleovoulou G. (2001) Classification of rainfall variability by using artificial neural networks. International Journal of Climatology **21**, 1401–1414

29. Cofiño A.S., Cano R., Sordo C., Gutiérrez J.M. (2002) Bayesian networks for probabilistic weather prediction. In Proceedings of the 15th European Conference on Artificial Intelligence, 695–700

30. Peña J.M., Lozano J.A., Larrañaga P. (2001) Performance evaluation of compromise conditional Gaussian networks for data clustering. International Journal of Approximate Reasoning **28**, 23–50

31. Murphy K.P. (2001) The Bayes net toolbox for Matlab. Computing Science and Statistics **33**. http://www.cs.berkeley.edu/ murphyk/Bayes/bnt.html